Lecture Notes in Physics

Edited by H. Araki, Kyoto, J. Ehlers, München, K. Hepp, Zürich
R. Kippenhahn, München, D. Ruelle, Bures-sur-Yvette
H. A. Weidenmüller, Heidelberg, J. Wess, Karlsruhe and J. Zittartz, Köln

Managing Editor: W. Beiglböck

342

M. Balabane P. Lochak C. Sulem (Eds.)

Integrable Systems and Applications

Proceedings of a Workshop
Held at Oléron, France
June 20–24, 1988

Springer-Verlag
Berlin Heidelberg New York London Paris Tokyo Hong Kong

Editors

M. Balabane
P. Lochak
C. Sulem
Ecole Normale Supérieure, Centre de Mathématiques Appliquées
45 rue d'Ulm, F-75230 Paris Cédex 05, France

ISBN 3-540-51615-8 Springer-Verlag Berlin Heidelberg New York
ISBN 0-387-51615-8 Springer-Verlag New York Berlin Heidelberg

Printing: Druckhaus Beltz, Hemsbach/Bergstr.
Bookbinding: J. Schäffer GmbH & Co. KG., Grünstadt
2158/3140-543210 – Printed on acid-free paper

PREFACE

This volume contains articles on most of the lectures presented at the workshop "Integrable Systems and Applications" held at the Ile d'Oléron (France), June 20 – 24, 1988. It was organized in the context of a Special Year on Nonlinear Phenomena.

The principal aim was to collect new material on nonlinear systems related to integrable systems. There were lectures on a variety of topics, reflecting different aspects of the subject, including the relevance and application of integrable systems to the description of natural phenomena, the development of perturbation theories, and the statistical mechanics of ensembles of objects obeying integrable equations.

One may notice that a large proportion of the physics-oriented lectures were centered around the three paradigmatic equations Korteweg de Vries, sine-Gordon and nonlinear Schrödinger, especially the latter. For a long time, these have proved of great mathematical interest and have also exhibited some "universality", which places them among the most frequently encountered integrable equations in the description of physical systems. Tidal waves, optical fibers, and laser beams were among the topics discussed. Some lectures were also devoted to multidimensional solitons, the integrability of Hamiltonian systems of ODEs and dissipative systems of PDEs.

We wish to express our thanks to the lecturers for their simulating presentations and to all the participants who contributed to the success of the meeting. We thank the CEA, CNRS, MRES, DRET, the Université Paris-Nord and the Ecole Normale Supérieure (CMA) for their financial support. Last but not least, we thank the "village du CAES, la vieille Perrotine" (St Pierre d'Oléron) for its kind hospitality.

Ile d'Oléron (France)
June 20 – 24, 1988

Mikhael Balabane
Pierre Lochak
Catherine Sulem

CONTENTS

*In the case of papers written by several authors, the author marked with a * presented the lecture.*

GLOBAL SOLUTIONS TO NON LINEAR DIRAC EQUATIONS IN MINKOWSKI SPACE

Alain BACHELOT
Département de Mathématiques Appliquées
Université de Bordeaux I
351, Cours de la Libération
33405 Talence Cedex

INTRODUCTION

The purpose of this paper is to expose the recent results of global existence for non linear Dirac equations or Dirac-Klein-Gordon systems in both cases where blow up can generally occur : 1) if the order of the non linearities is critic with respect to the space dimension ; 2) if the Cauchy data is large. More precisely let's consider an hyperbolic symetric system

$$\mathscr{L}(\partial_t,\partial_x)\psi = f(\psi),\ x\in\mathbb{R}^n \tag{1}$$

where

$$|f(\psi)| = O(|\psi|^d),\ |\psi| \to 0 .$$

If we want to obtain global asymptotically free solutions of (1) for small initial data, we expect that the energy contribution of $f(\psi)$ is finite, i.e.

$$\int_{-\infty}^{+\infty} \|f(\psi(t))\|_{L^2(\mathbb{R}^n_x)} dt<+\infty . \tag{2}$$

We test estimate (2) with regular wave packet free solutions of $\mathscr{L}(\partial_t,\partial_x)\psi_0=0$ which satisfy

$$\|\psi_0(t)\|_{L^2(\mathbb{R}^n_x)}=cst,\ |\psi_0(t)|_{L^\infty(\mathbb{R}^n_x)}=O(|t|^{-(n-1)/2})\ ;$$

therefore,

$$\|f(\psi_0(t))\|_{L^2(\mathbb{R}^n_x)}=O(|t|^{-(d-1)(n-1)/2})$$

then the case of quadratic non linearities in Minkowski space $\mathbb{R}^{3+1}$ is critic and we know that the local solution can blow up [10]. So it is interesting to find bilinear interaction f such that

$$\|f(\psi_0(t))\|_{L^2(\mathbb{R}^3_x)} \in L^1(\mathbb{R}_t). \tag{3}$$

We know that a sufficient condition for (3) is that f is null on the kernel of the symbol of $\mathcal{L}$: it is the algebraic condition of *compatibility* of f with $\mathcal{L}$ introduced by B. Hanouzet and J.L. Joly [9] (and so [1]) which is related to the Lorentz invariance and the *Null Condition* of S. Klainerman [13]. We study these notions in part I and apply them in part II to solve the global Cauchy problem with small initial data for the Dirac-Klein-Gordon systems

$$\begin{aligned} -i\gamma^\mu\partial_\mu\psi+M\psi &= f(\varphi,\psi), \quad M\neq 0 \\ \Box\varphi+m^2\varphi &= g(\varphi,\psi), \qquad m\geqslant 0 \end{aligned} \tag{4}$$

Now, if we take large initial data, we cannot expect that (2) is verified and in fact all cases can occur : global existence, blow up or stationary solutions, even for d=3, n=3 ; for instance, M. Balabane, Th. Cazenave, A. Douady, F. Merle prove in [5] the existence of infinitely many stationary states for

$$-i\gamma^\mu\partial_\mu\psi+M\psi = (\overline{\psi}.\psi)\psi \quad , \; M\neq 0 \tag{5}$$

and we establish in part IV the existence of global asymptotically free solutions to systems generelising (4) and (5) for arbitrarly large initial data under two algebraic assumptions on the non linearities and on the polarization of initial data : on the one hand we suppose that the systems are Lorentz invariant, and on the other hand, we consider initial data satisfying an approximated Lochak-Majorana condition. We study in part III this last condition which implies that the chiral invariant is small ; then, the Lorentz-invariance implies that the non linearities are small, and (2) is true again and we obtain the global existence and the asymptotic freedom. The complete demonstrations are published in [2], [3], [4].

I - COMPATIBILITY - LORENTZ-INVARIANCE - NULL CONDITION

We specify some notations $g^{\mu,\nu}=\mathrm{diag}(1,-1,-1,-1)$ is the Lorentz metric on Minkowski space $\mathbb{R}^4=\mathbb{R}_t\times\mathbb{R}^3_x$, $x^0=t, (x^1,x^2,x^3)=x$. The Dirac matrices γ^μ, $0\leqslant\mu\leqslant3$, satisfy the relations

$$\gamma^\mu\gamma^\nu+\gamma^\nu\gamma^\mu=2g^{\mu,\nu}I \quad , \quad \tilde{\gamma}^\mu=g^{\mu,\mu}\gamma^\mu$$

where $\tilde{A}$ notes the conjugate transposate of matrix A. We introduce so the matrix $\gamma^5=-i\gamma^0\gamma^1\gamma^2\gamma^3$.

We consider the generators $(\Gamma_\sigma)_{1\leqslant\sigma\leqslant10}$ of Poincaré group

$$(\Gamma_\sigma)_{1\leqslant\sigma\leqslant10} = (\partial_\mu=\partial/\partial x^\mu,\ \Omega_{\mu,\nu}=x_\mu\partial_\nu-x_\nu\partial_\mu)_{0\leqslant\mu,\nu\leqslant3}$$

The Lorentz invariance of the wave equation is expressed by the commuting relations

$$[\Gamma_\sigma, \Box] = 0 .$$

To study the Dirac system we introduce

$$(\hat{\Gamma}_\sigma)_{1\leq\sigma\leq 10} = (\partial_\mu, \Omega_{\mu,\nu} + \tfrac{1}{2}\gamma_\mu\gamma_\nu)_{0\leq\mu,\nu\leq 3}$$

which satisfy

$$[\hat{\Gamma}_\sigma, -i\gamma^\mu\partial_\mu] = 0 .$$

In the case of massless fields, we can use the scaling invariance with the radiation operator Γ_0 :

$$\Gamma_0 = x^\mu\partial_\mu,\quad [\Box,\Gamma_0] = 2\Box,\quad [-i\gamma^\mu\partial_\mu, \Gamma_0] = -i\gamma^\mu\partial_\mu .$$

We recall some definitions :

i) A sesquilinear form f on $\mathbb{C}^N$ is said compatible with a first order differential system

$$A(\partial) = \sum_{j=1}^{n} A_j \frac{\partial}{\partial x^j} + iB$$

where A_j and B are P×N matrices with constant coefficients if

$$\forall\xi\in\mathbb{R}^n\setminus\{0\},\ V\in \mathrm{Ker}\Big(\sum_{j=1}^{n}\xi_j A_j + B\Big) \Rightarrow f(V,V) = 0 .$$

ii) A sesquilinear form $N(\psi_1', \psi_2')$ on $\mathbb{C}^{16}\times\mathbb{C}^{16}$ satisfies the null condition if $\forall\psi_i\in C^1(\mathbb{R}^4,\mathbb{C}^4)$, ${}^t\psi_i = (\psi_i^1, \psi_i^2, \psi_i^3, \psi_i^4)$, $\psi_i' = (\partial_\mu\psi_i)_{0\leq\mu\leq 3}$,

$$\forall (X_\mu)\in\mathbb{R}^4,\ g^{\mu,\nu}X_\mu X_\nu = 0 \Rightarrow \forall h,k,\ \sum_{0\leq\mu,\nu\leq 3} \frac{\partial^2 N}{\partial(\partial_\mu\tilde{\psi}_1^h)\partial(\partial_\nu\psi_2^k)} X_\mu X_\nu = 0 .$$

iii) The spinorial representation of the whole Lorentz group O(3,1) is the mapping

$$L = (L^\mu_\nu)\in O(3,1) \rightarrow \Lambda\in SL(4,\mathbb{C})/\{-1,1\},$$

defined by

$$L^\mu_\nu\gamma^\nu = \Lambda^{-1}\gamma^\mu\Lambda .$$

We note DO(3,1) the orthochroneous proper Lorentz subgroup.
The compatible forms for massless Dirac system are described by the following :

THEOREM I.1 - Let f be a sesquilinear form on $\mathbb{C}^4$; the following assertions are equivalent :

1) $\forall\psi_i \in \mathbb{C}^4$, $\forall L \in DO(3,1)$, $f(\Lambda\psi_1,\Lambda\psi_2)=f(\psi_1,\psi_2)$.
2) f is compatible with the massless Dirac system $\mathscr{L}_0=-i\gamma^\mu\partial_\mu$.
3) $N(\psi_1',\psi_2')=f(\mathscr{L}_0\psi_1,\mathscr{L}_0\psi_2)$ satisfies the null condition.
4) $\exists C>0$, $\forall\psi_i \in C^1(\mathbb{R}^4,\mathbb{C}^4)$,

$$|N(\psi_1',\psi_2')|\leqslant C(1+|t|+|x|)^{-1}\sup_{0\leqslant\sigma,\tau\leqslant 10}|\Gamma_\sigma\psi_1(t,x)||\Gamma_\tau\psi_2(t,x)| \qquad (6)$$

5) $\exists\alpha,\beta\in\mathbb{C}/f(\psi_1,\psi_2)=\tilde{\psi}_1(\alpha\gamma^0+\beta\gamma^0\gamma^5)\psi_2$.

If the mass is non null we must avoid Γ_0 and the is only one compatible form.

THEOREM I.2 - Let f be a sesquilinear form on $\mathbb{C}^4$; the following assertions are equivalent :

1) $\forall\psi_i \in \mathbb{C}^4$, $\forall L \in O(3,1)$, $f(\Lambda\psi_1,\Lambda\psi_2)=(\det L)\, f(\psi_1,\psi_2)$.
2) f is compatible with the mass Dirac system $\mathscr{L}_M=-i\gamma^\mu\partial_\mu+M$, $M\neq 0$.
3) $N(\psi_1',\psi_2')=f(\mathscr{L}_0\psi_1,\mathscr{L}_0\psi_2)+g^{\mu,\nu}f(\partial_\mu\psi_1,\partial_\nu\psi_2)$ satisfies the null condition and

$$|N(\psi_1',\psi_2')|\leqslant C(1+|t|+|x|)^{-1}\sup_{1\leqslant\sigma,\tau\leqslant 10}|\Gamma_\sigma\psi_1(t,x)|\Gamma_\tau\psi_2(t,x)|. \qquad (7)$$

4) $\exists\alpha\in\mathbb{C}/f(\psi_1,\psi_2)=\alpha\tilde{\psi}_1\gamma^0\gamma^5\psi_2$.

It is crucial that the radiation operator does not appear in estimate (7). The estimates (6)(7) play a fundamental role in our study : if ψ_j are regular wave packet free solutions (6)(7) imply

$$\|N(\psi_1',\psi_2')(t)\|_{L^2(\mathbb{R}^3_x)}=O(|t|^{-\alpha}),\alpha=2 \text{ if } M=0,\ \alpha=5/2 \text{ if } M\neq 0\ ,$$

instead of $\alpha=1$ if $M=0$, $\alpha=3/2$ if $M\neq 0$, for ordinary product.

II - GLOBAL SOLUTIONS FOR MASS AND MASSLESS FIELDS INTERACTING.

We consider system (4) with quadratic nonlinearities

$$-i\gamma^\mu\partial_\mu\psi+M\psi=\varphi V\psi, \qquad (8.1)$$

$$\Box\varphi+m^2\varphi=\tilde{\psi}F\psi \qquad (8.2)$$

where V and F are 4×4 matrices with constant coefficients. If both masses M and m are non null, the uniform decay is fast enough to assure the global existence [12]. If both masses M and m are null, the conformal invariance allows us to use Penrose transform and to obtain global solutions [7]. The interesting case is :

$$M\neq 0\ ,\ m=0 \qquad (9)$$

We make two algebraic assumptions on the nonlinearities :

$$\tilde{V}\gamma^0=\gamma^0 V, \quad \tilde{F}=F, \tag{10}$$

$$F=ig\gamma^0\gamma^5, \quad g\in\mathbb{R} \tag{11}$$

(10) implies the conservation of the spinorial charge and (11) has two equivalent interpretations according to theorem I.2 : φ is a pseudoscalar Lorentz invariant field ; F is compatible with the mass Dirac system. The main example is the Yukawa model of nuclear forces with the interaction lagrangien $ig\varphi\tilde{\psi}\gamma^0\gamma^5\psi$.
We choose small very regular initial data :

$$\psi(0,x)=\varepsilon\psi_0(x), \varphi(0,x)=\varepsilon\varphi_0(x), \partial_t\varphi(0,x)=\varepsilon\varphi_1(x),$$
$$\psi_0\in\mathcal{D}(\mathbb{R}^3_x,\mathbb{C}^4), \varphi_j\in\mathcal{D}(\mathbb{R}^3_x,\mathbb{R}), \quad 0<\varepsilon \ . \tag{12}$$

THEOREM II.1 - Under hypotheses (9)(10)(11), there exists $\varepsilon_0>0$ such that for $\varepsilon\in]0,\varepsilon_0[$, the Cauchy problem (8)(12) has a unique solution $(\psi,\varphi)\in C^\infty(\mathbb{R}^4)$. Moreover (ψ,φ) is asymptoticaly free : there exist $\psi^\pm$ $\varphi^\pm$ satisfying

$$-i\gamma^\mu\partial_\mu\psi^\pm+M\psi^\pm=0, \quad \Box\varphi^\pm=0$$

$$\lim_{t\to\pm\infty}\left(\|\psi(t)-\psi^\pm(t)\|_{L^2(\mathbb{R}^3_x)}+\sum_{\mu=0}^{3}\|\partial_\mu\varphi(t)-\partial_\mu\varphi^\pm(t)\|_{L^2(\mathbb{R}^3_x)}\right)=0 \ .$$

In fact φ has even a nice behaviour in L^2-norm :

$$\varphi,\varphi^\pm\in C^0(\mathbb{R},L^2(\mathbb{R}^3_x)), \|\varphi(t)-\varphi^\pm(t)\|_{L^2(\mathbb{R}^3_x)}\longrightarrow 0, \quad t\to\pm\infty. \tag{13}$$

The key of the proof of theorem II.1 is the compatibility of F, (11) which allows to transform the nonlinear wave equation (8.2) into a better wave equation

$$\Box(\varphi+(2M)^{-2}\tilde{\psi}F\psi)=N(\psi',\psi')+\mathcal{R}$$

where N satisfies the null condition and (7) and $\mathcal{R}$ is cubic.
An interesting question is the necessity of (11) to solve the global Cauchy problem. We show that we can replace (11) by

$$V=ig\gamma^5, \quad g\in\mathbb{R} \ . \tag{14}$$

We point out that, if system (8) comes from a lagrangian, then (11) and (14) are equivalent and we find the Yukawa model again.

THEOREM II.2 - Under hypotheses (9)(10)(14), the conclusions of theorem II.1 hold again.

But in this case φ has not necessarly a good behaviour in L^2-norm ;

we prove only $\|\varphi(t)\|_{L^2(\mathbb{R}^3_x)}=O(t^{1/2})$. This time, algebraic condition (14) is used to transform equation (8.1) :

$$(\Box+M^2)\psi=-g(\partial_\mu\varphi)\gamma^\mu\gamma^5\psi+\text{cubic}$$

and we note that in the quadratic part, φ appears only with the first derivatives which are easily estimated in L^2.

III - CHIRAL INVARIANT AND MAJORANA CONDITION

To estimate a Lorentz-invariant non linearity $F(\tilde{\psi}\gamma^0\psi, i\tilde{\psi}\gamma^0\gamma^5\psi)$, we introduce according to G. Lochak [14] the *chiral invariant* of ψ, $\rho(\psi)$ defined by

$$\rho^2=|\tilde{\psi}\gamma^0\psi|^2+|\tilde{\psi}\gamma^0\gamma^5\psi|^2.$$

We are concerned by the spinors ψ for which the chiral invariant is null. In the case of a free solution of the linear Dirac equation, a necessary and sufficient condition to $\rho=0$ is the Majorana condition generalized :

$$\exists z\in\mathbb{C}\ ,\ |z|=1,\ \psi=z\gamma^2\psi^+,$$

where ψ^+ is the complex conjugate of ψ and we have choosen the Dirac matrices in order to γ^2 is symetric :

$$\gamma^2=\begin{pmatrix}0 & \sigma^2\\ -\sigma^2 & 0\end{pmatrix},\quad \sigma^2=\begin{pmatrix}0 & -i\\ i & 0\end{pmatrix}.$$

We have the same result for the Dirac systems with a time dependent potential A

$$-i\gamma^\mu\partial_\mu\psi=A\psi \tag{15}$$

where A satisfies

$$A,\partial_\mu A\in L^\infty_{loc}(\mathbb{R}_t;L^\infty(\mathbb{R}^3_x;\mathbb{R}Id+i\mathbb{R}\gamma^5)). \tag{16}$$

PROPOSITION III.1 - Let ψ be a solution of (15) and $\psi\in C^0(\mathbb{R}_t,(L^2(\mathbb{R}^3_x))^4)$ $\psi(0,x)=\psi_0(x)$. Then the following assertions are equivalent :

i) $\exists z\in\mathbb{C},\ |z|=1,\ \psi_0=z\gamma^2\psi_0^+$;

ii) $\forall x\in\mathbb{R}^3,\ \rho(\psi_0(x))=0$;

iii) $\forall(t,x)\in\mathbb{R}^4,\ \rho(\psi(t,x))=0$;

iv) $\forall(t,x)\in\mathbb{R}^4,\ \exists z\in\mathbb{C},\ |z|=1,\ \psi(t,x)=z\gamma^2\psi^+(t,x)$.

For z=1, the implication i) $\Rightarrow$ iv) was proved by J. Chadam and R. Glassey [6].

IV - GLOBAL EXISTENCE OF LARGE AMPLITUDE SOLUTIONS

First, we consider the mass Dirac-Klein-Gordon system in Minkowski space $\mathbb{R}^{3+1}$:

$$\begin{cases} -i\gamma^{\mu}\partial_{\mu}\psi+M\psi=\varphi V\psi+F(\overline{\psi}\psi,i\overline{\psi}\gamma^{5}\psi)\psi, \\ \Box\varphi+m^{2}\varphi=G(\overline{\psi}\psi,i\overline{\psi}\gamma^{5}\psi)-k\varphi^{3} \end{cases} \quad (17)$$

where the masses M and m are non null

$$M\neq 0,\ m\neq 0. \quad (18)$$

We define the vector space $\mathcal{M}$ of 4×4 matrices

$$\mathcal{M}=\{\alpha I+i\beta\gamma^{5},\ \alpha,\beta\in\mathbb{R}\}.$$

The hypotheses on the non linearities are following :

$$V\in\mathcal{M} \quad (19)$$
$$F\in C^{\infty}(\mathbb{R}^{2},\mathcal{M}),\ \ |F(u,v)|=O(|u|+|v|),|u|+|v|\to 0 \quad (20)$$
$$G\in C^{\infty}(\mathbb{R}^{2},\mathbb{R}),\ \ |G(u,v)|=O(|u|+|v|),|u|+|v|\to 0 \quad (21)$$
$$k\geqslant 0 \quad (22)$$

$\overline{\psi}$ notes the usual Dirac conjugate

$$\overline{\psi}=\tilde{\psi}\gamma^{0}\ . \quad (23)$$

Many models of the relativistic fields theory satisfy these hypotheses ; the scalar and pseudo-scalar Yukawa models of the nuclear forces, the interactions of Heisenberg, Federbusch, the magnetic monopole of G. Lochak.
We choose arbitrarly large Cauchy data in a neighboorhood of decoupling data for which the chiral invariant is null

$$\psi\Big|_{t=0}=\Psi_{0}+\varepsilon\chi_{0}\ ,\ 0<\varepsilon, \quad (24)$$
$$\Psi_{0},\ \chi_{0}\in\mathcal{D}(\mathbb{R}^{3}_{x},\mathbb{C}^{4}), \quad (25)$$

$$\Psi_{0}=z\gamma^{2}\Psi_{0}^{*}\ ,\ z\in\mathbb{C}\ ,\ |z|=1, \quad (26)$$

$$\varphi\Big|_{t=0}=\varphi_{0}\ ,\ \partial_{t}\varphi\Big|_{t=0}=\varphi_{1} \quad (27)$$
$$\varphi_{0},\varphi_{1}\in\mathcal{D}(\mathbb{R}^{3}_{x},\mathbb{R}). \quad (28)$$

According to proposition III.1, assumption (26) implies that Ψ_0 is a decoupling data : for $\varepsilon=0$, the scalar field φ does not depend on ψ and ψ satisfies a linear equation. So we solve the Cauchy problem in a neighbourhood of such a solution. Remark Ψ_0 and scalar field φ can be as large as we want.

THEOREM IV.1 - There exists ε_0 such that for any $0\leqslant\varepsilon\leqslant\varepsilon_0$ the Cauchy problem (17) to (28) has a unique solution (ψ,φ) in $C^\infty(\mathbb{R}^4)$. Moreover, this solution is asymptotically free : there exist $\psi^\pm$, $\varphi^\pm$ satisfying

$$\psi^\pm \in C^0(\mathbb{R}_t,(L^2(\mathbb{R}^3_x))^4),\ -i\gamma^\mu\partial_\mu\psi^\pm+M\psi^\pm=0,$$
$$\varphi^\pm\in C^0(\mathbb{R}_t,H^1(\mathbb{R}^3_x))\cap C^1(\mathbb{R}_t,L^2(\mathbb{R}^3_x)),\Box\varphi^\pm+m^2\varphi^\pm=0,$$
$$\lim_{t\to\pm\infty}\|\psi(t)-\psi^\pm(t)\|_{L^2}+\|\varphi(t)-\varphi^\pm(t)\|_{H^1}+\|\partial_t\varphi(t)-\partial_t\varphi^\pm(t)\|_{L^2}=0.$$

This result shows how complicated is the problem of large amplitude solutions : indeed, we have obtained large solutions asymptotically free but we know that there exist so stationnary solutions for V=0, $F(\overline{\psi}\psi,\overline{\psi}\gamma^5\psi)=\overline{\psi}\psi$, [5].
Now we consider the non linear massless Dirac system

$$-i\gamma^\mu\partial_\mu\psi=F(\overline{\psi}\psi,i\overline{\psi}\gamma^5\psi)\psi, \tag{29}$$

and F verifies (20).
The global Cauchy problem for small initial data was solved by J.P. Dias, M. Figueira [8]. Here we choose larga data satisfying (24) (25) (26).

THEOREM IV.2 - There exists $\varepsilon_0>0$ such that for any $0\leqslant\varepsilon\leqslant\varepsilon_0$, the Cauchy problem (29)(24)(25)(26) has a unique solution $\psi\in C^\infty(\mathbb{R}^4)$ which is asymptotically free : there exists $\psi^\pm$ verifying

$$\psi^\pm\in C^0(\mathbb{R}_t,(L^2(\mathbb{R}^3_x))^4),\ -i\gamma^\mu\partial_\mu\psi^\pm=0,$$
$$\lim_{t\to\pm\infty}\|\psi(t)-\psi^\pm(t)\|_{L^2}=0\ .$$

Moreover ψ and the free solution have the same decay inside the light cone : for $0\leqslant C<1$

$$|\psi(t)|_{L^\infty(\{|x|\leqslant C|t|\})}=O(|t|^{-2}). \tag{30}$$

(30) is a consequence of the important fact following : let's consider Φ defined by

$$\Phi=x_\mu\gamma^\mu\psi\ . \tag{31}$$

Then Φ is solution of a non linear wave equation which allows us to make convenient estimates and we have

$$|\Phi(t)|_{L^\infty(\mathbb{R}^3_x)}=O(|t|^{-1}) \tag{32}$$

that proves (30).
Now we want point out the remarkable properties of asymptotic behaviour of relativistic quantities $\overline{\psi}\psi$ and $\overline{\psi}\gamma^3\psi$. We use another characterization of the compatibility of a sesquilinear form with the massless Dirac system : the factorization formula of B. Hanouzet and J.L. Joly [9] :

PROPOSITION IV.1 - Let f be a sesquilinear form on $\mathbb{C}^4$; the following assertions are equivalent :

i) f is compatible with the massless Dirac system $-i\gamma^\mu\partial_\mu$;

ii) there exist 4×4 matrices $P(x^\mu)$, $Q(x^\mu)$, homogeneous of order -1 such that

$$f(\psi,\psi)=\tilde{\Phi}P\psi+\tilde{\psi}Q\Phi, \tag{33}$$

where Φ is defined by (31).

Theorem I.1 and (32)(33) allow to obtain for non linear system (29) a strong result of equipartition of energy which is well known in the linear case [1] [11] :

THEOREM IV.2 - Let ψ be the solution of (29) given by theorem IV.1. Then the relativistic quantities satisfy :

$$|\overline{\psi}(t)\psi(t)|_{L^1(\mathbb{R}^3_x)}+|\overline{\psi}(t)\gamma^5\psi(t)|_{L^1(\mathbb{R}^3_x)}=O(|t|^{-1}),$$

$$|\overline{\psi}(t)\psi(t)|_{L^\infty(\mathbb{R}^3_x)}+|\overline{\psi}(t)\gamma^5\psi(t)|_{L^\infty(\mathbb{R}^3_x)}=O(|t|^{-3}).$$

We can prove by the same methods the existence of global solutions for massless Dirac equation with cubic relativistic nonlinearity in two space dimension.

CONCLUSION

If we consider mass and massless fields interacting in Minkowski space, there are two difficulties : on the one hand the mass breaks the conformal invariance and we cannot transform the global Cauchy problem in $\mathbb{R}^{3+1}$ into a local Cauchy problem on $S^3\times\mathbb{R}$ by using Penrose transform. On the other hand, the uniform decay of massless field is a priori only t^{-1} and the energy of quadratic nonlinearities does not decay fast enough to assure the global existence.

Nevertheless, we have proved that the global Cauchy problem with small data is well posed for the mass Dirac system quadratically coupled with a massless scalar field if the nonlinearities satisfy the algebraic property of *compatibility* with the Dirac system, related to the Lorentz invariance and the null condition.

If the initial data are large, some blow-up can occur or certain stationnary solutions can exist. In the case of the mass Dirac-Klein-Gordon system with quadratic nonlinearities or the massless Dirac system with cubic nonlinearity, we have proved the existence of global solutions asymptotically free, for arbitrarly large Cauchy data under two assumptions : on the one hand the nonlinearities are Lorentz-invariant, on the other hand the polarization of initial spinor satisfies a generalized Majorana condition ; then the nonlinearities are small again and decay more fast than an ordinary product.

Therefore we constat that in critic cases, quadratic nonlinearities in three space dimension, or large initial data, we can obtain global asymptotically free solutions under algebraic hypotheses on the nonlinearities and the Cauchy data.

REFERENCES

[1] A. Bachelot, *Equipartition de l'énergie pour les systèmes hyperboliques et formes compatibles*. Ann. Inst. Henri Poincaré, Physique Théorique, vol.46, n°1, 1987, pp.45-76.

[2] A. Bachelot, *Problème de Cauchy global pour des systèmes de Dirac-Klein-Gordon*. Ann. Inst. Henri Poincaré, Physique Théorique, vol.48, n°4, 1988, pp.387-422.

[3] A. Bachelot, *Global existence of large amplitude solutions for Dirac-Klein-Gordon systems in Minkowski space*. Colloque international "Problèmes hyperboliques non linéaires", Bordeaux, Juin 1988, to appear in Lecture Notes in Math, Springer-Verlag.

[4] A. Bachelot, *Global existence of large amplitude solutions for nonlinear massless Dirac equation*. To appear in Portugaliae Math.

[5] M. Balabane, T. Cazenave, A. Douady, F. Merle, *Existence of excited states for a nonlinear Dirac field*. To appear.

[6] J. Chadam, R. Glassey, *On certain global solutions of the Cauchy problem for the (classical) coupled Klein-Gordon-Dirac equations in one and three space dimensions*. Arch. Rat. Mech. Anal. 54, 1974, pp.223-237.

[7] Y. Choquet-Bruhat, D. Christodoulou, *Existence of global solutions of the Yang-Mills, Higgs and spinor field equations in 3+1 dimensions*. Ann. Scient. Ec. Norm. Sup. 4ème série, t.14, 1981, pp.481-500.

[8] J.P. Dias, M. Figueira, *Sur l'existence d'une solution globale pour une équation de Dirac non linéaire avec masse nulle*. C.R. Acad. Sci. Paris, série I, t.305, 1987, pp.469-472.

[9] B. Hanouzet, J.L. Joly, *Applications bilinéaires sur certains sous-espaces de type Sobolev*. C.R. Acad. Sci. Paris, série I, t.294, 1982, pp.745-747.

[10] B. Hanouzet, J.L. Joly, *Explosion pour des problèmes hyperboliques semi-linéaires avec second membre non compatible*. C.R. Acad. Sci. Paris, t.301, n°11, 1985, pp.581-584.

[11] B. Hanouzet, J.L. Joly, *Applications bilinéaires compatibles avec un système hyperbolique*. Ann. Inst. Henri Poincaré, Analyse non linéaire, t.4, n°4, 1987, pp.357-376.

[12] S. Klainerman, *Global existence of small amplitude solutions to nonlinear Klein-Gordon equations in four space-time dimensions*. Comm. Pure and Appl. Math., t.38, 1985, pp.631-641.

[13] S. Klainerman, *The null condition and global existence to nonlinear wave equations*. Lectures in Appl. Math., t.23, 1986, pp.293-326.

[14] G. Lochak, *Wave equation for a magnetic monopole*. Int. J. Theor. Phys. 24, n°10, 1985, pp.1019-1050.

Statistical Mechanics of the NLS Models andTheir Avatars

R. K. Bullough†, Yu–zhong Chen†, S. Olafsson†#) and J. Timonen‡

†) Department of Mathematics, UMIST, PO Box 88, Manchester M60 1QD, UK

‡) Department of Physics, University of Jyväskylä, SF–40100, Jyväskylä, Finland

#) Currently at Department of Mathematics, University of Southampton

Abstract "In Vishnuland what avatar? Or who in Moscow (Leningrad) towards the czar [1]". The different manifestations (avatars) of the Nonlinear Schrödinger equation (NLS models) are described including both classical and quantum integrable cases. For reasons explained the sinh-Gordon and sine-Gordon models, which can be interpreted as covariant manifestations of the 'repulsive' and 'attractive' NLS models, respectively, are chosen as generic models for the statistical mechanics. It is shown in the text how the quantum and classical free energies can be calculated by a method of functional integration which uses the classical action-angle variables on the real line with decaying boundary conditions, even though we define the finite density thermodynamic limit by imposing periodic boundary conditions. Both this method and an equivalent method of 'generalised Bethe ansatz' exploit the classical complete integrability of the models, and in quantum form are manifestly fermi-bose equivalent. The state of current knowledge of quantum and classical integrability is reviewed. Algebraic manifestations of the NLS models as supersymmetric NLS models are described. The fermi-bose equivalence of the quantum models has a natural algebraic origin.

1 Introduction: the Avatars of the NLS Models

The classical nonlinear Schrödinger equation (NLS model) in one space and one time (1+1) dimension is

$$-i\phi_t = \phi_{xx} - 2c\phi^*\phi^2, \tag{1}$$

where $\phi(x,t) \in \mathcal{C}, x \in \mathcal{R}$, and ϕ^* is complex conjugate of ϕ. It is integrable, i.e., it can be solved by the inverse scattering or inverse spectral transform (ST) method [2,3]. It is Hamiltonian with

$$H[\phi] = \int dx \left[\phi_x^*\phi_x + c(\phi^*)^2\phi^2\right] \tag{2}$$

and Poisson bracket $\{\phi, \phi^*\} = i\delta(x-x')$. Hamilton's equations $\phi_t = \{H, \phi\}$ are equation (1). It is 'completely integrable' [2,4,5], i.e. it has a complete set of independent constants commuting under the bracket. Because $\phi(x,t)$ is a field, a continuous infinity of such constants must be found: they have been — action-angle type variables $P(k)$ can be found and the $P(k)$, $k \in \mathcal{R}$, commute under the bracket [2,4].

The number $c \in \mathcal{R}$ and is a classical coupling constant. It determines three different manifestations ('avatars' [6]) of the NLS models — the case $c>0$ is the 'repulsive' NLS, and the case $c<0$ is 'attractive'. The terminology is clarified by the quantum models below. The case $c = 0$ is the linear Schrödinger equation (LS): it has no self-potential and is the equation of a free non-relativistic particle.

The quantum forms of the NLS models use operators $\hat{\phi}, \hat{\phi}^{+}$, and $\hat{H}$. They are normally ordered so that, for quantum repulsive NLS anyway, there is a well defined ground state. Heisenberg's equations of motion and commutator are

$$-i\hat{\phi}_t = \hat{\phi}_{xx} - 2c\hat{\phi}^{+}\hat{\phi}^{2} \quad ; \quad [\hat{\phi}, \hat{\phi}^{+}] = \delta(x - x'). \tag{3}$$

This is the second quantised form, with $\hat{N} = \int dx \hat{\phi}^{+}\hat{\phi}$, of the Schrödinger wave mechanical problem

$$\sum_{i=1}^{N}[-\frac{\partial^2}{\partial x_i^2} + c\sum_{j\neq i}\delta(x_i - x_j)]\Psi(\{x_i\}) = E\Psi(\{x_i\}). \tag{4}$$

The wave mechanical problem is solved by the *Bethe ansatz* (BA) wave functions [7,8]

$$\Psi(\{x_j\}) = \sum_{P} a(P)\ e^{i\sum_{j=1}^{N} k_{P_j} x_j} \quad ; \quad x_1<x_2<\ldots<x_N, \tag{5a}$$

in which $a(P) = \Pi e^{i\Delta_{jj'}}$ and $e^{i\Delta_{jj'}}$ is the 2-body S-matrix phase shift for each of the transpositions $j \leftrightarrow j'$ making up the permutation P. Bose symmetry is imposed so that $\Psi(\{x_j\})$ is also defined for $x_2<x_1<\ldots<x_N$, etc. . The phase shifts

$$e^{i\Delta_{jj'}} = -\frac{c - i(k_j - k_{j'})}{c + i(k_j - k_{j'})} \quad ; \quad \Delta_{ij} = -2\tan^{-1}\frac{c}{k_i - k_j}. \tag{5b}$$

The smooth branch $-2\pi<\Delta_{ij}<0$ is taken — see below. Periodicity on $x_j \to x_j + L$ then requires

$$Lk_i = 2\pi n_i + \sum_{j\neq i}\Delta_{ij} \quad ; \quad i = 1, \ldots, N \quad , \tag{6}$$

and the n_i are integers.

The wave numbers k_i solving (6) are the only wave numbers (N of them) allowed. For $L \to \infty$ and $N \to \infty$ with $NL^{-1}>0$, a *finite density* limit, both the k_i and the wave numbers $2\pi n_i L^{-1}$ are dense and distinct. When $c<0$, additional 'string' solutions arise generalising (6) [7]. When $c>0$, (3) or (4) is the problem of N bosons on a line with repulsive δ-function interactions of strength c [9,10], hence the terminology above. The BA wave functions (5) solve (4) exactly [9]: $E = \sum_{i=1}^{N} k_i^2$ and the k_i label this eigenstate. Yang and Yang [10] used the same BA method to solve the quantum statistical mechanics (SM). Evidently the conditions (5) are crucial to both of these calculations.

With this as background the purpose of the present paper is partly to describe in brief the two new and equivalent methods we have devised recently to solve for the quantum *and* classical free energies. They apparently apply to "all" of the integrable models [5,11–13]. Our method of functional integration is actually sketched here, and the method of 'generalised BA' [5,11–14] is only briefly mentioned for comparisons. Both methods have an explicit bose-fermi equivalence [7,15], and we show how the classical SM derives naturally from the boson form.

The word 'avatar' appeared in the prolegomena to this meeting and, shortly, we make a small digression into Hindu mythology to explain it. All the *many* 'avatars' of the NLS models are completely integrable and the main purpose of this paper is really to exhibit some of these avatars put into the contexts of 'classical integrability' and 'quantum integrability'. The authors' current understanding of these two properties is exhibited in the Figure 1 called 'Solitons'.

Our functional integration route [5,11,12] to the quantum and classical partition functions Z of the integrable models in 1+1 is shown in Fig. 1 (top left to top right and down). We use §§ 2 and 3 to fill in some details of this route. We shall use a *covariant* manifestation of the NLS models as 'generic' for the route.

Three classical covariant models are the s-G model [11], the sinh-G model [12], and the classical massive Thirring model (MTM) with commutative bracket [14,16]. These models have equations of motion and Hamiltonians:

(i)Sine-Gordon model

$$H[\phi]=\frac{1}{\gamma_0}\int dx\,[\frac{1}{2}\gamma_0^2\Pi^2+\frac{1}{2}\phi_x^2+m^2(1-\cos\phi)];\quad \phi_{xx}-\phi_{tt}=m^2\sin\phi, \tag{7}$$

where real positive γ_0 is a coupling constant, and $\{\pi,\phi\}=\delta(x-x')$.

(ii) Sinh-Gordon (sinh-G) model

$$H[\phi]=\frac{1}{\gamma_0}\int dx[\frac{1}{2}\gamma_0^2\Pi^2+\frac{1}{2}\phi_x^2+m^2(\cosh\phi-1)];\quad \phi_{xx}-\phi_{tt}=m^2\sinh\phi, \tag{8}$$

with the same bracket as s-G. By canonical transformation $\Pi\to\Pi\gamma_0^{-\frac{1}{2}}$, $\phi\to\phi\gamma_0^{1/2}$ and analytic continuation $\gamma_0\to-\gamma_0$ sinh-G is found from s-G. Indeed sinh-G is 'repulsive' and s-G 'attractive' in the sense of the two NLS models with $c>0, c<0$.

(iii) The classical MTM model with commutative Poisson bracket is [14,16]

$$(-i\partial_\nu\gamma^\nu+m)\phi+g^2\gamma^\nu\phi(\bar\phi\gamma_\nu\phi)=0\ , \tag{9a}$$

$$\phi=\begin{bmatrix}\phi_1\\ \phi_2\end{bmatrix}\ ,\quad \delta_{\mu\nu}=\begin{bmatrix}1&0\\0&-1\end{bmatrix}\ ,\quad \gamma^0=\begin{bmatrix}0&1\\1&0\end{bmatrix}\ ,\quad \gamma^1=\begin{bmatrix}0&-1\\1&0\end{bmatrix} \tag{9b}$$

with $\gamma_\mu=\delta_{\mu\nu}\gamma^\nu$; $\mu,\nu=0,1$. The field $\phi\in C$ and $\bar\phi=\phi^*\gamma^0$; g is real and g^2 is a coupling constant. It has a soliton solution which is 'breather like' and very similar to the soliton solution of the attractive NLS [2,14,16]. The Hamiltonian and bracket of this MTM are

$$H[\phi]=\int dx\left[-\frac{1}{2}\bar\phi\gamma^1\overleftrightarrow{\partial}_x\phi+m\bar\phi\phi+\frac{1}{2}g^2(\bar\phi\gamma_\nu\phi)(\bar\phi\gamma^\nu\phi)\right];\ \{\phi_i,\bar\phi_j\}=\delta_{ij}\delta(x-x'). \tag{10}$$

with $i,j=0,1$. When negative energy particles are ignored [14] H is positive definite for $\frac{1}{2}g^2=-c>0$ and its action-angle variables then pass smoothly to those of the attractive NLS in non-relativistic limit. Reference to the classical action-angle variables of sinh-G and s-G shows these two models have properties similar to those of the 'repulsive' and 'attractive' NLS, respectively. However, it is the rest mass m of the covariant models which makes their classical SM tractable [17]. In this sense the covariant models become generic.

With reference to classical integrability and the Lax pair (Fig. 1) it is unnecessary to say anything here. The reader is referred to [2–4] and the recent [18,19]. For the Sklyanin bracket one may refer to [4,5,20].

In our functional integral method we work with the classical action, do not introduce operators, and appear to use only the classical r-matrix. However, the quantum R-matrix and the Yang–Baxter relations [5,21] underlie the theory: they define YBZF algebras or 'quantum groups' [22] emphasizing the algebraic basis of the theory.

We now turn briefly aside to define 'avatar' and thereby introduce still other 'avatars' of the NLS models. According to [6] (the square brackets denote our insertions) 'Avatar' means: The advent to earth of a deity in visible form. The ten *avăta'ras* of Vishnu (evidently the NSL model) in Hindu mythology are by far the most celebrated. 1st advent in the form of a fish [LS model?]; 2nd advent, in that of a tortoise [classical attractive NLS, for it was arguably the 2nd advent amongst the integrable models [2]]; 3rd of a hog [repulsive NLS]; 4th, of a monster, half man and half lion, to destroy the giant Iranian [quantum spin$-\frac{1}{2}$ XYZ model, see below]; 5th in the form of a dwarf [quantum attractive NLS, hard to give it its full structure!]; 6th in human form, under the name of Rama [classical sinh-G, very human in its SM]; 7th, under the same figure and name, to slay the thousand-armed giant Cartasuciriargunan [classical s-G, $-\gamma_0 \to \gamma_0$; the giant destroyed is given in [11]]; 8th, as a child named Krishna, who performed numerous miracles (this is the most memorable of all the advents) [quantum repulsive NLS; miracles are in [7–9]]; 9th, under the form of Buddha [classical Landau-Lifshitz (L-L') model — see below: if one does not mind parametrisation by Jacobian elliptic functions, the classical L-L' model *contains* all of s-G, sinh-G and the NLS models [4,14,20]: Sklyanin first introduced the r-matrix in [20]: it should indeed be worshipped; its classical SM is in [14]]. These are all past. The 10th advent will be in the form of a white horse (Kalki) with wings, to destroy the earth [the assignment of Kalki must remain open!].

These remarks, humourous or not, introduce further avatars of the NLS models: spin$-\frac{1}{2}$ XYZ is

$$H = -\sum_{\alpha,n} J_\alpha S_n^\alpha S_{n+1}^\alpha \quad ; \quad \alpha = x, y, z \tag{11}$$

with $[S_n^\alpha, S_m^\beta] = \delta_{nm}\epsilon^{\alpha\beta\gamma}S_n^\gamma$. This H commutes with the row-to-row transfer matrix of the solvable 8-vertex model in 2+0 dimensions ([23], Fig. 1). It also maps to quantum MTM by Jordan-Wigner transformation and continuum limit [2]. This model is fermi-bose equivalent to quantum s-G [2]. The classical and continuum limit of spin$-\frac{1}{2}$ XYZ is the L-L' model

$$\vec{S}_t = \vec{S} \times \vec{S}_{xx} + \vec{S} \times J \cdot \vec{S} \quad , \quad \vec{S} = (S^1, S^2, S^3) \quad , \quad |\vec{S}| = 1 \tag{12}$$

with $J = \mathrm{diag}(J_1, J_2, J_3)$; $J_1 < J_2 < J_3$. This reduces to the continuous classical Heisenberg ferromagnet in the isotropic case $J_1 = J_2 = J_3$; and this is related to attractive NLS by gauge transformation [24].

Before the SM we make two further remarks. The integrable NLS in 2+1 dimensions is the Davey-Stewartson (D-S) equations ([3] and Fig. 1)

$$i\phi_t + \frac{1}{2}(\phi_{yy} + \alpha^2\phi_{xx}) = -\alpha^2\beta|\phi|^2\phi + w\phi$$

$$w_{xx} - \alpha^2 w_{yy} = 2\alpha^2\beta(|\phi|^2)_{xx} \tag{13}$$

with α=1 or i and $\beta = \pm 1$. This system is solved by an inverse ST through a non-local Riemann-Hilbert or 'D-bar' problem [3,18,25]. When $w_{yy} = \phi_{yy} = 0, w = 2\alpha^2\beta|\phi|^2$: with $\frac{1}{2}\alpha^2 t \to t$ this is the NLS model (1) with $c = \beta$. Reference [26] reports action-angle variables for the D-S system (13) in 2+1. However, we have not found a non-trivial classical or quantum SM for this model yet. Other models in 2+1 [26] offer the same difficulty.

The second remark is that Fig. 1 shows bose-fermi equivalence inherent from the big infinite dimensional loop algebra gl(∞). This algebra has two realisations, one in the Lie algebra of bilinear free fermion operator products, the other in terms of vertex operators on a bosonic space [15]: Fermi-bose equivalence is made *manifest* in the SM of repulsive NLS and sinh-G and in a different way in the equivalence of quantum MTM and s-G [2]. This feature indicates the importance of the underlying algebras to the theory.

2 General Strategy for Free Energies

We take the quantum partition function Z as the functional integral

$$Z = Tr \int \mathcal{D}\Pi\mathcal{D}\phi \exp S[\phi], \quad S[\phi] = i\int_0^\beta d\tau \int dx \,[\Pi(x,\tau)\phi_\tau(x,\tau) - H[\phi]]\,. \tag{14}$$

For an integrable lattice this becomes an infinite product. $S[\phi]$ is the Wick rotated classical action evaluated on the space-time torus $-\frac{1}{2}L \leq x < \frac{1}{2}L,\ 0 \leq \tau < \beta,\ \beta^{-1} = T =$ temperature; Π, ϕ are classical canonical variables. The Tr in (14) requires periodicity in τ, the thermodynamic limit is achieved by periodic b.c.s in x, period $L \to \infty$. Z in (14) is then the Wick rotated Feynman propagator in Hamiltonian form: it is defined on a symplectic manifold (a phase space). Integral (14) embraces Feynman's integral in terms of the Lagrangian: if $H[\phi]$ has a density in the form $\alpha\Pi^2 + V[\phi(x,\tau)]$, then $Z = \int \mathcal{D}\phi \exp S[\phi]$ with $S[\phi] = \int_0^\beta d\tau \int dx(-\frac{1}{2}\phi_\tau^2 + V[\phi])$ on a space-time torus, and the measure $\mathcal{D}\phi$ is Feynman's measure. The measure $\mathcal{D}\Pi\mathcal{D}\phi$ in (14) is actually easier to use than Feynman's.

Our line of calculation is to exploit the complete integrability of the classical integrable models and rewrite (14) in the form

$$Z = \int \mathcal{D}\mu \exp S[p]. \tag{15}$$

Here $S[p]$ is the classical action expressed in terms of the classical action-angle variables. We choose these to be the action-angle variables for the model defined on the real line with vanishing b.c.s at $\pm\infty$ for these are relatively easy to find [4,27,28]. The proper finite density thermodynamic limit can still be found by choosing the measure $\mathcal{D}\mu$ appropriately. That this is so can be proved by discretizing the functional integral (14) to an (integrable) lattice, finding the action-angle variables for it under periodic b.c.s, connecting these with action-angle variables on the real line as $L \to \infty$ and determining

$\mathcal{D}\mu$ in the process. All these steps have been carried out for s-G, sinh-G, NLS and L-L' models [5,11–14]; currently there are problems with the Toda lattice in taking the thermodynamic limit [17].

We quote here the now well known [2,4,5,27,28] action-angle variables for s-G, sinh-G and the two NLS models on the real line with vanishing b.c.s: for s-G

$$H[p] = \sum_{i=1}^{N_K+N_{\bar K}} (M^2 + p_i^2)^{1/2} + \sum_{j=1}^{N_B} (4M^2 \sin^2\theta_j + \hat p_j^2)^{\frac{1}{2}} + \int_{-\infty}^{\infty} dk\omega(k)P(k); \tag{16}$$

$M = 8m\gamma_0^{-1}$, $\omega(k) = (m^2 + k^2)^{\frac{1}{2}}$. The brackets are $\{p_i, q_i\} = \delta_{ij}$, etc.; $\{4\gamma_0^{-1}\theta_j, \Phi_{j'}\} = \delta_{jj'}$; $\{P(k), Q(k')\} = \delta(k - k')$. For sinh-G

$$H[p] = \int_{-\infty}^{\infty} dk\omega(k)P(k), \tag{17}$$

in which $\omega(k)$, $P(k)$ and $Q(k)$ are as for s-G. For NLS, $c<0$,

$$H[p] = \sum_{j=1}^{N} (c\hat p_j^2 \theta_j^{-1} - \frac{1}{12c}\theta_j^3) + \int_{-\infty}^{\infty} dk\omega(k)P(k) \tag{18}$$

and $\omega(k) = k^2$. The brackets are $\{\hat p_j, \hat q_{j'}\} = \delta_{jj'}$, $c^{-1}\{\theta_j, \Phi_{j'}\} = \delta_{jj'}$, $\{P(k), Q(k')\} = \delta(k - k')$. For NLS, $c>0$, the Hamiltonian is (17) with $\omega(k) = k^2$. There are still other avatars of the NLS models, covariant or not. For $H[p]$ for repulsive NLS is the non-relativistic limit of (17) for sinh-G when $m = \frac{1}{2}$ (in appropriate units). Then (18) for attractive NLS is the non-relativistic limit of (16) for s-G providing the kink and antikink contributions (sum to $N_K+N_{\bar K}$) are simply discarded. It is a feature of the MTM with commutative bracket that it has in many respects a more natural non-relativistic limit which is the attractive NLS [14]:

$$H[p] = \sum_{+,-} \int_{-\infty}^{\infty} dk\omega_\pm(k)P_\pm(k) + \sum_{j=1}^{N_B} E_j \tag{19}$$

with $\omega_\pm(k) = \pm(k^2+m^2)^{\frac{1}{2}}$, and this is bounded below if, and only if the negative energy phonons are rejected. Then since $E_j = (\hat p_j^2 + M_j^2)^{\frac{1}{2}}$ and $M_j = 2mg^{-2}\sin(\frac{1}{2}g^2\Pi_j)$, in which Π_j is an action variable, one finds [14], under the condition $\theta_j \ll m$, that $H[p]$ is (18) with $\theta_j = 2cm\Pi_j$: the phase shifts pass smoothly to those of the attractive NLS under the same conditions [14]. The condition $\theta_j \ll m$ puts a lower bound on the Hamiltonian (18) and it is possible to compute a partition function and free energy: the result, in classical limit, is our final result (26) given below [14].

In classical limit the partition function is (14) but $S[\phi] = \int_0^\beta d\tau(-H[\phi]) = -\beta H[\phi]$ (for each x $\int_0^\beta d\tau \Pi\phi_\tau = \hbar^{-1}\int^{\beta\hbar} d\tau\Pi\phi_\tau \to 0$ as $\hbar \to 0$). The *classical* functional integral (14) can be evaluated by the transfer integral method (TIM) [8,11,12]. For s-G the problem reduces to finding the smallest eigenvalue of a Mathieu equation. Our result for this eigenvalue, found by matched asymptotic expansion methods, yields the free energy [11,13]

$$FL^{-1} = F^{(1)} + F^{(2)} + F^{(3)} + \cdots + F_{KG} \quad , \tag{20}$$

where, with $M = 8m\gamma_0^{-1}$ and $t = (M\beta)^{-1}$, $\beta = T^{-1}$,

$$\begin{aligned} F^{(1)} = & -\beta^{-1}m(8/\pi t)^{\frac{1}{2}}e^{-1/t}\big[1 - \frac{7}{8}t - \frac{59}{128}t^2 - \frac{897}{1024}t^3 - \cdots\big] \\ & -\beta^{-1}m\big[\frac{1}{4}t + \frac{1}{8}t^2 + \frac{3}{16}t^3 + \frac{53}{128}t^4 + \frac{594}{2^9}t^5 + \cdots\big] \ , \end{aligned} \tag{21a}$$

$$F^{(2)} = \frac{8m}{\pi}Me^{-2/t}\{\ell n(\frac{4C}{t}) - \frac{5}{4}t\big[\ell n(\frac{4C}{t}) + 1\big] - \cdots\} \ . \tag{21b}$$

Here ℓnC=0.5771... (Euler's constant), $F^{(q)} = O\left(e^{-q/t}\right)$, $q = 3, 4, \ldots$, and $F_{KG} = \beta^{-1}a_0^{-1}\left[\ell n(\beta a_0^{-1}) + \frac{1}{2}ma_0\right]$. Note that because of the choice of classical SM there is the classical ultraviolet divergence: lim $a_0 \to 0$ cannot be taken in F_{KG} (a_0 is a lattice parameter introduced in order to discretize Z, thus to define it, and then to calculate it).

The same problem arises for the sinh-G model which is otherwise much simpler, for there are no soliton contributions. Our result by the TIM finds the smallest eigenvalue of the Sturm-Liouville problem for the sinh-G which is the same as that for the s-G with $(1 - \cos\phi)$ replaced by $(\cosh\phi - 1)$. From this we find [12]

$$FL^{-1} = -\beta^{-1}m\left[\frac{1}{4}t - \frac{1}{8}t^2 + \frac{3}{16}t^3 - \frac{53}{128}t^4 + \cdots\right] + F_{KG}. \tag{22}$$

Note that this is precisely the analytical continuation $\gamma_0 \to -\gamma_0$ from (21), (22).

The application of the TIM to the repulsive NLS models raises problems not yet completely solved [30]. This is why we choose the covariant models as generic.

3 Functional Integral Method Using Action-Angle Variables

We have to evaluate Z in the form (15). Despite the results in § 2 it is still worth doing this in classical limit in order ultimately to get connection with the quantum theories: we need $Z = \int \mathcal{D}\mu \exp(-\beta H[p])$.

Crucial results found by Floquet theory applied to a classical integrable sinh-Gordon *lattice* with spacing a_0 under periodic b.c.s period $L = Na_0$ are [31], with $\omega(k) = (m^2 + k^2)^{\frac{1}{2}}$,

$$L^{-1}H[p] = L^{-1}\sum_{n=-\frac{1}{2}N}^{\frac{1}{2}N}\omega(\tilde{k}_n)P_n + O(L^{-1}) \tag{23}$$

which means that $L^{-1}H[p] = O(1) + O(L^{-1})$ as $L \to \infty$ and $P_n = O(1)$;

$$\tilde{k}_n = k_n - L^{-1} \sum_{m \neq n} \Delta_c(\tilde{k}_n, \tilde{k}_m) P_m, \tag{24}$$

where $k_n = 2\pi n L^{-1}$, $\Delta_c(k,k') = -\frac{1}{4}\gamma_0 m^2 \left(k\omega(k') - k'\omega(k)\right)^{-1}$. Note how (24) generalises (6) to the classical problem. The action variables relate directly to the action variables $P(k)$ of sinh-G (see below).

For both sinh-G and repulsive NLS integrable lattice theories show that the natural measure is [12] $\mathcal{D}\mu = \lim_{N\to\infty} (2\pi)^{-N-1} \prod_{n=-\frac{1}{2}N}^{\frac{1}{2}N} dP_n dQ_n$. Since $\hbar = 1$, $h = 2\pi$: there is no normalisation problem with this measure as there is for the Feynman measure.

Note that $\omega(\tilde{k}_n)$ in $H[p]$, (23), depends on the P_n through (24). Thus, from this viewpoint, the oscillators making up (23) are *nonlinear* oscillators (from the viewpoint of our generalised BA they are large amplitude phonons [32]). With (24) Z can be evaluated by iterating $\omega(\tilde{k}_n) = \omega(k_n)$ + terms in P_n. Apparently this iteration is formal only for classical repulsive NLS; there are no problems for the classical sinh-G model [12]. The result of this iteration is that [12]

$$\frac{F}{L} = \frac{1}{2\pi\beta} \int dk \ell n\left(\beta\omega(k)\right) - \left(\frac{1}{2\pi\beta}\right)^2 \int \frac{dq}{\omega(q)} \int dk \Delta_c(k,q) \frac{d}{dk} \ell n \omega(k) - \cdots. \tag{25}$$

Here as in what follows all integrals are from $-\infty$ to ∞ unless marked otherwise. But (25) is exactly the iteration of

$$\epsilon(k) = \omega(k) + \frac{1}{2\pi\beta} \int \frac{d\Delta_c}{dk} \ell n(\beta\epsilon(k')) dk' \ , \quad \frac{F}{L} = \frac{1}{2\pi\beta} \int \ell n(\beta\epsilon(k)) dk \tag{26}$$

with $\Delta_c = \Delta_c(k,k')$. Moreover one can evaluate all the integrals in (25) and regain the TIM result (22) exactly term by term. Thus the strictly asymptotic expansion (22) is summable to (26) — despite the problem of the u.v. divergence: the system (26) is defined by formally evaluating the first iterate for FL^{-1} with $a_0 > 0$ and this yields F_{KG}: all other terms can be found exactly and agree with (22) term by term when $a_0 \to 0$!

There is now a pair of bose-fermi equivalent forms for the quantum theories — namely (bose) [12]

$$\epsilon(k) = \omega(k) + \frac{1}{2\pi\beta} \int \frac{d\Delta_b}{dk} \ell n(1 - e^{-\beta\epsilon(k')}) dk', \quad \frac{F}{L} = \frac{1}{2\pi\beta} \int \ell n(1 - e^{-\beta\epsilon(k)}) dk. \tag{27}$$

This can be formed directly from the quantum functional integral with quantisation conditions $P_n = 0, 1, 2, \cdots$ or by our method of 'generalised BA' [12]. The 'generalised BA' follows [10] and defines an entropy and free energy and minimises this [12].

The fermi form of (27) is [12]

$$\frac{F}{L} = \mu\bar{n} - \frac{1}{2\pi\beta} \int \ell n \left(1 + e^{-\beta\bar{\epsilon}(k)}\right) dk \tag{28a}$$

$$\bar{\epsilon}(k) = \omega(k) - \mu - \frac{1}{2\pi\beta} \int \frac{d}{dk} \Delta_f(k,k') \ell n \left(1 + e^{-\beta\bar{\epsilon}(k')}\right) dk' \tag{28b}$$

in which μ is a chemical potential and $\bar{n} = \lim_{L\to\infty} NL^{-1}>0$. This result can be found directly from the quantum functional integral with quantisation conditions $P_n = 0, 1$ or by our method of 'generalised BA' using a fermion description. In this case except that we are treating sinh-G, not repulsive NLS, the argument follows Yang and Yang [10]. Thus for sinh-G (and NLS) the generalised BA generalises from the fermion to boson description and to the classical problem and results (26). Note that when $P_n = 0, 1$ (24) is (6) provided the classical phase shift Δ_c is replaced by the fermi 2-body S-matrix phase shift Δ_f. For sinh-G

$$\Delta_f = -2\tan^{-1}\left\{m^2 \sin\left[\frac{1}{8}\gamma_0''/(k\omega(k') - k'\omega(k))\right]\right\} \tag{29}$$

with [12] $\gamma_0'' = \gamma_0/(1+\gamma_0/8\pi)$. The smooth branch $-2\pi<\Delta_f<0$ is taken. This relates to the bose phase shift Δ_b through

$$\Delta_b = \Delta_f + 2\pi\theta(k'-k), \tag{30}$$

where $\theta(x)$ is the unit step function. The relation (30) is exactly that needed to relate (28) (with $\mu = 0$) to (27): (27) and (28) are then equivalent and together display bose-fermi equivalence. When $\gamma_0 \to 0$ (classical limit) the bose form Δ_b (with a singularity at $k = k'$) becomes the classical phase shift Δ_c given in (24). The classical results (26) derive from the bose form in the same way. Note that the bose phase shift Δ_b and the classical phase shift Δ_c both have the property $\Delta \to 0$ as $|k-k'| \to \infty$: Δ_f does not have this property.

We must comment on the quantisation procedure. The classical action $S[\phi]$ in (14) becomes

$$S[p] = i\int_0^\beta d\tau \int dk\,[P(k)Q_\tau(k,\tau) - H[p]] \tag{31}$$

in (15). For real free energy (real ℓnZ) it is more-or-less necessary that $\int_0^\beta d\tau P_n Q_{n,\tau} = \oint P_n dQ_n = 2\pi m_n$ with m_n an integer. The action-angle variables P_n, Q_n used already in (23) and (24) relate to $P(k)$, $Q(k)$ as $P_n \leftrightarrow P(\tilde{k}_n)d\tilde{k}_n$, $Q_n = Q(\tilde{k}_n)$ so $P_n = m_n$. Since $d\tilde{k}_n = O(L^{-1})$ and P_n is $O(1)$ $P(\tilde{k}_n) = O(L)$ as $L \to \infty$. This is the crucial feature of the thermodynamic limit: the $P(k)$ in (17) for sinh-G are bounded for vanishing b.c.s on the real line. They are not bounded in the thermodynamic limit taken through periodic b.c.s. Thus the former is at zero density: the latter is at finite density. Indeed in the method of generalised BA $L^{-1}P(k)$ defines a finite particle density $\rho(k)>0$ [12].

The quantisation above is Bohr quantisation. But it is not trivial quantisation. The R-matrix and the Yang-Baxter relations determine the quantum 2-body S-matrix phase shifts Δ_f or Δ_b [12]: These then enter into (24) instead of Δ_c with the choices $P_n = 0, 1$ or $P_n = 0, 1, 2, \cdots$ for the fermi or bose descriptions. Thus quantum integrability is transferred from the quantum functional integral to the condition (24) imposed on the functional integral (and the role of the functional integral becomes one of minimising the quantum free energy — as is plain in a comparison with our generalised BA method).

This completes the sketch of calculating quantum and classical Z's by the functional integral method for sinh-G. For repulsive NLS $\omega(k) = k^2$, and $\Delta_f = -2\tan^{-1}(c/(k-k'))$ as in (5b): then Δ_b is defined through (30) and $\Delta_c = -2c/(k-k')$, the usual limit $c \to 0$

of Δ_b. Although all three systems of integral equations identical to (26), (27) and (28), respectively, apparently apply (with $\omega(k) = k^2$ and the appropriate Δ's), the actual iteration of the classical system (26) offers difficulties for repulsive NLS. This is why sinh-G is chosen as generic for all cases.

Both methods functional integral and generalised BA extend to quantum and classical s-G, and the other integrable models. The generalised BA results are given in [14] and these and other results will be reported at length.

One interesting result is the classical SM for the ferromagnetic ($J_1 = J_2 = J_3 = J$) in (12). We find [14,33]

$$\begin{aligned} \frac{F}{L} &= \frac{1}{2\pi\beta}\int dk\ \ell n\,(\beta^*\epsilon(k)) - J(1+h) \\ \epsilon(k) &= k^2 + h + \frac{1}{2\pi\beta}\int \frac{d}{dk}\Delta_c(k,k')\ell n(\beta^*\epsilon(k'))dk' \end{aligned} \tag{32}$$

with $\beta^* = J\beta$, h a transverse magnetic field, and

$$\Delta_c(k,k') = J^{-1}kk'(k-k')^{-1} \quad , \tag{33}$$

the classical magnon-magnon phase shift (it is a feature of our analysis that all of the classical phase shifts like (33) emerge through the analytical properties of the classical spectral data and (33) is found this way: note that (33) has coupling constant J^{-1} and does not depend on $(k - k')$ alone). The system (32) iterates. With the scaled temperature $s = (\beta J h^{\frac{1}{2}})^{-1}$ (32) yields the low temperature asymptotic expansion

$$FL^{-1} = \beta^{-1}a_0^{-1}\left[\ell n(\pi^2 J\beta a_0^{-2}) - 2\right] - J(1+h) + Jh(s - \frac{1}{4}s^2 - \frac{1}{64}s^3 + \frac{5}{512}s^4 + \cdots) \tag{34}$$

which agrees with, and extends, the result [34] found by the TIM. The expressions (32) agree in form with those we find for the *attractive* NLS [14]. These results are [14] expressions (26) with the phase shift Δ_c

$$\Delta_c(k,k') = 2c(k-k')^{-1}, \quad c>0, \tag{35}$$

which depends on $k - k'$ alone. The results (26) follow formally from (32) for $h \to 0$ (with a shift of energy): however Δ_c, given by (35), for the attractive NLS model is not the same as (33) for the ferromagnet. The asymptotic expansion of (32), namely (34), is not defined for $h \to 0$ since s is not defined. The system (26) for the attractive NLS does not iterate and we have not found any low temperature asymptotic expansion (compare [30]). These particular considerations apart the ferromagnet continues to play its role as an avatar of the attractive NLS model.

Note that (26) is exactly what we find for the classical SM of the *repulsive* NLS except that $c \to -c$ in the phase shift Δ_c Eqn. (35). This classical system also does not iterate and we have no simple low temperature asymptotic expansion. It is remarkable that both attractive NLS with breather-like solitons and phonons and repulsive NLS with no solitons and only phonons (see (17) and (18)) have formally equivalent classical SM. A similar situation obtains for s-G [32] where breathers act like interacting phonons in the classical SM (and the original phonons are eliminated [32]). The analysis of the classical SM for the NLS models is given in part in [14] and will be reported elsewhere.

4 Algebraic Manifestations of the NLS Models

We get another view of the classical integrability and find still another avatar of the NLS model by following the route shown in the Fig. 1 from the box 'loop algebras' through 'KSA theorem' and back to the classical 'integrable model(s) (=NEE)'. The 'NEE' means 'nonlinear evolution equation'. These are of the form (in the field $\phi(x,t)$) $\phi_t = K[\phi]$ where $K[\phi]$ is a nonlinear functional of ϕ: the NLS is of this form; s-G (in light-cone coordinates) is e.g. $\phi_t = \int_{-\infty}^{x} \sin\phi dx$. The general form is [3]

$$U_t - V_x + [U,V] = 0 \tag{36}$$

with U, V in the algebra $sl(N, \mathcal{C})$ [3].

We appeal first of all to the KSA (Kostant-Symes-Adler) theorem, given in its original form [35,19] for a finite number of degrees of freedom, and to the observation of Kirillov [19] that the coadjoint orbits of a simple Lie group G form a symplectic manifold endowed with a bracket. KSA [35] show that it is a direct sum splitting of the algebra g (dual g^*) of G which is responsible for complete integrability of finite dimensional systems associated with g. We extend to the infinite dimensionsal loop algebras $\hat{g} = g \otimes \mathcal{C}[\lambda, \lambda^{-1}]$, centre free Kac-Moody Lie algebras [36] (e.g. $\hat{g} = s\ell(N,\lambda)$), with dual $\hat{g}^* = g^* \otimes \mathcal{C}[\lambda, \lambda^{-1}]$. Several authors have made this connection with the classical integrable models [37]. The approach sketched here uncovers (we believe) some new aspects. Let $\hat{g} = \sum g^{(j)}$, $g^{(j)} = g \otimes \lambda^j$: $\hat{g}$ has elements $X(\lambda) = \sum X_j \lambda^j$. Different gradings can be defined on $\hat{g}$: we use only $[X(\lambda), Y(\lambda)] = \sum_{i,j,} [X_i, X_j] \lambda^{i+j}$. Split $\hat{g} = g_0^+ + g^-$, where $g_0^+ = \sum_{j \geq 0} g^{(j)}$, $g^- = \sum_{j<0} g^{(j)}$: these are subalgebras in $\hat{g}$; $[g_0^+, g_0^+] \subseteq g_0^+$, $[g^-, g^-] \subseteq g^-$. The infinite dimensional algebra $\hat{g}$ is equipped with a non-degenerate bilinear form $(X,Y) = \mathrm{Res}(<X,Y>\lambda^{-1})$, where $<,>$ is an extension of the Killing form on g to $\hat{g}$ [38]: Res means 'residue' ($<X,Y>$ is a Laurent polynomial in λ). With respect to this form one can identify $(g_0^+)^*$, the dual of g_0^+, with $g_0^- = \sum_{j \leq 0} g^{(j)}$: g_0^- is equipped with a Poisson bracket structure [37]

$$\{f,g\}(X) = (X, [P \bigtriangledown f, P \bigtriangledown g]) \quad ; \quad f, g \in \mathcal{C}(g_0^-), \tag{37}$$

where P is the projection $P: \ \hat{g} \to g^-$ [36,37].

The KSA theorem may be stated as follows:– Let A be a ring of adinvariant polynomials on g_0^-. Then every element $\phi \in A$ gives rise [35,39] to a vector field of the form $\Delta_\phi(X) = [\bigtriangledown\phi(X), X]$: $\bigtriangledown$ is defined through (,) by $d\phi(X) = (\bigtriangledown\phi(X), dX)$. Consider the countably infinite set of functions defined on g_0^- as $\rho_k(X) = -\mathrm{Res}(\lambda^{-k-1}<X,X>)$. A simple calculation [19] shows they are in involution with respect to the bracket (37) and define an infinite set of commuting flows. Moreover, because they are adinvariant on g_0^-, we find that equivalently $[\bigtriangledown\rho_k, \bigtriangledown\rho_\ell] = 0$. The adinvariance also means by the KSA theorem that they define an infinite set of vector fields of the form ($k \in \mathcal{Z}_+$)

$$\Delta_k(X) = [P \bigtriangledown \rho_k(X), X], \text{ or } \partial_k X = [P_k, X] \tag{38}$$

where $P_k = P \bigtriangledown \rho_k = -\sum_{j \geq k+1} X_j \lambda^{k-j} \in g^-$, and $\partial_k = \partial_{t_k}$: the parameter t_k is associated with the vector field Δ_k. We find further that

$$\partial_n P_k - \partial_k P_n + [P_k, P_n] = 0 \quad ; \quad n, k \in Z_+ \ . \tag{39}$$

This vanishing curvature condition evidently generalises (36): $t_1 \leftrightarrow x$, $t_n \leftrightarrow t$; $P_k, P_n \in g^- \leftrightarrow U, V \in s\ell(N, C)$. Thus (38), second expression, represents a hierarchy of integrable nonlinear flows in 1+1 in the usual Lax pair form [2] and (39) is the Zakharov-Shabat form [3].

To gain further insight into these NEEs we can expand both sides of Eqns. (38) in powers of λ. In this way we eventually find that the hierarchy of NEEs in the infinite number of times t_n for $\hat{X}_1$ are

$$\partial_n \hat{X}_1 = \sum_{j=0}^{n} D_x^j [\hat{X}_1, \tilde{X}_{n-1}], \tag{40}$$

where $D_x = \partial_x - adX_1$, x is identified with t_1, and $\tilde{X}_j$ and $\hat{X}_j$ are the diagonal and off-diagonal parts of X_j, respectively: $X = \sum_{j \geq 0} X_j \lambda^{-j} \in g_0^-$. To find the fields contained in these NEEs we have to specify the basis for g. Let it be E_a, F_a, H_a; $a = 1, \ldots,$ rank(g) : $[H_a, E_b] = K_{ab} E_b$, $[H_a, F_b] = -K_{ab} F_b$, $[E_a, F_b] = \delta_{ab} H_a$, and K_{ab} is the Cartan matrix for g. With $\hat{X}_j = e_j^a E_a + f_j^a F_a$, $\tilde{X}_j = h_j^a H_a$, we find for the flows and recursion relations

$$e_n^a = \partial_x e_{n-1}^a + e_1^a h_{n-1}^b K_{ab} = \partial_{n-1} e_1^a \tag{41a}$$

$$f_n^a = \partial_x f_{n-1}^a - f_1^a h_{n-1}^b K_{ab} = \partial_{n-1} f_1^a \tag{41b}$$

$$h_n^a = I_x \left(e_1^a \partial_x f_{n-1}^a - f_1^a \partial_x e_{n-1}^a \right) . \tag{41c}$$

By using the recursion relations one can reduce the flows to hierarchies of NEEs in 2·rank(g) variables e_1^a, f_1^a. By working out the hamiltonian structure [19] we find the Hamiltonian densities for the hierarchies (41a,b) are given by $h_k = \sum_{a,b} h_0^a K_{ab} h_k^b$. Thus, e.g. if $g = A_1 (= s\ell(2, C))$, $\partial_{k-2} e = \{H_k, e\}$, $\partial_{k-2} f = \{H_k, f\}$ with new Hamiltonians $H_k = i \int_{-\infty}^{\infty} h_k dx$ and bracket $\{e(x),\ f(x')\} = i\delta(x - x')$: $\{H_k, H_\ell\} = 0$; $e^a, f^a = e_1^a, f_1^a$. When $k = 4$, in particular we regain the NLS equation (1). Indeed as expected [2] $g = s\ell(2, C)$ yields all of the AKNS systems [2,3]. If we use the Cartan-Weyl basis for g we find other hierarchies in dim$(g)-$rank(g) variables more complicated than (41) and these also include the AKNS systems for $g = s\ell(2, C)$. Thus the loop algebra $s\ell(2, \lambda) \otimes C[\lambda, \lambda^{-1}]$ is an algebraic manifestation of the NLS model in this rather comprehensive sense.

Moreover the scheme developed here can be extended [40,41] to include several super-algebras (as Fig. 1, boxes SUSY, indicates). In Ref. [40] it is found that the equations related to the super-algebra $B(0, 1)$ lead to super versions of the NLS equation [42] and of the Korteweg-de Vries (KdV) and modified KdV (mKdV) equations [43].

The equations related to the SUSY algebra $s\ell(2, 1)$ contain [40] a generalisation of the super-NLS system with cubic nonderivative interactions in two complex spinor fields. The system of equations *reduces* to e.g.

$$\begin{aligned} -ie_t &= ae_{xx} - 2aeee^* - c\chi\chi^* e + b\psi\psi^* e - e\psi^*\chi_x - b\psi^*_x\chi \\ -i\psi_t &= b\psi_{xx} + aee^*\psi + c\chi\chi^*\psi - ce^*\chi_x - ae^*_x\chi \\ -i\chi_t &= c\chi_{xx} - aee^*\chi + b\psi\psi^*\chi - be\psi_x + a\psi e_x \end{aligned} \tag{42}$$

where a, b, c are constants and the χ, ψ are anticommuting fields. For $\chi = \psi = 0$ and $a = 1$, (42) is the NLS model (1) with $c = 1$.

The algebraic manifestations of the NLS models show that Hindu mythology scarcely does justice to the avăta'ras of Vishnu. There are considerably more than 10! [44].

References

[1.] Browning R. Poems

[2.] Bullough RK, Caudrey PJ. Solitons. (Springer, Heidelberg 1980), and references

[3.] Caudrey PJ. Physica 6D:51 (1982); In: Fordy AP (ed.) Soliton Theory a Survey of Results. (Manchester UP, Manchester 1989), and references

[4.] Faddeev LD, Takhtajan LA. Hamiltonian Methods in the Theory of Solitons. (Springer, Berlin 1987)

[5.] Bullough RK, Pilling DJ, Timonen J. In: Lakshmanan M (ed.) Solitons. (Springer, Heidelberg 1988)

[6.] Brewer, The Reverend Ebenezer Cobham, 1810–1897. The Dictionary of Phrase and Fable, p. 77. (Avenel Books, New York 1978)

[7.] Thacker HB. Rev.Mod.Phys. 53:253 (1981), and references

[8.] Bullough RK, Pilling DJ, Timonen J. In: Claro F (ed.) Nonlinear Phenomena in Physics. (Springer, Berlin 1985)

[9.] Lieb EH, Liniger W. Phys.Rev. 130:1605 (1963)

[10.] Yang CN, Yang CP. J.Math.Phys. 10:1115 (1969)

[11.] Timonen J, Stirland M, Pilling DJ, Cheng Yi, Bullough RK. Phys.Rev.Lett. 56:2233 (1986)

[12.] Bullough RK, Pilling DJ, Timonen J. J.Phys. A19:L955 (1986)

[13.] Timonen J, Bullough RK, Pilling DJ. Phys.Rev. B34:6525 (1986)

[14.] Chen Yu-zhong. Ph.D. Thesis. University of Manchester (submitted March 1989), and references

[15.] Jimbo M, Miwa T. In: D'Ariano CM, Montorsi S, Rasetti MG (eds.) Integrable Systems in Statistical Mechanics. (World Scientific, Singapore 1985)

[16.] Kuznetsov EA, Mikhailov AV. Teor.Mat.Fiz. 30:303 (1977)

[17.] Bullough RK, Pilling DJ, Cheng Yi, Chen YZ, Timonen J. In: Fordy AP, Degasperis A (eds.) Nonlinear Evolution Equations: Integrability and Spectral Methods. (Manchester UP, Manchester 1989)

[18.] Bullough RK, Olafsson S. In: Truman A, Davies B (eds.), Proc. of the 9th Congress of the International Association of Mathematical Physicists, Swansea 1988. (Adam Hilger, Bristol 1989)

[19.] Bullough RK, Olafsson S. In: Solomon A (ed.) Proc. of the 17th International Conference on Differential Geometric Methods in Theoretical Physics, Chester 1988. (World Scientific, Singapore 1989)

[20.] Sklyanin E. LOMI Preprint E–3–1979 (1979)

[21.] Izergin AG, Koregin VE. Lett.Math.Phys. 5:199 (1981)

[22.] de Vega HJ. Lectures given at the Sao Paolo School on Strings, Integrable and Conformal Theories, (Brazil 1987)

[23.] Baxter RJ. Exactly Solved Models in Statistical Mechanics. (Academic Press, New York 1982)

[24.] Takhtajan LA. Phys.Lett. 64A:235 (1977); Lakshmanan M. Phys.Lett. 61A:53 (1977)

[25.] Bullough RK, Olafsson S. To be published

[26.] Jiang Zhuhan. Ph.D. Thesis, University of Manchester (1987)

[27.] Dodd RK, Bullough RK. Physica Scripta 20:514 (1979)

[28.] Tahtajan LA, Faddeev LD. Theor.Mat.Fiz. 21:160 (1979)

[29.] Stirland M, Bullough RK. To be published

[30.] Bullough RK, Pilling DJ, Timonen J. In: Takeno S (ed.) Dynamical Problems in Soliton Systems. (Springer, Berlin 1985) p. 105

[31.] Cheng Yi. Ph.D. Thesis, University of Manchester (1987)

[32.] Timonen J, Chen YZ, Bullough RK. Nucl.Phys. B, 5A:58 (1988)

[33.] Chen YZ, Bullough RK, Timonen J, Tognetti V, Vaia R. To be published

[34.] Nakamura K, Sasada T. J.Phys. C11:L171 (1978)

[35.] Kostant B. Advances in Mathematics 34:195 (1979); Symes W. Inventiones Math. 59:13 (1980); Adler M. Inventiones Math. 50:219 (1979)

[36.] Kac V. Infinite Dimensional Lie Algebras — An Introduction. (Cambridge UP, Cambridge 1985)

[37.] Reyman AG, Semenov-Tian-Shansky MA. Inventiones Math. 54:81 (1979); 63:423 (1981); Flaschka H, Newell AC, Ratin J. Physica 9D:300 (1983)

[38.] Frenkel IB, Kac VG. Inventiones Math. 62:23 (1980)

[39.] Adler M, van Moerbecke P. Advances in Mathematics 38:267 (1980)

[40.] Olafsson S. J.Phys. A22:157–167 (1989)

[41.] Olafsson S. Loop Algebras and Their Relation to the Conformal Structure of Integrable Systems. To be published

[42.] Kulish PP. ICTP Preprint IC/85/38 (1985)

[43.] Kuperschmidt BA. Phys.Lett. 102A:213 (1984)

[44.] Radhakrishnan S. Bhagavadgita 7 edn. (Blackie & Son, Bombay 1982)

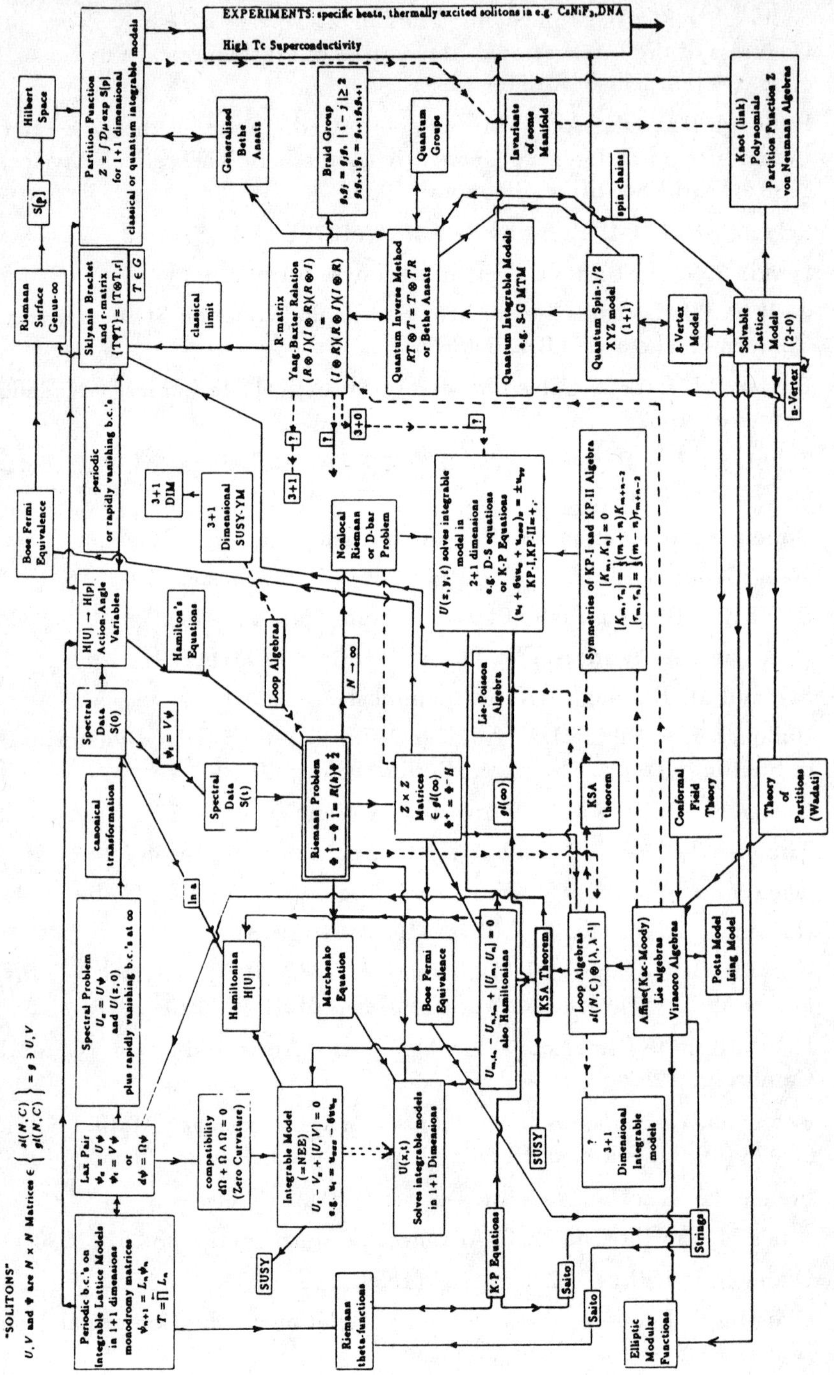

Fig. 1. Overview of generalised 'Soliton' theory as of date (August, 1988). A hard arrow indicates that a minimal connection (at least) between the 'boxes' is already achieved: dashed arrows indicate expectation by the authors that such connection can be achieved, or better. Note the *experimental* output from the partition function Z, top right!

ASYMPTOTIC BEHAVIOR OF SOLUTIONS AND UNIVERSAL ATTRACTORS FOR A SYSTEM OF NONLINEAR HYPERBOLIC EQUATIONS

Piotr Biler

Mathematical Institute, University of Wrocław, Poland
and (1987/88)
Laboratoire d'Analyse Numérique, Université de Paris-Sud

The system of two semilinear Schrödinger and Klein-Gordon equations

$$i\psi_t + \Delta\psi + i\varepsilon\psi = -\psi\phi + F$$

$$\phi_{tt} + \delta\phi_t - \Delta\phi = |\psi|^2 + G$$

defined on a bounded domain in $\mathbb{R}^n$, $n=1,2,3$, describes the dynamics of a complex nucleon field ψ and a real meson field ϕ coupled through the Yukawa interaction. The small dissipation is introduced to this system by the zeroth order terms $i\varepsilon\psi$ and $\delta\phi_t$, $\varepsilon, \delta > 0$ and F, G are the exterior forces. We consider the initial-boundary value problem for the system above with the initial data $\langle\psi, \phi, \phi_t\rangle(0) \in V = H^1_o \times H^1_o \times L^2$ or in the space $W = (H^2 \cap H^1_o)\times(H^2 \cap H^1_o)\times H^1_o$, $\psi=0$, $\phi=0$ on the boundary.

The classical Hamiltonian system with $\varepsilon=\delta=0$, $F=G=0$ was studied in [F-T] and [H-W]. These authors have proved, using Galerkin and integral equations methods respectively, the global in time solvability of the Cauchy problem. However their approach did not give any significant information on the asymptotics of the solutions in the space W. Our techniques based on energy inequalities and nonlinear interpolation theory lead to the following result

PROPOSITION 1. The unique global in time solution of the conservative homogeneous system with the initial data in V is bounded in V: $\| \langle\psi(t), \phi(t), \phi_t(t)\rangle \|_V = O(1)$ as t tends to ∞. For more regular data in W $\| \langle\psi(t), \phi(t), \phi_t(t)\rangle \|_W = O(t^2)$ if $n=3$, $=O(t^{1+c})$ for any $c>0$ if $n=2$ and $=O(t)$ if $n=1$.

The next case in our study is the damped homogeneous system with $\varepsilon, \delta>0$ and $F=G=0$. Here the solutions converge to zero exponentially and the optimal rates of decay are obtained following the methods developed in [B1].

PROPOSITION 2. For the homogeneous system with sufficiently small damping parameter δ in the Klein-Gordon equation ($\delta^2/4 <$ the first eigenvalue of $-\Delta$) $\|\psi(t)\|_{H^1} = O(\exp(-\varepsilon t))$, $\|\phi(t)\|_{H^1} = O(\exp(-\min(2\varepsilon, \delta/2)t))$ for $t \to +\infty$ except for the resonance case $2\varepsilon = \delta/2$ where $\|\phi(t)\|_{H^1} = O(t\exp(-\delta t/2))$.

The main subject is a study of the trajectories of the dynamical system $S(t)$ in V or in W associated to the full system of equations. These trajectories are captured by a bounded set A in V. More precisely $\lim \operatorname{dist}(S(t)B,A) = 0$ for each set B of initial data bounded in V, as $t \to +\infty$. The universal attractor A constructed as $\bigcap_{s\geq 0} \overline{\bigcup_{t\geq s} S(t)B_R}$, where B_R is a sufficiently large ball in V, consists of all globally ($t\to\pm\infty$) bounded trajectories. A is $S(t)$-invariant and the property described above justifies the name "attractor" . The most important geometric property of A is given in the following

THEOREM. The attractor A is compact in V and its fractal dimension is finite.

These results may be interpreted as follows: the dynamics of the dynamical system associated to those equations is governed by a finite number of modes as t tends to $+\infty$. The estimate of the dimension of A is given in terms of the uniform Lyapunov exponents. This technique presented in [CFT] was generalized to more difficult situations in [G-T] and [G1], [G2]. Our system has intermediate properties between two important physical models: systems of two Klein-Gordon equations and of two Schrödinger equations. Our results and techniques have also an intermediate character: first the attractor is constructed in the weak topology of V (like for damped nonlinear Schrödinger equations) and then it is shown to be bounded in W, hence compact in V (like for damped nonlinear Klein-Gordon equations).

Finally we consider the singular perturbation problem for the system

$$i\psi_t + \Delta\psi + i\varepsilon\psi = -\psi\phi + F$$

$$\beta^2\phi_{tt} + \beta\phi_t - \Delta\phi = |\psi|^2 + G$$

where ε, $\beta \to 0$, $\beta \leq \varepsilon$, $\varepsilon = O(\beta)$, $\|F\| = O(\varepsilon)$, $\|G\| = O(\beta)$. Physically this corresponds to the infinite limit of the velocity of propagation of disturbances in the Klein-Gordon equation, so to an instantaneous response of ϕ to variations of ψ. It is shown that the solutions of the system converge in a weak sense to the solutions of the non-linear Schrödinger equation

$$i\psi_t + \Delta\psi = \psi\Delta^{-1}(|\psi|^2).$$

The proof is similar to that in [S-W]. The estimates of dimension of the corresponding attractors $A(\varepsilon,\beta)$ grow to infinity as $\varepsilon,\beta \to 0$ but the sets $A(\varepsilon,\beta)$ stay in a bounded subset of the phase space V. It would be interesting to reveal some relations between the attractors $A(\varepsilon,\beta)$ of the slightly damped system and the time asymptotics of the solutions to the limit equation. Possibly $A(\varepsilon,\beta)$ approximate in a certain sense an invariant measure on the energy levels of the conservative equation.

Some other results on the asymptotic behavior of solutions are given for the models with higher order Yukawa interaction terms $-p|\psi|^{2p-2}\psi\phi$, $|\psi|^{2p}$, $p \geq 1$, and stronger dissipation terms of Ginzburg-Landau type $-i\mu\Delta\psi$ and on the regular exponential decay of the solutions of the corresponding homogeneous system with $F=G=0$.

The crucial role in obtaining all energy estimates for the damped system is played by functionals corresponding to the invariants (integrals) of the conservative system, cf [B2].

REFERENCES

[B1] P. Biler, Exponential decay of solutions of damped nonlinear hyperbolic equations, Nonlinear Analysis 11, 1987, 841-849.

[B2] P. Biler, Attractors for the system of Schrödinger and Klein-Gordon equations with Yukawa coupling, Prépublications Université de Paris-Sud, 88-01, 1-40.

[CFT] P. Constantin, C. Foiaş, R. Temam, Attractors representing turbulent flows, Memoirs of A.M.S. vol. 53, 1985, 314.

[F-T] I. Fukuda, M. Tsutsumi, On coupled Klein-Gordon-Schrödinger equations II, J. Math. Anal. Appl. 66, 1978, 358-378.

[G1] J.-M. Ghidaglia, Finite dimensional behavior for weakly damped driven Schrödinger equations, Ann. IHP, Analyse Non Linéaire 5, 1988, 365-405.

[G2] J.-M. Ghidaglia, Upper bounds on the Lyapunov exponents for dissipative perturbations of infinite dimensional Hamiltonian systems, this volume.

[G-T] J.-M. Ghidaglia, R. Temam, Attractors for damped nonlinear hyperbolic equations, J. Math. pures appl. 66, 1987, 273-319.

[H-W] N. Hayashi, W. von Wahl, On the global strong solutions of coupled Klein-Gordon-Schrödinger equations, J. Math. Soc. Japan 39, 1987, 489-497.

[S-W] S.H. Schochet, M.I. Weinstein, The nonlinear Schrödinger limit of the Zakharov equations governing Langmuir turbulence, Comm. Math. Phys. 106, 1986, 569-580.

SOLITONS IN TWO DIMENSIONS

M. Boiti*, J.JP. Leon, L. Martina*, F. Pempinelli*
Laboratoire de Physique Mathématique, USTL - F 34060 Montpellier

We display here the soliton solutions to the following nonlinear evolution equations in two dimensions:

i) the Davey Stewartson equation [1]

$$iQ_t = -\frac{1}{2}\sigma_3(Q_{xx}+Q_{yy}) + \sigma_3 Q^3 + [Q, A]$$

$$(\partial_x + \sigma_3\partial_y)A = \partial_y(Q^2) \quad , \tag{1}$$

ii) the third order equation [2][3]

$$Q_t = \frac{1}{4}Q_{xxx} + \frac{3}{4}Q_{xyy} - \frac{3}{2}Q^2Q_x - \frac{3}{2}\sigma_3[Q, A]_x - \frac{3}{4}\{Q_y, A\} + \frac{3}{4}[Q, B]$$

$$(\partial_x + \sigma_3\partial_y)A = \partial_y(Q^2) \quad ,$$

$$(\partial_x + \sigma_3\partial_y)B = -\sigma_3[Q_{xy}, Q] \quad . \tag{2}$$

The 2 x 2 matrix field Q is off-diagonal

$$Q = \begin{pmatrix} 0 & q(x,y,t) \\ r(x,y,t) & 0 \end{pmatrix} \quad . \tag{3}$$

The spectral transform solution of these equations proceeds through the two dimensional Zakharov Shabat problem solved in [4]:

$$(\partial_x + \alpha\sigma_3\partial_y + Q)\psi = 0 \quad ; \quad Q \to 0 \ , \ x^2 + y^2 \to \infty \quad ; \tag{4}$$

$$\psi \exp[-ik(\sigma_3 x - y)] \to \mathbb{1} + O(\tfrac{1}{k}) \quad , \quad |k| \to \infty \tag{5}$$

We have considered this problem in [3] on the ground that the "zero order" solution ψ_0 of

$$(\partial_x + \sigma_3\partial_y)\psi_0 = 0 \tag{6}$$

can be chosen as

$$\psi_0 = \begin{pmatrix} \alpha_1(x-y) & 0 \\ 0 & \alpha_2(x+y) \end{pmatrix} e^{+ik(\sigma_3 x - y)} \tag{7}$$

---

* Permanent address: Dipartimento di Fisica, Università di Lecce, 7310 Lecce, Italy.

where α_1 and α_2 are two arbitrary functions of their arguments going to 1 as $|k|$ goes to infinity.

We display here the soliton solutions to (1) and (2) on the basis of the Bäcklund transform method [5].

The simplest gauge transformation $B = B(Q', Q)$ which generates the Bäcklund transform Q' of Q (see [6]) is given by the equation

$$B(Q', Q) = a\,\partial_y - \frac{1}{2}\,\sigma_3(Q'a - aQ) - \\ - \frac{1}{2}\,\sigma_3\, a\, J(Q'^2 - Q^2) + b \quad . \tag{8}$$

The operator $J = (\partial_x + \sigma_3\,\partial_y)^{-1}$ is defined with the boundary requirement that, in a given direction, $\lim JM = 0$ for any diagonal (well behaved at infinity) matrix $M = M(x, y)$. a, b are 2×2 constant diagonal matrices.

For $a = \begin{pmatrix} 1 & 0 \\ 0 & 0 \end{pmatrix}$, $b = \begin{pmatrix} \lambda & 0 \\ 0 & 1 \end{pmatrix}$ we get the so-called elementary Bäcklund gauge of first kind $B^{(I)}(Q', Q\,;\,\lambda)$, for $a = \begin{pmatrix} 0 & 0 \\ 0 & 1 \end{pmatrix}$, $b = \begin{pmatrix} 1 & 0 \\ 0 & \mu \end{pmatrix}$ the elementary Bäcklund gauge of second kind $B^{(II)}(Q', Q\,;\,\mu)$ and for $a = \begin{pmatrix} 1 & 0 \\ 0 & 1 \end{pmatrix}$, $b = \begin{pmatrix} \lambda & 0 \\ 0 & \mu \end{pmatrix}$ the Bäcklund gauge $B(Q', Q\,;\,\lambda, \mu)$ which generates the soliton.

We are interested in the solution Q that can be derived by comparing two elementary Bäcklund gauges and by starting from the zero solution. The main property to be used is the commutativity of Bäcklund gauges which is given in the specific case by the formula

$$B(Q, 0\,;\,\lambda, \mu) = B^{(II)}(Q, Q^{(I)}\,;\,\mu)\, B^{(I)}(Q^{(I)}, 0\,;\lambda) = \\ = B^{(I)}(Q, Q^{(II)}\,;\,\lambda)\, B^{(II)}(Q^{(II)}, 0\,;\,\mu) \tag{9}$$

It results that

$$Q^{(I)} = \begin{pmatrix} 0 & 0 \\ r^{(I)} & 0 \end{pmatrix} \quad , \quad Q^{(II)} = \begin{pmatrix} 0 & q^{(II)} \\ 0 & 0 \end{pmatrix} \quad , \quad Q = \begin{pmatrix} 0 & q \\ r & 0 \end{pmatrix} \tag{10}$$

where

$$r^{(I)} = \rho(x+y\,,\,t)\, e^{-\lambda(x-y)} \quad , \quad q^{(II)} = \eta(x-y, t)\, e^{\mu(x+y)} \tag{11}$$

$$q = \frac{q_y^{(II)} + (\lambda - \mu)\, q^{(II)}}{1 + \frac{1}{4} r^{(I)} q^{(II)}} \quad , \quad r = \frac{r_y^{(I)} - (\lambda - \mu)\, r^{(I)}}{1 + \frac{1}{4} r^{(I)} q^{(II)}} \tag{12}$$

with

$$\rho(x+y\,,\,t) = \iint d\ell \wedge d\bar{\ell}\; e^{-i\ell(x+y)}\, \tilde{\rho}(\ell\,,\,t)$$

$$\eta(x-y,t) = \iint d\ell \wedge d\bar{\ell}\, e^{i\ell(x-y)}\,\tilde{\eta}(\ell,t) \tag{13}$$

The time evolution is given by the equations

$$\tilde{\rho}(\ell,t) = \tilde{\rho}(\ell,o)\exp[\omega(\ell)t + \omega(-i\lambda)t]$$

$$\tilde{\eta}(\ell,t) = \tilde{\eta}(\ell,o)\exp[-\omega(\ell)t - \omega(-i\mu)t] \tag{14}$$

where $\tilde{\rho}(\ell,o)$ and $\tilde{\eta}(\ell,o)$ are arbitrary functions and where $\omega(k)$ is the dispersion relation: $\omega = ik^2$ for eq. (1) and $\omega = ik^3$ for eq. (2). The explicit form of $B(Q,0;\lambda,\mu)$ is

$$B(Q,0;\lambda,\mu) = \mathbb{I}\,\partial_y + B^{(1)}(Q,0;\lambda,\mu) \quad , \tag{15}$$

$$B^{(1)}(Q,0;\lambda,\mu) = \begin{bmatrix} \lambda - \frac{1}{4}r^{(I)}q & -\frac{1}{2}q \\ \frac{1}{2}r & \mu - \frac{1}{4}q^{(II)}r \end{bmatrix}$$

By applying the gauge operator $B(Q,0;\lambda,\mu)$ to $e^{ik(\sigma_3 x - y)}$ we get the following eigenfunction relative to Q in (10)

$$\psi(x,y,k) = \left\{ \mathbb{I} - \frac{i}{4}\begin{bmatrix} \dfrac{r^{(I)}q}{k+i\lambda} & \dfrac{2q}{k+i\mu} \\ -\dfrac{2r}{k+i\lambda} & \dfrac{q^{(II)}r}{k+i\mu} \end{bmatrix} \right\} \exp[ik(\sigma_3 x - y)] \tag{16}$$

The spectral transform of Q can be explicitly computed [6]:

$$R(k,\ell) = i\pi \begin{bmatrix} 0 & \delta(k+i\mu)\,\tilde{\eta}(\ell,t)(\ell+i\lambda) \\ -\delta(k+i\lambda)\,\tilde{\rho}(\ell,t)(\ell+i\mu) & 0 \end{bmatrix} \tag{17}$$

and , consequently, the solution Q is indeed related to discrete eigenvalues of the spectral problem.

In order to get from the general solution Q a soliton solution we have to choose the arbitrary functions $\rho(x+y,0)$ and $\eta(x-y,0)$ in such a way that Q is localized in the space at any time. This can be easily done in the reduced case

$$r = \varepsilon\bar{q} \quad , \quad \varepsilon \in \mathbb{R} \ . \tag{18}$$

If we introduce the following functions

$$\rho(x+y,0) = 2\exp[-\mu(x+y)]\,S(\sigma)$$

$$\eta(x-y,0) = 2\exp[\lambda(x-y)]\,T(\tau) \tag{19}$$

with

$$\sigma = \frac{1}{\mu + \bar{\mu}} \exp[(\mu + \bar{\mu})(x + y)]$$

$$\tau = \frac{\varepsilon}{\lambda + \bar{\lambda}} \exp[-(\lambda + \bar{\lambda})(x - y)] \qquad (20)$$

the reduction condition (18) at $t = 0$ becomes the following functional equation

$$[1 + \bar{S}(\sigma)\,\bar{T}(\tau)]\,\frac{dS}{d\sigma}(\sigma) = [1 + S(\sigma)\,T(\tau)]\,\frac{dT}{d\tau}(\tau) \qquad (21)$$

It admits the following four different solutions

$$\begin{cases} S = a\sigma + b \quad , \quad b\bar{a} - \bar{b}a = 0 \\ T = \bar{a}\tau + c \quad , \quad ca - \bar{c}\bar{a} = 0 \end{cases} \qquad (22)$$

$$\begin{cases} S = a\sigma + b \\ T = \dfrac{1}{a\bar{b} - \bar{a}b}\,(\bar{a} + \bar{c}\exp[(\bar{a}b - a\bar{b})\tau]) \\ |c|^2 = |a|^2 \end{cases} \qquad (23)$$

$$\begin{cases} S = \dfrac{1}{\bar{a}b - a\bar{b}}\,(a + c\exp[(a\bar{b} - \bar{a}b)\sigma]) \\ T = \bar{a}\,\tau + \bar{b} \\ |c|^2 = |a|^2 \end{cases} \qquad (24)$$

$$\begin{cases} S = \dfrac{b}{a}\exp[-a\,\sigma] \quad , \quad a + \bar{a} = 0 \\ T = c\,|c|^2\,a\exp[-\dfrac{1}{a|c|^2}\,\tau] \\ |c|^2 = |b|^2 \end{cases} \qquad (25)$$

with a, b and c arbitrary complex constants subjected to the indicated conditions. Only in the first case, with a convenient choice of the constants, one gets a localized solution.

Therefore in the following we choose

$$\tilde{\rho}(\ell, o) = \rho[\delta(\ell + i\mu) + \delta(\ell - i\bar{\mu})]$$

$$\tilde{\eta}(\ell, o) = \eta[\delta(\ell + i\lambda) + \delta(\ell - i\bar{\lambda})] \qquad (26)$$

with η and ρ arbitrary real constants.

We get for the one-soliton solution

$$q = \frac{2\lambda_R\,\eta\exp[i\varphi]}{D}$$

$$r = \frac{2\mu_R\, \rho \exp[-i\varphi]}{D}$$

$$D = \gamma \exp(\xi_1) + \gamma \exp(-\xi_1) + (1+\gamma) \exp(\xi_2) + \gamma \exp(-\xi_2)$$

$$\gamma = \frac{1}{4}\eta\rho$$

$$\varphi = (\mu_I + \lambda_I)x + (\mu_I - \lambda_I)y - [\omega_I(-i\lambda) + \omega_I(-i\mu)]t$$

$$\xi_1 = (\lambda_R + \mu_R)x - (\lambda_R - \mu_R)y - [\omega_R(-i\lambda) + \omega_R(-i\mu)]t$$

$$\xi_2 = (\lambda_R - \mu_R)x - (\lambda_R + \mu_R)y - [\omega_R(-i\lambda) - \omega_R(-i\mu)]t \tag{27}$$

For $\lambda_R \mu_R \neq 0$ and $\gamma(1+\gamma) > 0$ the above formula defines (up to the phase factor $\exp[\pm i\varphi]$ in the numerator) a 2-dimensional bell-shaped solution, exponentially decreasing in all directions of the (x, y)-plane, moving without deformation with velocity $\vec{v} = (v_x, v_y)$

$$v_x = \frac{1}{2\lambda_R \mu_R}[\mu_R \omega_R(-i\lambda) + \lambda_R \omega_R(-i\mu)]$$

$$v_y = \frac{1}{2\lambda_R \mu_R}[-\mu_R \omega_R(-i\lambda) + \lambda_R \omega_R(-i\mu)] \tag{28}$$

In general the soliton is the envelope of the plane wave $\exp[\pm i\varphi]$. The initial position of the soliton can be moved arbitrary by the translation

$$x \to x - x_o \quad , \quad y \to y - y_o \quad , \quad x_o, y_o \in R \tag{29}$$

In general ψ solves the following integral equations

$$\psi^{(+)}(k,x,y) = \psi_o - \int_{-\infty}^{x} dx' \begin{pmatrix} q\psi_{21}^{(+)} & q\psi_{22}^{(+)} \\ 0 & r\psi_{12}^{(+)} \end{pmatrix} (x', y-(x-x')\sigma_3) +$$

$$+ \int_{x}^{\infty} dx' \begin{pmatrix} 0 & 0 \\ r\psi_{11}^{(+)} & 0 \end{pmatrix} (x', y-(x-x')\sigma_3) =$$

$$\doteq \psi_o + G_{xy}^{(+)}\psi^{(+)}(k,.,.) \quad , \quad \operatorname{Im} k > 0 \quad , \tag{30}$$

$$\psi^{(-)}(k,x,y) = \psi_o - \int_{-\infty}^{x} dx' \begin{pmatrix} q\psi_{21}^{(-)} & 0 \\ r\psi_{11}^{(-)} & r\psi_{12}^{(-)} \end{pmatrix} (x', y-(x-x')\sigma_3)$$

$$+ \int_{x}^{\infty} dx' \begin{pmatrix} 0 & q\psi_{22}^{(-)} \\ 0 & 0 \end{pmatrix} (x', y-(x-x')\sigma_3) =$$

$$\div \quad \phi_o + G^{(-)}_{xy}\, \phi^{(-)}(k,.,.) \quad , \qquad \mathrm{Im}\, k < 0 \quad , \tag{31}$$

where ϕ_o is defined in (7).

The eigenfunction corresponding to the soliton solution (27) solves the above equations with ϕ_o given by (7) and with

$$\alpha_1 = 1 - \frac{2\, i\, \lambda_R \gamma}{k + i\lambda} \; \frac{1}{\gamma + (\theta(\mu_R) + \gamma)\exp[\xi_1 + \xi_2]}$$

$$\alpha_2 = 1 - \frac{2\, i \mu_R \gamma}{k + i\mu} \; \frac{1}{\gamma + (\theta(-\lambda_R) + \gamma)\exp[\xi_2 - \xi_1]} \tag{32}$$

(θ is the step function).

The DSI soliton is obtained for $\omega(k) = ik^2$, while the 2DMKdVI soliton is obtained for $\omega(k) = ik^3$.

The auxiliary functions A for the DSI equation and A, B for the 2DMKdVI equation can be uniquely determined by using the so-called t-component of the Bäcklund transformation, which can be written by using the Bäcklund gauge in (15) as follows

$$T_2(Q)\, B(Q, 0; \lambda, \mu) - B(Q, 0; \lambda, \mu)\, T_2(0) = 0 \tag{33}$$

By equating to zero the coefficients of the powers of the differential operator ∂_y one gets for the DSI equation

$$A = \frac{1}{2}\,(\partial_x - \sigma_3 \partial_y)\, \partial_y \log Q^2 \tag{34}$$

and for the 2DMKdVI equation

$$A = \frac{1}{2}\,(\partial_x - \sigma_3 \partial_y)\, \partial_y \log Q^2$$

$$B = 0 \tag{35}$$

The auxiliary functions A can be obtained also by applying to $(Q^2)_y$ the following inverse operator $(\partial_x + \sigma_3 \partial_y)^{-1}$

$$(\partial_x + \partial_y)^{-1} f(x, y) = \int_{-\infty}^{x} dx'\, f(x', y - x + x') + \frac{4\mu_R \gamma}{\gamma + (\theta(\mu_R) + \gamma)\exp[\xi_1 + \xi_2]}$$

$$(\partial_x - \partial_y)^{-1} f(x, y) = \int_{-\infty}^{x} dx'\, f(x', y + x - x') + \frac{4\mu_R \gamma}{\gamma + (\theta(-\lambda_R) + \gamma)\exp[\xi_2 - \xi_1]}$$

We may now superpose two elementary solitons Q_1 and Q_2 by using the commutativity of the Bäcklund transformations [2]:

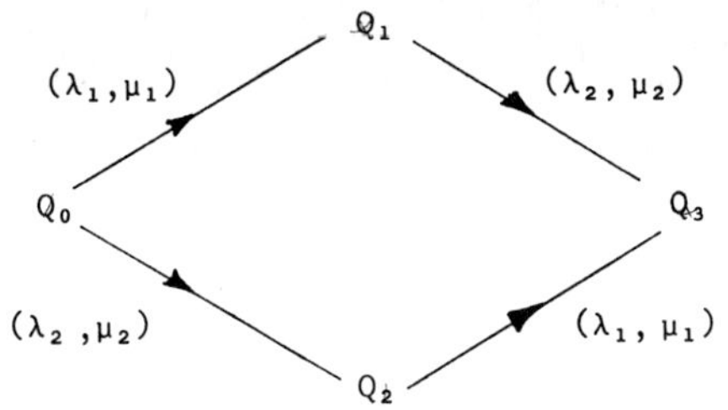

or equivalently for the Bäcklund gauges

$$B(Q_3, Q_1; \lambda_2, \mu_2)\, B(Q_1, Q_0; \lambda_1, \mu_1) =$$

$$= B(Q_3, Q_2; \lambda_1, \mu_1)\, B(Q_2, Q_0; \lambda_2, \mu_2) \tag{36}$$

for $Q_0 = 0$. The general full Bäcklund gauge is given in (15).
From the nonlinear superposition formula (36), which can be more conveniently rewritten with self explanatory notations

$$(\mathbb{1}\,\partial_y + B^{(1)}_{31})(\mathbb{1}\,\partial_y + B^{(1)}_{10}) = (\mathbb{1}\,\partial_y + B^{(1)}_{32})(\mathbb{1}\,\partial_y + B^{(1)}_{20}) \quad , \tag{37}$$

and the formula

$$[\sigma_3, B^{(1)}_{ij}] = Q_j - Q_i \tag{38}$$

one easily derives

$$Q_3 = Q_0 - [\sigma_3, B^{(1)}_{10} + B^{(1)}_{31}] \quad ,$$

$$B^{(1)}_{31} = \{ - (B^{(1)}_{10} - B^{(1)}_{20})_y + (B^{(1)}_{10} - B^{(1)}_{20})\, B^{(1)}_{20} \}\, (B^{(1)}_{10} - B^{(1)}_{20})^{-1} \quad . \tag{39}$$

Equation (39) furnishes the nonlinear superposition formula we are looking for. For $Q_0 = 0$, $B^{(1)}_{i0}$ $(i = 1, 2)$ given in eq. (15) and a convenient choice of λ_i, μ_i $(i = 1, 2)$ we get a solution Q_3 which describes the interaction of the two solitons Q_1 and Q_2 . In fact, in 2+1 dimensions, the two interacting solitons described by Q_3 can be different from the starting fields Q_1 and Q_2 , not only because their relative position in the plane is shifted by the interaction, but also because their amplitude can be renormalized by the interaction.
Rather than studying the properties of this Q_3 "bi-soliton" solution, it seems more convenient here to give a few pictures describing the interaction for some choices of the parameters.

FIGURE CAPTION

Figure 1a, b. Plots of the two-soliton solution Q_3 with Q_1 and Q_2 for the following choice of parameters:

$$\mu_1 = \lambda_1 = 1\,(v_1 = .5)\,, \quad \rho_1 = \eta_1 = 1\,, \quad x_{01} = 10\,, \quad y_{01} = 9\,,$$

$$\mu_2 = \lambda_2 = 4\,(v_2 = .5)\,, \quad \rho_2 = \eta_2 = .8\,, \quad x_{02} = 10\,, \quad y_{02} = 14\,.$$

This choice produces matrices Q_i proportional to $\begin{pmatrix} 0 & 1 \\ 1 & 0 \end{pmatrix}$. The grid spacing is $\Delta x = \Delta y = .4$. Time runs from -8 to +8 ($\Delta t = 2$).

Figure 2a, b, c, d. The same parameters as in fig. 1 are used to show the effect of the interaction in the rest frame of one soliton (the other soliton is not visible in these pictures).

Figure 3a, b. These two stes of plots show the pattern of the level contours at different times for the modulus of y-soliton, in its own frame of reference, during the interaction with a x-soliton, not visible.

The parameters of the two solitons are:

$\lambda_1 = \mu_1 = 1.0 \quad v_1 = 1.0 \quad \rho_1 = \eta_1 = 1. \quad (x_{10}, y_{10}) = (-4.0, 15.0)$

(for the x-moving soliton)

$\lambda_2 = .9\,, \quad \mu_2 = .8 \quad v_2 = 0.0 \quad \rho_2 = \eta_2 = 4.0 \quad (x_{20}, y_{20}) = (10.0, 5.0)$

(for the y-moving soliton).

The time runs from -10.0 up to 30.0 in our unity of time ($\Delta t = 5$). The corresponding plots go from the left to the right and from the top to the bottom.

The lowest level plotted contour is 0.05 with steps of 0.10 for all plots.

The star indicates the position of the relative maxima. Composing the plot at $t = -10.0$ with that one at $t = 30.0$, the reader gets an estimate of the x- and y-shift.

Figure 4a. This plot represents, in the coordinates (ξ_1, ξ_2), the pattern of the level contours of the associated field A to the DSI eq., in the case of the single soliton solution. The following parameters are used:

$\lambda = \mu = 1.0 \quad \gamma = \frac{1}{4}\eta\rho = 1.0 \quad (\xi_{10}, \xi_{20}) = (0, 0)$

The function is real and bounded between -8.15 and 0.00 .

The smallest closed contour represents the lowest level at -8.0 . Increments of 1.5 are drawn.

Figure 4b. As in the previous plot, the level contour pattern is given for the single soliton associated field A in the DSI eq., where the following choice of parameter is used:

$\lambda = -\mu = 1.0 \quad \gamma = 1.0 \quad (\xi_{10}, \xi_{20}) = (0, 0).$

The function is real and bounded between -1.66 and 1.66, and it is an odd function by $\xi_2 \rightarrow -\xi_2$, with negative values in the upper half plane.

The inner contour in the upper plane represents the lowest level at -1.6 . Increments of 0.64 are drawn.

REFERENCES

[1] A.DAVEY, K. STEWARTSON, Proc. Roy. Soc. Lond. A 338, 101 (1974).

[2] M. BOITI, J.JP. LEON, L. MARTINA, F. PEMPINELLI, Phys. Lett. A .

[3] M. BOITI, J. JP. LEON, F. PEMPINELLI, "Multidimensional solitons and their spectral transforms", Preprint PM/88-44 Montpellier (1988), Sub. to J. Math. Phys.

[4] A.S. FOKAS, Phys. Rev. Lett. 51, 3 (1983); A.S. FOKAS, M.J. ABLOWITZ, J. Math. Phys. 25, 2494 (1984).

[5] M. BOITI, B.G. KONOPELCHENKO, F. PEMPINELLI, Inv. Problems 1, 33 (1985).

[6] M. BOITI, J. JP. LEON, L. MARTINA, F. PEMPINELLI, J. Phys. A 21, (1988).

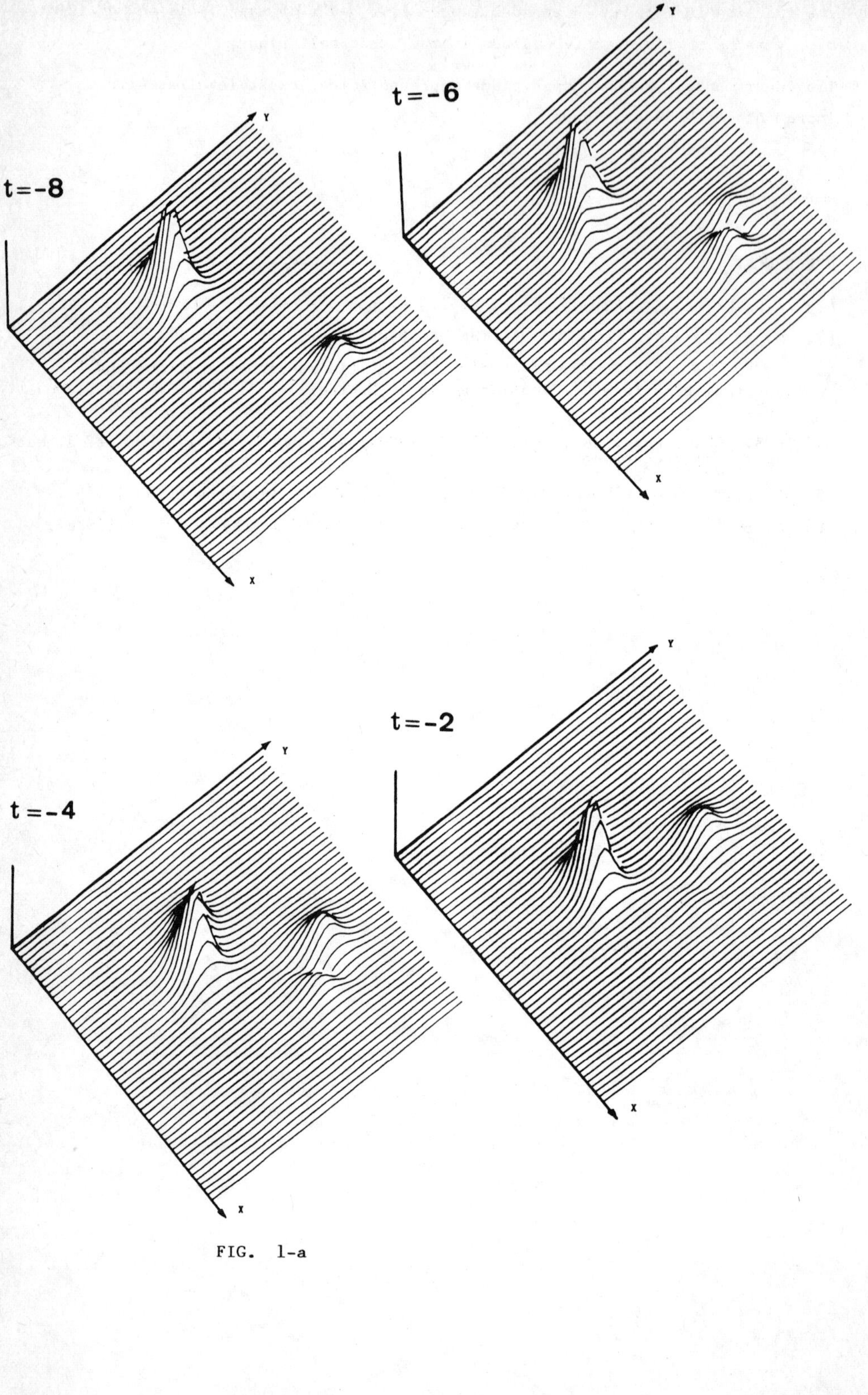

FIG. 1-a

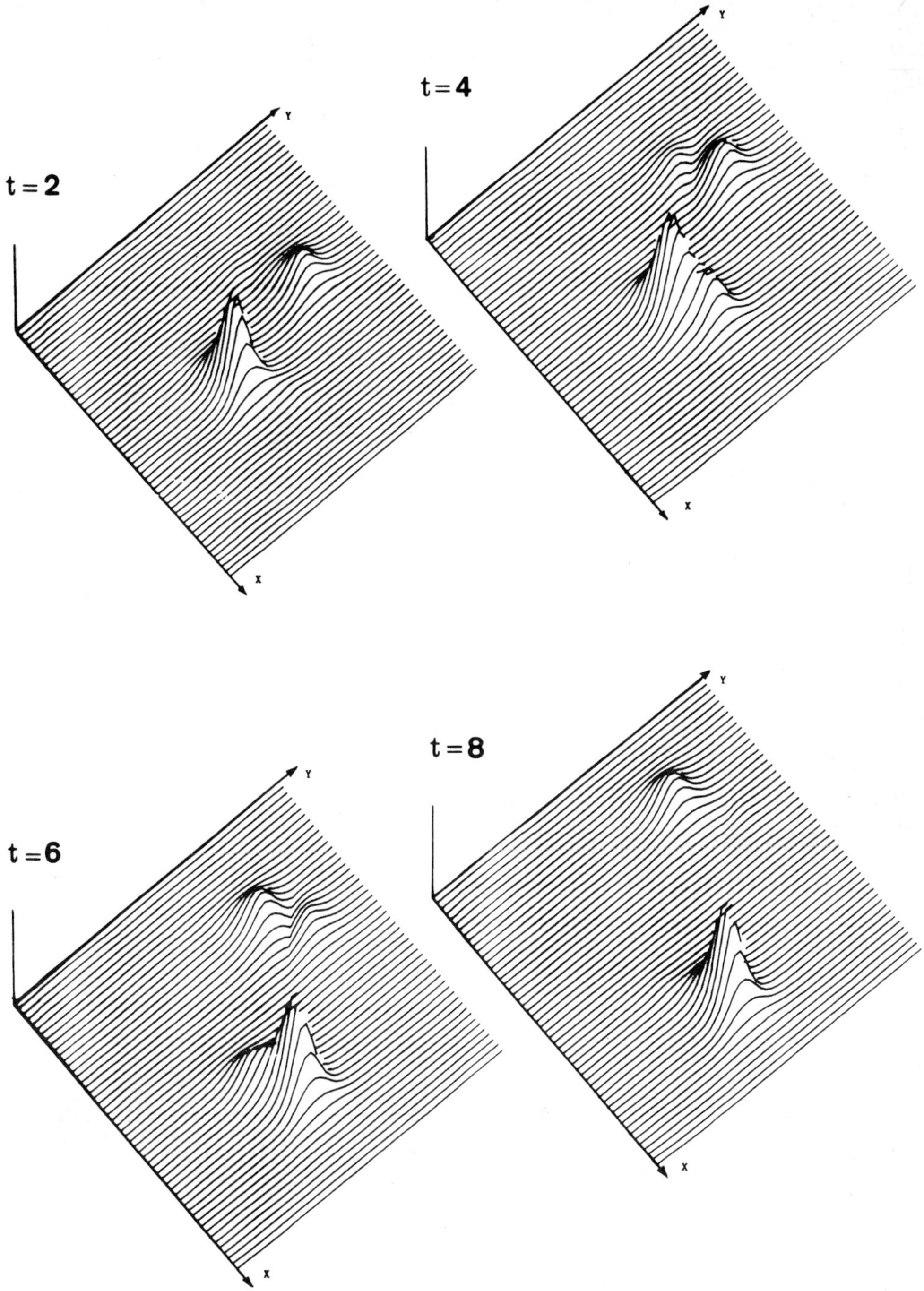

FIG. 1-b

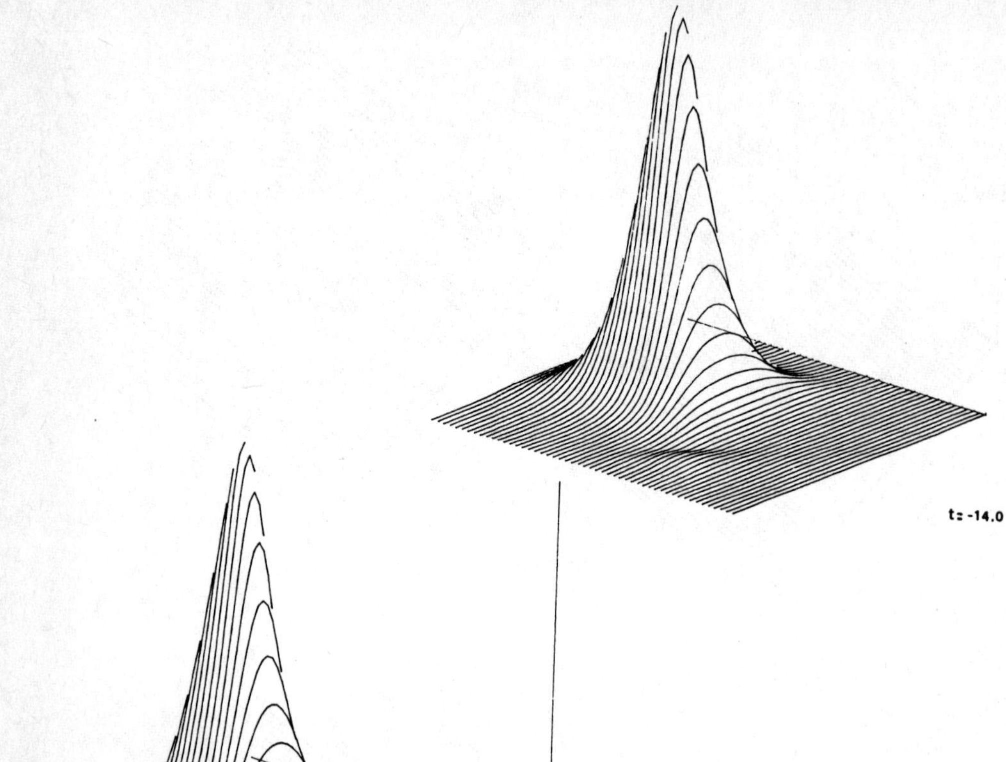

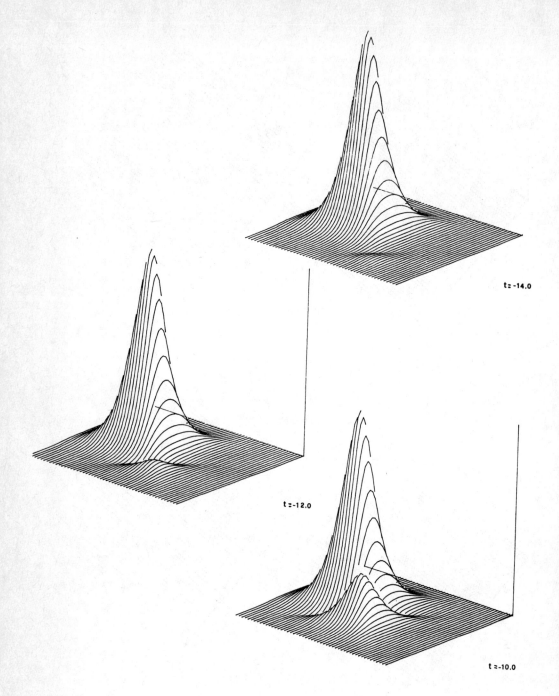

FIG. 2-a

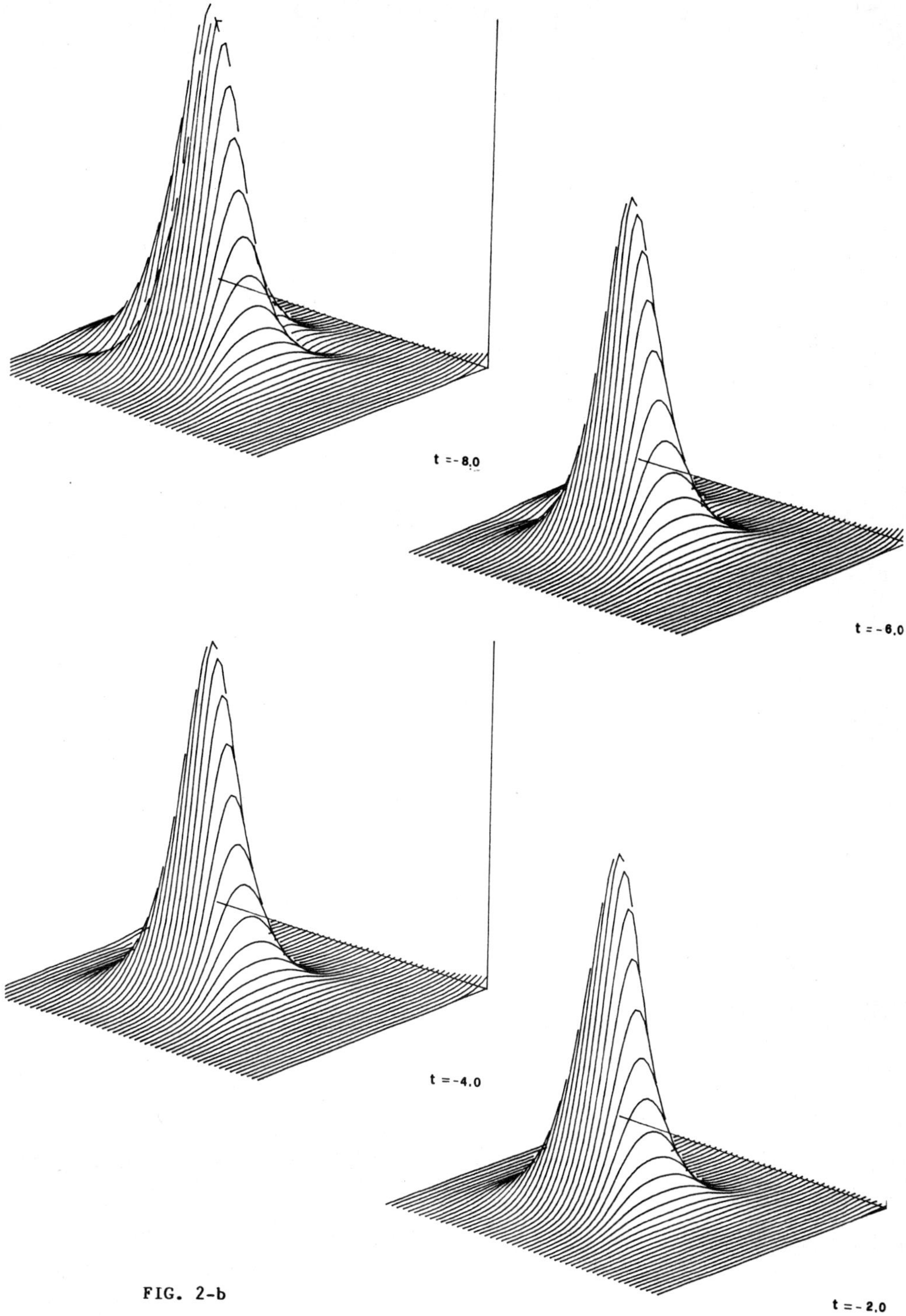

FIG. 2-b

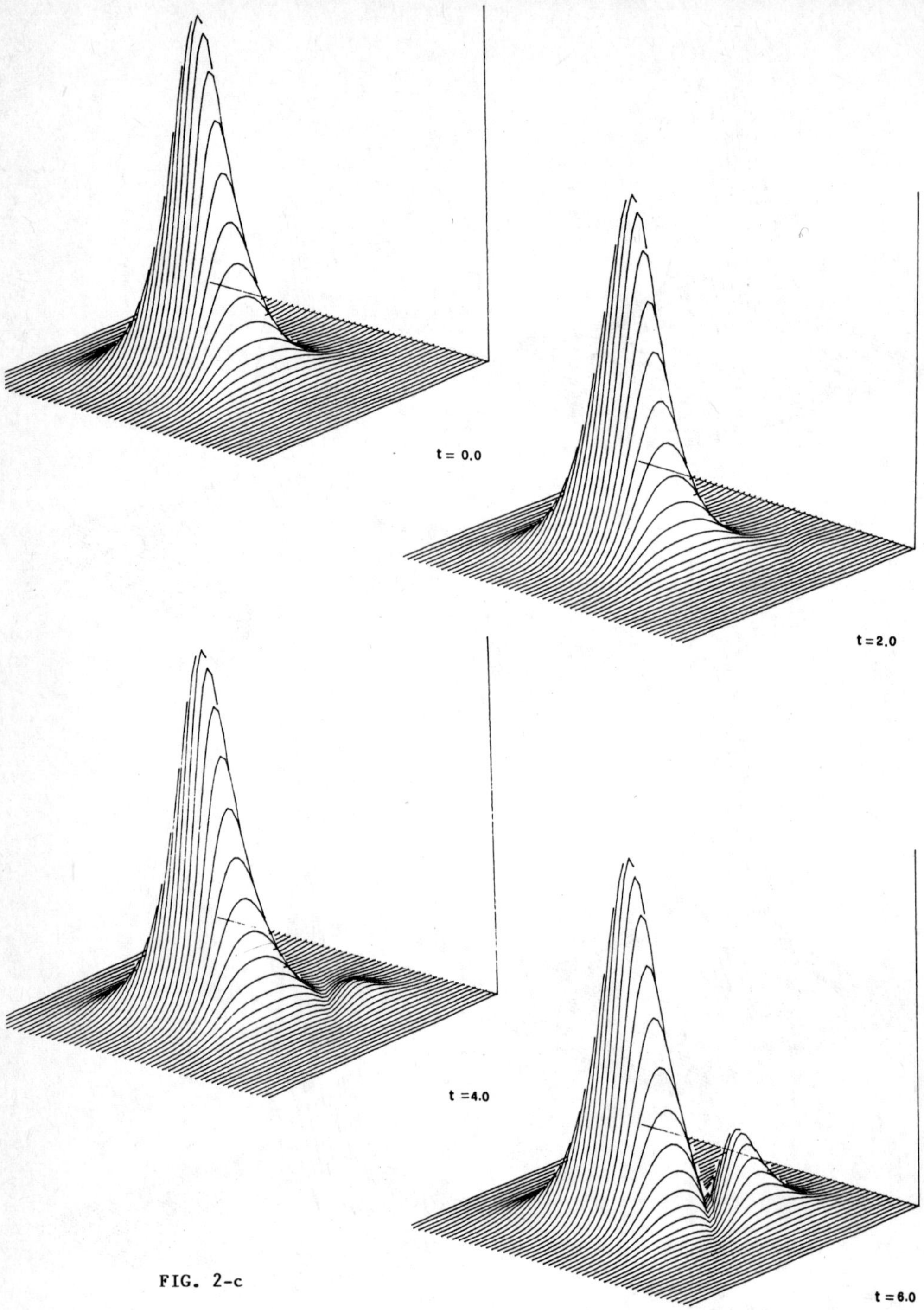

FIG. 2-c

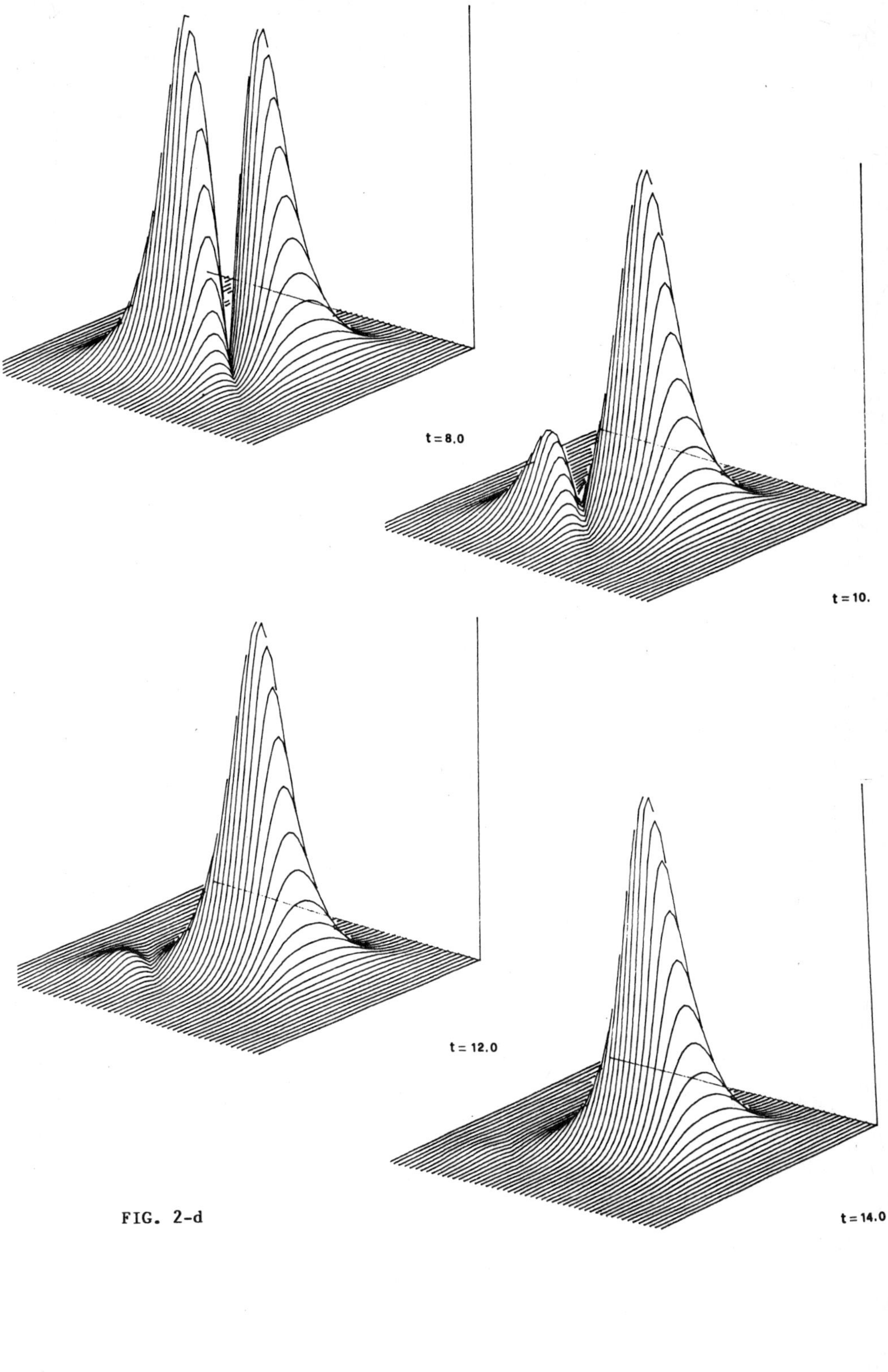

FIG. 2-d

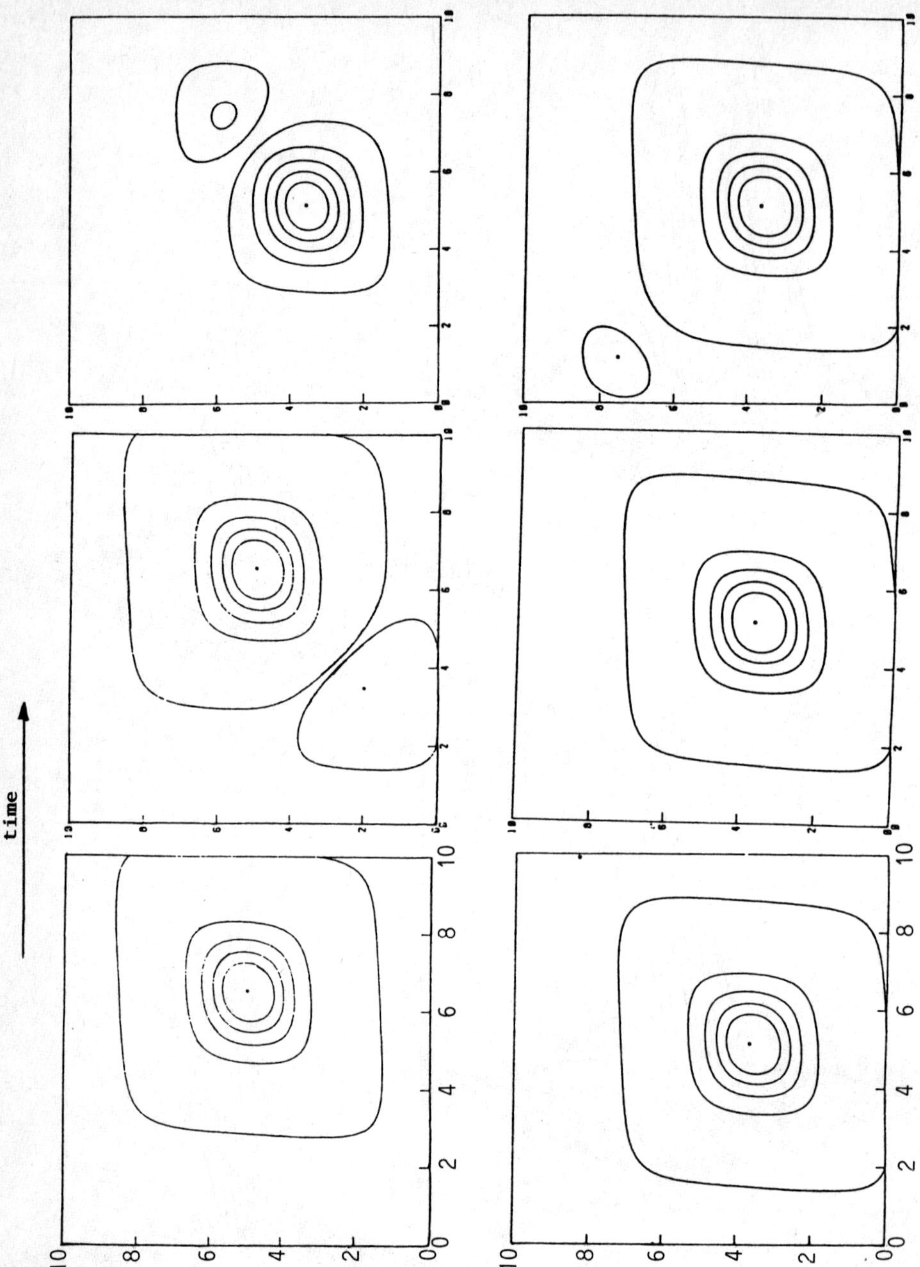

FIG. 3-a

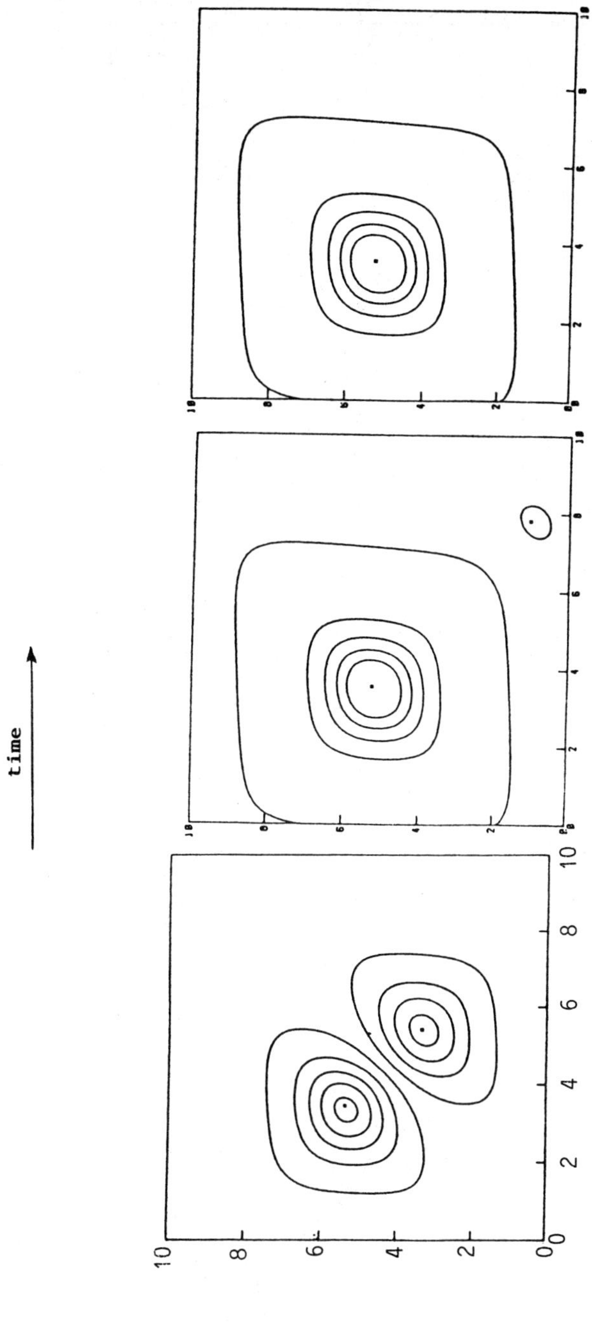

FIG. 3-b

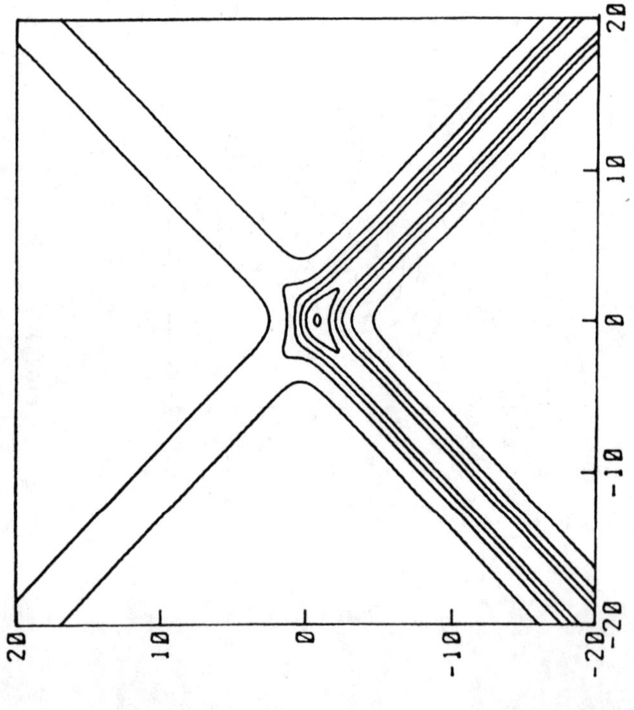

FIG. 4-a

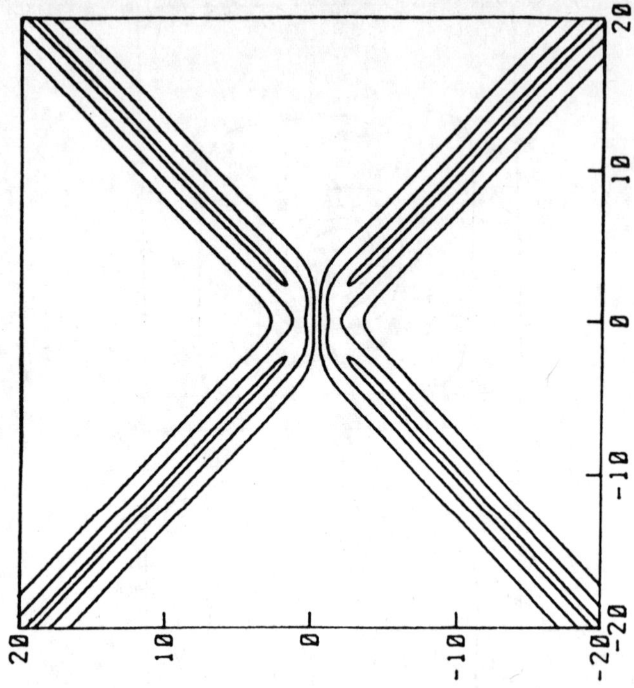

FIG. 4-b

NONLINEAR WAVE PROPAGATION THROUGH A RANDOM MEDIUM AND SOLITON TUNNELING

J. G.Caputo, A. C.Newell and M. Shelley
Department of Mathematics
University of Arizona
Tucson, Az. 85721

Abstract

We have studied the propagation of non-linear waves across a random medium, using the nonlinear Schrödinger equation with a random potential as a model. By simulating a scattering experiment, we show that non-linearity leads to an improvement of the transmission only when it contributes to create pulses. The propagation properties of these pulses can be described by an equivalent particle theory. Numerical experiments show that this is an approximation: two situations are presented depending on the two ratios of the amplitude and velocity to the perturbation. Concluding remarks link the time-dependent and time-independent regimes.

I. Introduction

We have chosen the nonlinear Schrödinger equation

$$i\,[\partial A/\partial t] + [\partial^2 A/\partial x^2] + 2\beta|A|^2 A = \sigma V(x)A \tag{1}$$

as a model for the study of the influence of non-linearity on wave propagation accross a random medium because it has the simplest possible non-linearity and also because of its canonical character in the description of weakly non-linear effects.

In the linear regime ($\beta = 0$) time and space become decoupled, one can write $A(x,t) = \varphi(x)\;\; e^{-i\omega t}$ so that (1) reduces to

$$\varphi_{xx} + (\omega - \sigma V(x))\varphi = 0 \tag{2}$$

On a discretized version of (2) it can be shown that φ grows exponentially with x. In terms of a scattering experiment on the slab of random medium of width L where a left incoming wave Ae^{ikx} is partially reflected and partially transmitted giving rise to $R(L)e^{-ikx}$ to the left of the medium and $T(L)e^{ikx}$ to the right, it has been shown by Anderson and Ishii [1,2] that $T(L)$ behaves like $T(L) = e^{-L/\delta}$ (δ is

the localisation length). Souillard and Rammal [3,4] have studied the non-linear situation assuming the ansatz giving rise to equation (3) is still valid. Because of bi-stability effects, the fixed input problem as described previously is now different from the fixed input problem in which T is imposed and not the incoming amplitude. For the physically interesting situation (fixed input) Rammal [4] finds that for small L $(= A^{2/3}/\delta)$, $R_1(L) < |R(L)| < R_2(L)$ with large oscillations. $R_1(L)$ and $R_2(L)$ behave like $\sqrt{L}$. For larger L $(> A^2/\delta)$, one recovers the exponential behavior.
For equation (1) these results are questionable because it is not clear how the time-independent non-linear solution evolves into the harmonic form given above. It is also unclear what is the mechanism for the increase of transport in the non-linear regime. To tackle these questions, we have performed the scattering experiment in the full time-dependent framework.

II. The scattering experiment: why we study soliton propagation

To integrate numerically equation (1) with boundary conditions A fixed at $x = \pm \infty$, we have used the following discrete approximation:

$$idA_n/dt + A_{n+1} - 2A_n + A_{n-1} + \beta|A_n|^2(A_{n+1} + A_{n-1}) = \sigma V_n A_n \qquad (3)$$

integrating the equations with a fourth order Adams-Moulton predictor corrector scheme. The particular form of the non-linear term was chosen so that the system is completely integrable when $V_n = 0$ everywhere and the boundary conditions are $q = 0$ at $x = \pm \infty$ [5]. This enables us to test the scheme very accurately and to have stable solutions when the wave-length is not very much bigger than 1.
For the scattering experiment, we write the solution $A_n(t)$ of (3) as the sum of a left incoming wave and an additional term U which we solve for, $A_n(t) = A\Delta_n e^{ikn-i\omega(k)t} + U_n(t)$. The amplitude coefficient Δ_n is 1. at the left of the slab and decreases to 0. at the boundary of it. The reflected and transmitted waves are absorbed by introducing a damping term in (3). The potential V_n is obtained by averaging a gaussian process V_{0_p} with an Orstein-Ulhenbeck factor:

$$U_n = \sum_{p=-\infty}^{n} V_{0_p} e^{-\gamma(n-p)}$$

γ^{-1} is the correlation length. Exactly as in [3,4] the non-linearity parameter β and V_n have same support.
We first have tested the scheme without randomness. U_n develops into the plane wave when $\beta = 0$, when $\beta > 0$ it decomposes into lumps (Benjamin-Feir instability). These can be related to multi-soliton solutions of the non-linear Schrödinger equation. When $\beta < 0$ we expect a superposition on non-linear plane-wave-like solutions but have not been able to characterise it.

The randomness was then turned on and γ^{-1} set to be close to k, to get maximum backscatter from the $V_n A_n$ term. For the runs presented below, the amplitude σ of the random medium was chosen to be $.25A$. The results we obtain are that in the non-linear focussing case $\beta > 0$ stable coherent pulses are created. These considerably increase the transport properties. The rate of decay of the modulus of the solution is much more important for both the non-linear defocussing case and the linear case. This rate is not significantly different for the last two cases. Figure 1 shows the logarithm of the modulus of U as a function of n for a given time. We have analysed the time dependance of $|U_n|$ for a given n. Figure 2 shows the real part of U_n for $n = 115$ as a function of time, the behavior is clearly harmonic. Figure 3 shows that the modulus $|U|$ decreases with time to tend to a stationary value. This is a confirmation that the harmonic ansatz makes sense for the linear case. As one moves more into the medium, fluctuations become more important. It takes more time for the solution to settle down to its stationary value. Note however that the amplitude decreases. Despite of that first indication, we have failed to extract the localisation length out of Figure 1. We should have averaged the field over a certain number of realisations in order to get a clear result; this was not possible due to the amount of computing time required. Instead, we estimated the localisation length using a transfer matrix procedure: if we write the solution of a discretisation of equation (2) as $e^{i\omega t} A_n$. A_{n+1} becomes then a linear function of A_n, A_{n-1}, V_n which we can calculate by putting the problem in a matrix form. We can compute the growth rate of the norm of a product of N such matrices T_n. We obtain the lyapunov exponent and its inverse the localisation length by averaging this growth rate over a certain number of realisations of the random medium. For our particular values of the parameters we obtain $l = 160$. The localisation process happens on the first two hundred nodes.

We repeated the scattering experiment for two non-linear media. All the other parameters remaining unchanged, we set $\beta = \pm 10$. There is not much difference between $\beta = -10$ and $\beta = 0$ as can be seen in figure 4. We reach a stationary state, and the values of the field are comparable for the two cases. Interestingly enough, the transfer matrix procedure shows no indication of a positive liapunov exponent. Even though the now non-linear transformation of the plane (A_{n+1}, A_n) is still area-preserving we cannot say anything about the sign of the growth-rate and its dependance on the distribution of the random potential.

On the contrary, for $\beta = 10$ the field does not decay as fast as for the cases mentioned above, as can be seen on figure 5. It does not tend towards a stationary state. There are pulses which get created and propagate through the medium as is shown in figure 6. From the comparison between the two values of the non-linearity considered, it is clear that these coherent structures are responsible for the increase in transport. Figure 7 shows one of these pulses in dotted lines and the fit that can be obtained using the exact solution of equation (3) (when $V = 0$)

and the parameters extracted from the numerics. This is good evidence for saying that these pulses are non-linear Schrödinger solitons. We have shown the bigger pulses but have seen quite a few smaller ones which can also be fitted with sech profiles. If we assume that the non-linear pulses of figure 6 are indeed solitons, as long as they are far apart, they should obey the following analytic expression:

$$A_s(x,t;\eta,v_e,x_0,\sigma_0) =$$

$$2\left[\eta/\sqrt{\beta}\right] Sech 2\eta(x - v_e t - x_0) exp(i\left[v_e/2\right]x + i(4\eta^2 - \left[v_e^2/4\right])t + i\sigma_0), \quad (4)$$

Except in the special case where the velocities v_e do not vary from pulse to pulse and are very large compared to the amplitudes η, we cannot have a field which is monochromatic. To get more insight into this we need to be able to predict the distribution of amplitudes and velocities when the randomness is present and strong. For a weak randomness the velocities are imposed by the phase velocity of the incoming plane wave. The amplitude probably corresponds to the most unstable wave-number. However the perturbation will affect these two parameters. Furthermore it is not clear how an initial velocity and amplitude distribution will evolve because of possible interactions between solitons due to the randomness. This can also be stated as the germination problem: where are pulses created in the random medium?, is there a well defined threshold?, are the time intervals between emissions regular or chaotic?..

We have wanted to understand how can coherent non-linear pulses travel more efficiently through a random medium. Therefore we have conducted a second type of numerical experiment, starting with a bounded initial condition and letting it propagate accross the medium.

III. The propagation of solitons

a) A simple theory

In the following the coefficient β has been taken equal to 1. The first idea for a theory for the propagation is an equivalent particle theory. We define averages using the solution A as in Quantum Mechanics by introducing the following quantities:

$$m = \int_{-\infty}^{+\infty} AA^* dx \ , \ m\bar{x} = \int_{-\infty}^{+\infty} xAA^* dx \ ,$$

$$mv_e = \int_{-\infty}^{+\infty} i(A\left[\partial A^*/\partial x\right] - A^*\left[\partial A/\partial x\right])dx \ , \quad (5)$$

Using equation (1) It can be shown exactly that:

$$\left[dm/dt\right] = 0 \ , \left[d\bar{x}/dt\right] = v_e \ , m\left[dv_e/dt\right] = F = -2\int_{-\infty}^{+\infty} \left[\partial V/\partial x\right] AA^* dx. \quad (6)$$

Formally, the field $A(x,t)$ behaves like a particle of mass m in a force field $F = -2\int_{-\infty}^{+\infty} [\partial V/\partial x]\, AA^* dx$. We now introduce an ansatz and assume that the field intensity AA^* moves collectively and is therefore a function of t only through $x - \bar{x}(t)$. In that case the force F is a function of the single coordinate $\bar{x}$ and the field behaves as a single particle. Moreover, we may use integration by parts twice and write $[1/m]\, F(\bar{x}) = -\,[\partial U(\bar{x})/\partial x]$ where the effective potential $U(\bar{x})$ is

$$U(\bar{x}) = 2\int_{-\infty}^{+\infty} [AA^*/m]\,(x-\bar{x})\cdot V(x)dx. \tag{7}$$

Thus the motion of a coherent pulse in a random potential $V(x)$ may be modelled by the motion of a single particle in an equivalent potential which is twice a weighted average of the original potential, the weighting being given by the shape of the normalized local intensity.

$$h(x-\bar{x}) = [1/m]\, AA^*. \tag{8}$$

It is clear that even though the original potential $V_0(x)$ is uncorrelated over neighboring sites and the smoothed potential $V(x)$ has correlation length γ^{-1}, the effective random potential $V(\bar{x})$ is highly correlated up to distances proportional to the width of the coherent pulse. The effect of averaging $V(x)$ is to increase the penetration distance because the effective potential $U(\bar{x})$ will only stop the particle where there is a sufficiently long sequence of positive values $V(x)$ over the width of the soliton to produce a sufficiently positive effective potential to trip the particle. Given that the correlation length γ^{-1} of the random medium is much less than the wavepacket width, a sequence of such values is an unlikely event and therefore the expected distance which the coherent pulse will travel into the medium before encountering such a sequence is large.

One might think in terms of an analogy of a train with many carriages riding over a sequence of random hills. If each carriage behaves individually, then that carriage will be stopped as soon as it reaches the first hill whose potential energy exceeds its kinetic energy. If, on the other hand, the train behaves collectively, it will take a large hill to exceed its total kinetic energy. Large fluctuations are unlikely but the probability of their occurrence can be calculated [6].

Because the weighting factor used in calculations of the effective potential $V(\bar{x})$ involves the normalized field intensity, the propagative characteristics of the pulse are independent of the magnitude of the nonlinearity. The place where the ratio of field intensity to noise level is crucial is in making the assumption that the coherent pulse moves as a collective unit. In the following numerical experiments we have tried to define how well this assumption holds.

b) **the numerical experiments**

The numerical model is (3) but the initial condition is a soliton and the boundary conditions are $A = 0$ at $x = \pm \infty$. We also kept the damping term to avoid unwanted reflections. We have considered pulses which are large with respect to the correlation length of the medium. In all cases, we have $\gamma \ width = 10$ so the averaging effect is important. The case of pulses traveling in a slowly varying medium has been adressed in the limit of small perturbation [7]. It seems that in this situation the soliton looses a lot of its wave-like features and behaves more like a particle. The choice of the velocity of the envelope of the pulse which is also the wave-length of the carrier wave was determined by the linear situation: it can be much larger than the correlation length and this will create small resonances or of the same order of magnitude for much larger resonnances. For these two cases, the "modus operandi" has been the same: we increased the standard deviation of the potential till we had a significant effect on the wave.

The slow pulse regime: long wave-length

The initial velocity has been chosen so that the the wave-length of the underlying carrier wave is large compared with the correlation length. We then increased the noise to amplitude ratio to about 10^{-3} so that the variations in velocity are significant. Figure 8 shows the raw potential V_n and the average potential computed with the soliton initial profile U_n vs. n. U_n has typical length scales of the size of the width of the soliton. Figure 9 shows the field $|A_n|$ as a function of n for a typical time-step as the pulse travels across the medium. It is leaving behind it a trail of smaller pulses. It is obvious in these conditions that the hypothesis of the particle picture breaks down. In fact, the soliton is slowing down as figure 10 displays it, comparing the position as a function of time for the soliton run and a newtonian particle evolving in U_n. This energy loss can be loosely correlated to the mass loss of the initial pulse. The mechanism of this dissipation is unclear but it probably lies in a resonnance occuring between the terms $4w^2$ and $v^2/4$.

Due to this dissipation mechanism, the soliton in the case studied will eventually encounter a potential barrier it cannot climb and go back, subsequently, it will come to a rest. In our numerical simulations, we have found that it remains remarquably stable then. The enveloppe oscillates but the pulse is still very well defined. In order to study this situation, we have done a series of numerical experiments starting the pulse with no initial velocity in a "well" of the average potential. If the standard deviation σ of the random potential is small, the envelope remains unchanged and the dynamics is on the phase. At first the phase is smooth only in the center of the pulse. Then progressively the oscillations damp out in the wings and one gets a wide region where the phase is uniform. It still oscillates due to the $-4\eta^2$ term in the field. If σ is increased, the envelope is affected. The soliton develops small bumps on each side and the value of its maximum increases . Nevertheless, the structure remains stable, it does not decay. The value of the maximum of the field is what can be expected from a simple scaling argument which would give a behavior in $\sigma^{1/2}$.

Fast pulse regime: short wave-length

We have chosen the velocity to be of order 1. The underlying carrier wave then has a wave-length which is very close to the correlation length of the random medium. Figure 12 shows $log(\eta)(t)$ vs. t where η is estimated from the numerics. It is apparent that the amplitude of the pulse is decaying exponentially as it makes progress through the medium. The speed is almost constant as can be infered from figure 13 showing the position of the maximum of $|A_n|^2$ as a function of time.

Link with the localisation theory

In order to make a connection with the localisation theory which is a steady-state theory, we need to be able to predict what is the mean free path of a pulse in a random medium. We need to fix better the parameter regions where the velocity or the amplitude of the pulse is irreversibly affected by the perturbation. It may be the case that out of the two parameters η and v_e only the former is modified by the perturbation. At this point we can infer that a high amplitude wave of small wave-length hitting a potential will generate fast solitons. These will have an exponentially vanishing amplitude. The mechanisms for the generation of pulses inside the medium are not known so that it is not clear whether a stationary state exists or not. In the slow regime, the pulses will settle down in potential wells. They will however continue to interact so neither here do we settle down to a stationary state.

REFERENCES

[1] P. W. Anderson
Phys. Rev. A **109**, 1492 (1958)

[2] K. Ishii
"localisation of eigenstates and transport phenomena in one dimensional disordered systems"
Prog. Theor. Phys. Supp. **53**, 77 (1973)

[3] P. Devillard and B. Souillard
"Polynomially decaying transmission for the non-linear Schrödinger equation in a random medium"
J. Stat. Phys. **43**, 423 (1986)

[4] B. Doucot and R. Rammal
"Invariant embedding approach to localisation." I and II
J. Phys. **48**, (1987)

[5] M. J. Ablowitz and J. F. Ladik
"Non-linear differential-difference equations and Fourier analysis"
J. Math. Phys. **17**,Number 6, 1011 (1976)

[6] J. G. Caputo, A. C. Newell and M. Shelley
to be published

[7] D. J. Kaup and A. C. Newell
"Solitons as particles, oscillators and in slowly changing media: a singular perturbation theory"
Proc. R. Soc. Lond. A **361**, 413-446 (1978)

FIGURES

1. scattering experiment for a linear wave:
$log_{10}|U_n|$ vs. n for a given time.
Parameters: wave amplitude $A = 4.10^{-2}, k = 10^{-1}$, length of experimental domain $N = 3000$, support of random medium $[65, 2940]$ $\sigma = 10^{-2}$, correlation length $\gamma^{-1} = 5$

2. scattering experiment for a linear wave:
real part of U_n for $n = 115$ as a function of time.
Parameters: wave amplitude $A = 4.10^{-2}, k = 10^{-1}$, length of experimental domain $N = 1500$, support of random medium $[65, 1440]$ $\sigma = 10^{-2}$, correlation length $\gamma^{-1} = 5$

3. scattering experiment for a linear wave:
modulus of U_n for $n = 115$ as a function of time.
same other parameters as in figure 2.

4. scattering experiment for a non-linear wave, $\beta = -10$:
$log_{10}|U_n|$ vs. n for a given time. On the same graph is plotted the curve of figure

1 to show that the two are very close.
Parameters: wave amplitude $A = 4.10^{-2}, k = 10^{-1}$, length of experimental domain $N = 3000$, support of random medium $[65, 2940]$ $\sigma = 10^{-2}$, correlation length $\gamma^{-1} = 5$

5. scattering experiment for a non-linear wave, comparison with the linear case. $\beta = +10$:
$log_{10}|U_n|$ vs. n for a given time.
Parameters: wave amplitude $A = 4.10^{-2}, k = 10^{-1}$, length of experimental domain $N = 3000$, time 1.2210^4, support of random medium $[65, 2940]$ $\sigma = 10^{-2}$, correlation length $\gamma^{-1} = 5$

6. scattering experiment for a non-linear wave, $\beta = +10$:
$|U_n|$ vs. n for a given time.
Parameters: wave amplitude $A = 4.10^{-2}, k = 10^{-1}$, length of experimental domain $N = 3000$, time= $1.22\ 10^4$, support of random medium $[65, 2940]$ $\sigma = 10^{-2}$, correlation length $\gamma^{-1} = 5$

7. scattering experiment for a non-linear wave, $\beta = +10$:
caracterisation of the pulses as sech profiles: the numerical solution is in dotted lines, the fit in full
Parameters: wave amplitude $A = 4.10^{-2}, k = 10^{-1}$, length of experimental domain $N = 3000$, time= $1.22\ 10^4$, support of random medium $[65, 2940]$ $\sigma = 10^{-2}$, correlation length $\gamma^{-1} = 5$

8. soliton propagation, raw potential
Parameters: amplitude $\eta = 2.10^{-2}$, length of experimental domain $N = 1300$, time intervals $= 1.8\ 10^2$, support of random medium $[40, 1300]$ $\sigma = 10^{-4}$, correlation length $\gamma^{-1} = 5$

9. soliton propagation, average potential:
Parameters: amplitude $\eta = 2.10^{-2}$, length of experimental domain $N = 1300$, time intervals $= 1.8\ 10^2$, support of random medium $[40, 1300]$ $\sigma = 10^{-4}$, correlation length $\gamma^{-1} = 5$

10. soliton propagation, slow pulse regime, non-linear case ($\beta = +1$):
$|A_n|$ vs. n for a given time
Parameters: amplitude $\eta = 2.10^{-2}$, velocity $v_e = 2.6\ 10^{-2}$, length of experimental domain $N = 1300$, time intervals $= 1.8\ 10^2$, support of random medium $[50, 1250]$ $\sigma = 10^{-4}$, correlation length $\gamma^{-1} = 5$

11 soliton propagation, slow pulse regime, non-linear case ($\beta = +1$):
$X(t)$ vs. t
Parameters: same as figure 10

12. fast pulse regime, non-linear case ($\beta = +1$):
η vs. t
Parameters: amplitude $\eta_0 = 2.10^{-2}$, velocity $v_e = 0.95$, length of experimental domain $N = 310^4$, time intervals $= 1.8\ 10^2$, support of random medium [50, 29950] $\sigma = 10^{-2}$, correlation length $\gamma^{-1} = 5$

13. fast pulse regime, linear case ($\beta = 0$):
position of maximum of $|A_n|^2$ vs. time
Parameters: same as figure 12

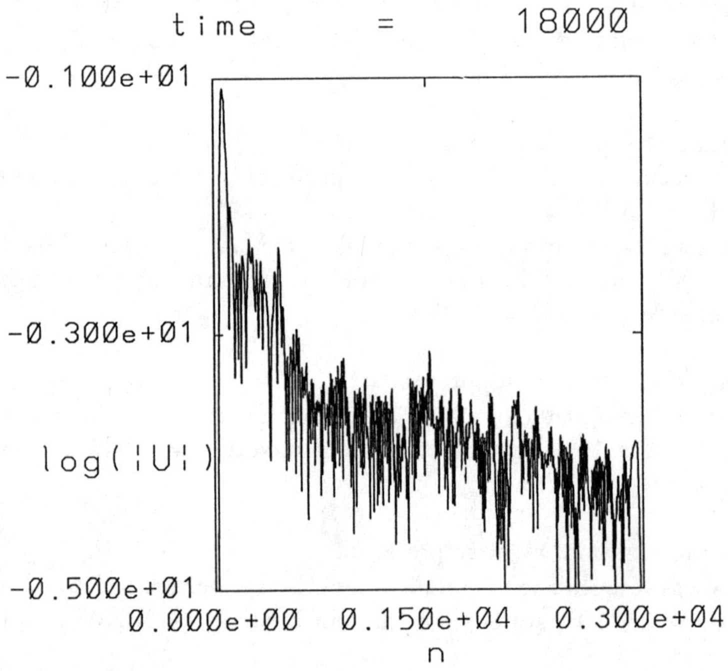

Fig. 1

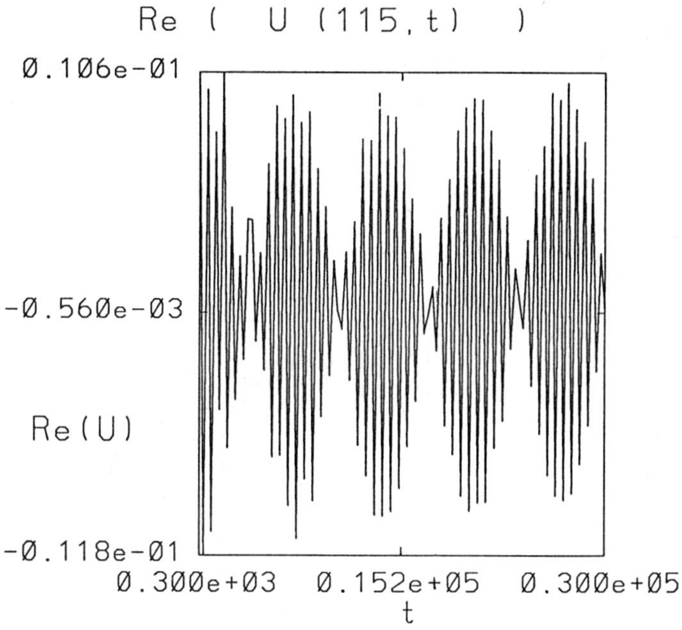
Re (U (115, t))
0.106e-01
-0.560e-03
Re(U)
-0.118e-01
0.300e+03
0.152e+05
0.300e+05
t

Fig. 2

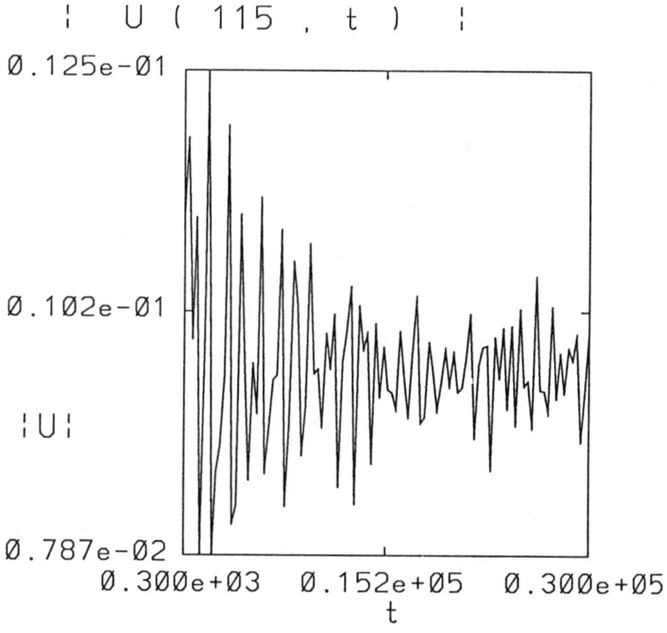
| U (115 , t) |
0.125e-01
0.102e-01
|U|
0.787e-02
0.300e+03
0.152e+05
0.300e+05
t

Fig. 3

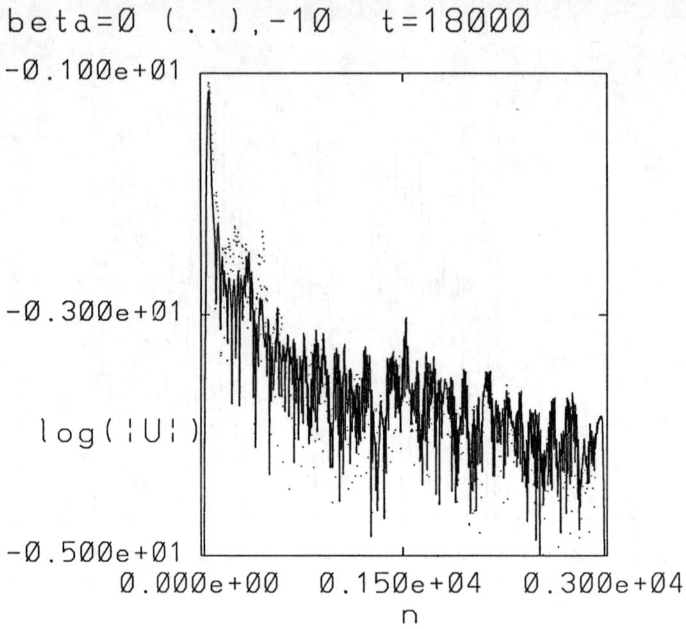
beta=Ø (..),-1Ø t=18ØØØ
-Ø.1ØØe+Ø1
-Ø.3ØØe+Ø1
log(|U|)
-Ø.5ØØe+Ø1
Ø.ØØØe+ØØ
Ø.15Øe+Ø4
Ø.3ØØe+Ø4
n

Fig. 4

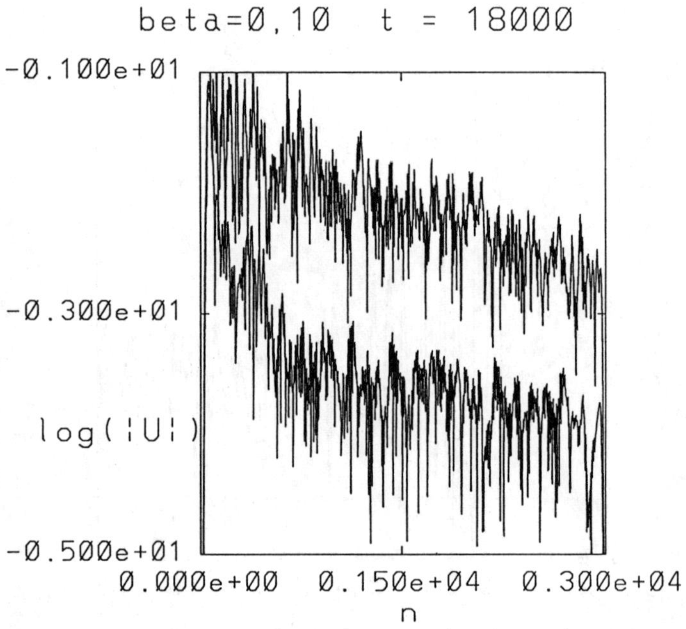
beta=Ø,1Ø t = 18ØØØ
-Ø.1ØØe+Ø1
-Ø.3ØØe+Ø1
log(|U|)
-Ø.5ØØe+Ø1
Ø.ØØØe+ØØ
Ø.15Øe+Ø4
Ø.3ØØe+Ø4
n

Fig. 5

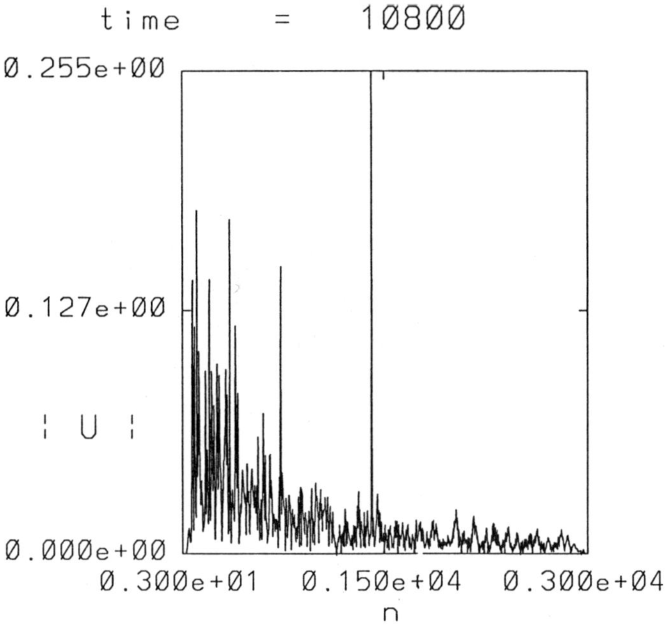
time = 10800
0.255e+00
0.127e+00
| U |
0.000e+00
0.300e+01
0.150e+04
0.300e+04
n

Fig. 6

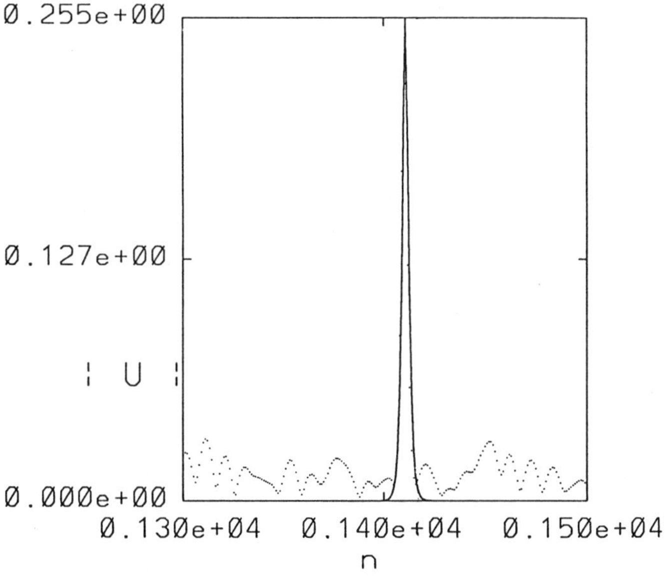
0.255e+00
0.127e+00
| U |
0.000e+00
0.130e+04
0.140e+04
0.150e+04
n

Fig. 7

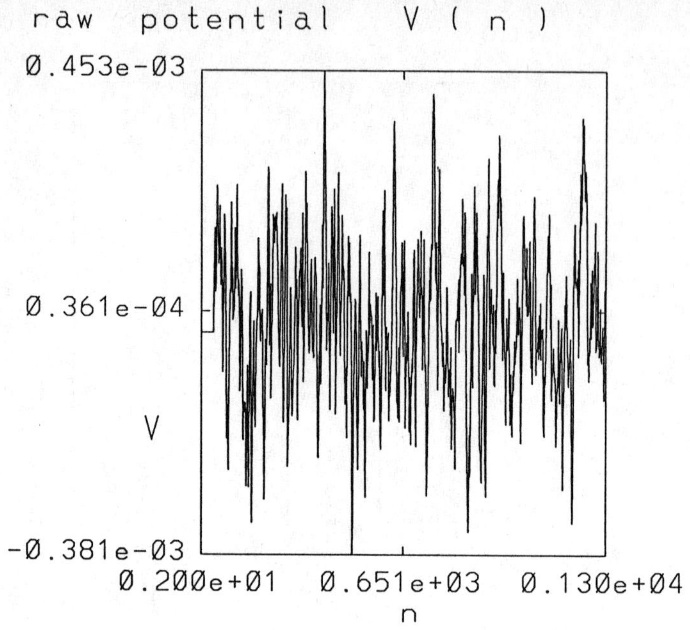
raw potential V (n)
Ø.453e-Ø3
Ø.361e-Ø4
V
-Ø.381e-Ø3
Ø.2ØØe+Ø1
Ø.651e+Ø3
Ø.13Øe+Ø4
n

Fig. 8

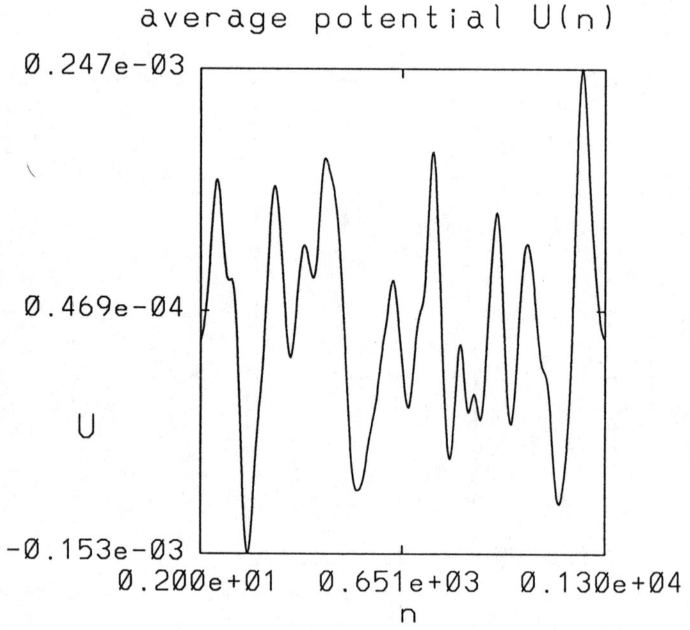
average potential U(n)
Ø.247e-Ø3
Ø.469e-Ø4
U
-Ø.153e-Ø3
Ø.2ØØe+Ø1
Ø.651e+Ø3
Ø.13Øe+Ø4
n

Fig. 9

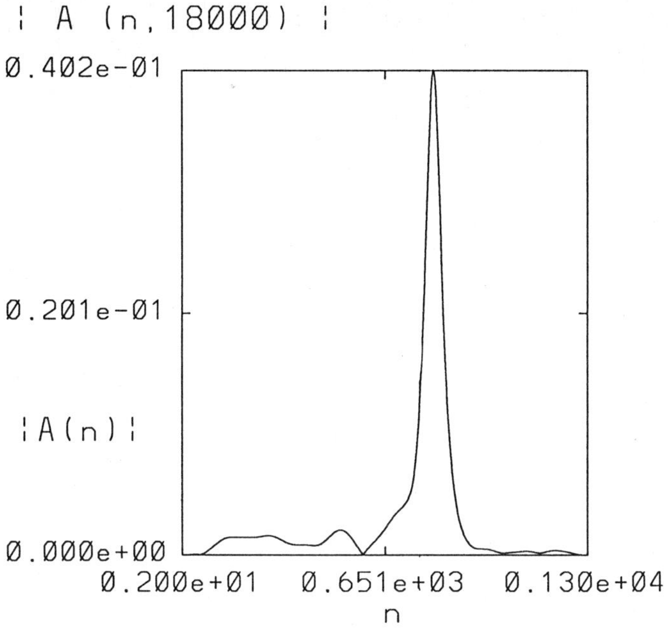
¦ A (n,18000) ¦
0.402e-01
0.201e-01
¦A(n)¦
0.000e+00
0.200e+01
0.651e+03
0.130e+04
n

Fig. 10

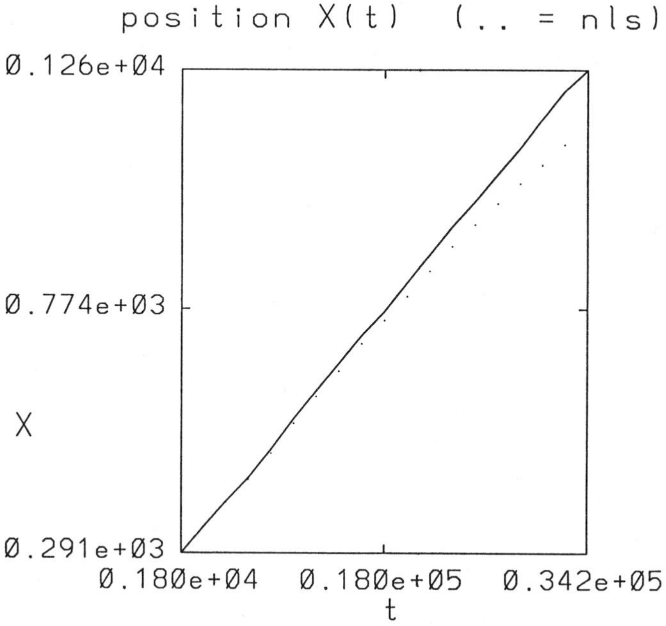
position X(t) (.. = nls)
0.126e+04
0.774e+03
X
0.291e+03
0.180e+04
0.180e+05
0.342e+05
t

Fig. 11

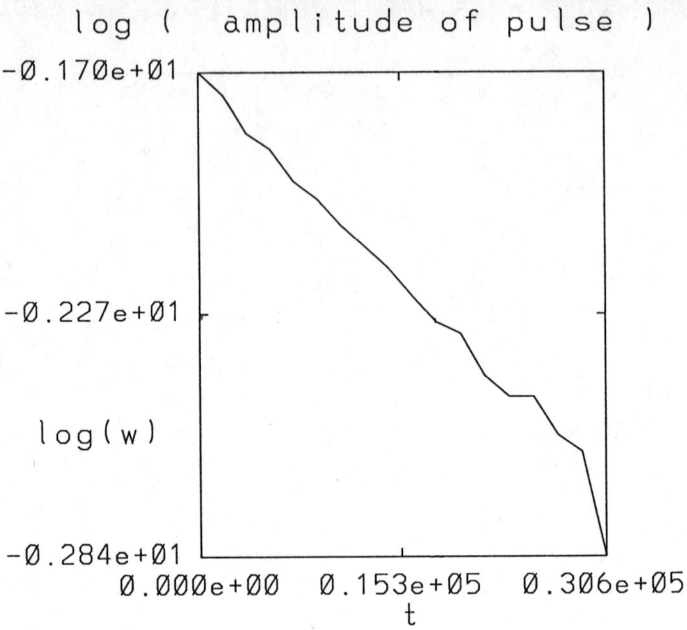

log (amplitude of pulse)
-Ø.17Øe+Ø1
-Ø.227e+Ø1
log(w)
-Ø.284e+Ø1
Ø.ØØØe+ØØ
Ø.153e+Ø5
Ø.3Ø6e+Ø5
t

Fig. 12

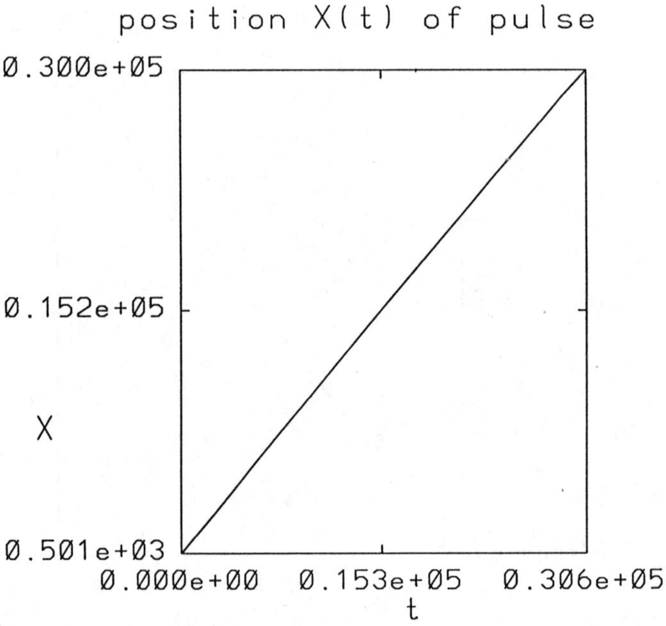

position X(t) of pulse
Ø.3ØØe+Ø5
Ø.152e+Ø5
X
Ø.5Ø1e+Ø3
Ø.ØØØe+ØØ
Ø.153e+Ø5
Ø.3Ø6e+Ø5
t

Fig. 13

Trace Formulae and Singular Spectra for the Schrödinger Operator

Walter Craig
Department of Mathematics
Brown University
Providence, RI 02912

§. 1. This paper addresses several questions in the spectral and inverse spectral theory of the one-dimensional Schrödinger operator. We consider the spectral problem

$$(1.1) \qquad L(q)\psi = (-\frac{d^2}{dx^2} + q(x))\psi = \lambda\psi, \quad -\infty < x < +\infty,$$

where $\lambda \in \mathbf{C}$ and $q(x) \in C(\mathbf{R})$, the set of bounded continuous functions on the line. When $q(x)$ is a periodic function, or if $q(x)$ belongs to a rather special class of hyperelliptic functions, the spectral and inverse spectral theory is complete [5][10][11]. In particular, it is known that the spectrum $\sigma(L(q))$ is entirely absolutely continuous spectrum; it consists of possible infinitely many bands, or zones of stability, with known asymptotics of the endpoints. The spectral class of these potentials is a torus, of possibly infinite dimension, which is associated with the compliment of $\sigma(L(q))$, the zones of instability. Finally, tangent to this torus there are infinitely many commuting flows, corresponding to the Korteweg deVries heirarchy of equations. These are used to solve any of the equations in the heirarchy, which are transparently isospectral flows by this

---

Research supported in part by the Alfred P. Sloan Foundation and the National Science Foundation.

method. In particular the first one is a flow of translation on the line, and is used to recover the potential from certain spectral data, that is, to solve the spectral inversion problem. The formula for recovery is the trace formula of the title.

This article describes a program to extend the above picture to a wider class of potentials. There have been approaches by several authors to this problem, most notibly Moser and Trubowitz [11] and Levitan [8]. More recently a connection has been found with the theory of Schrödinger operators with ergodic coefficients (Kotani [6]), making it relevant to obtain as broad an extended class of potentials as possible. This article describes an approach via the trace formula to solve the spectral and inverse spectral problems analogous to the periodic case. We are able to identify all potentials for which the trace formula holds. Furthermore, among those whose spectrum $\sigma(L(q))$ obeys certain geometrical conditions, the spectral class is identified as a torus associated with the spectral gaps. The vector fields for the flows of the Korteweg deVries heirarchy are obtained, and are shown to possess global solutions. Applying these results and their converses, we obtain several results on the spectral theory of the Schrödinger operator with an ergodic potential. In particular, we show that for a large class of ergodic potentials the Schrödinger operator has a non-zero component of singular spectrum.

§2. General Theory

For $\lambda \in \mathbb{C} - [\lambda_0, +\infty)$, where the ray $[\lambda_0, +\infty)$ contains the spectrum $\sigma(L(q))$, the resolvant $R(\lambda)$ for the operator $L(q)$ exists; it is an integral operator with integral kernel the Greens function $g(x,y;\lambda)$. Our analysis of the spectral and inverse spectral problems focuses on $g(x,x;\lambda)$, the diagonal of the Greens function. The following are well known facts about the diagonal.

<u>*Lemma 2.1:*</u> *The diagonal $g(x,x;\lambda)$ is analytic in $\mathbb{C} - [\lambda_0, +\infty)$ satisfies $g(x,x;\bar{\lambda}) = \overline{g(x,x;\lambda)}$, and when* im $\lambda > 0$ *then* im $g(x,x;\lambda) > 0$.

The full Greens function is recoverable from the diagonal via the formula

$$(2.1) \quad g(x,y;\lambda) = \sqrt{g(x,x;\lambda)\,g(y,y;\lambda)} \; \exp\left(-\frac{1}{2}\int_y^x \frac{1}{g(z,z;\lambda)}\,dz\right)$$

We have defined the Schrödinger operator on the line in (1.1). Consider an auxilliary operator given by imposing additional Dirichlet at $x = 0$;

$$(2.2) \quad L_D(q)\psi = \left(-\frac{d^2}{dx^2} + q(x)\right)\psi$$

$$\text{for } -\infty < x < 0 \text{ and } 0 < x < +\infty \quad \psi(0) = 0 \ .$$

It is well known that for $q(x) \in C(\mathbb{R})$, both $L(q)$ and $L_D(q)$ are Weyl limit-point case at $\pm\infty$, and their spectra are closed sets in $\mathbb{R}$ which are bounded below. The compliment of $\sigma(L(q))$ is thus a union of at most countably many disjoint open intervals. Label end points $\{\lambda_j\}_{j=0}^{\infty}$;

$$(2.3) \quad \mathbb{R} - \sigma(L(q)) = (-\infty, \lambda_0) \cup \bigcup_{j=1}^{\infty} \left(\lambda_{2j-1}, \lambda_{2j}\right)$$

The diagonal of the Greens functions at zero, $g(0,0;\lambda)$ is real in the spectral gaps, and as λ varies in each gap it has at most one zero, which is simple. By convention we label this zero μ_j; if $g(0,0;\lambda)$

has no zero in $(\lambda_{2j-1}, \lambda_{2j})$ set $\mu_j = \lambda_{2j-1}$ if $g > 0$ there, and $\mu_j = \lambda_{2j}$ if $g < 0$.

In the periodic case there is a beautiful formula relating the sets $\{\lambda_j\}_{j=0}^{\infty}$, $\{\mu_j\}_{j=1}^{\infty}$ and the potential; this is the trace formula.

$$(2.4) \qquad q(0) = \lambda_0 + \sum_{j=1}^{\infty} \lambda_{2j} + \lambda_{2j-1} - 2\mu_j \ .$$

The goal of this article is to examine to what extent this formula holds and then to use it analytically to discuss the inverse spectral problem and the initial value problem for the nonlinear evolution equations of the Korteweg deVries heirarchy.

A by-product of the analysis will be an existence theorem for singular spectra for Schrödinger operators with ergodic potentials. Roughly, an ergodic potential is one which possesses certain recurrance properties under translation. These include random potentials as well as periodic, almost periodic and many other classes of potentials which are bounded and continuous, but which do not decay to zero as $|x| \to \infty$. Formally, consider $C(\mathbb{R}) = \Omega$ as a probability space, with the topology of uniform convergence on compact sets. Let the translation flow $(\varphi_\xi q)(x) = q(x+\xi)$ be considered as a one parameter group on Ω. Then $q(x)$ is an ergodic potential if $q \in \operatorname{supp} P$, where P is a probability measure on Ω which is invariant and also ergodic with respect to φ_ξ. Random potentials have spectra which are entirely singular, in fact entirely dense pure point spectra, while periodic potentials have only absolutely continuous spectra. It is an outstanding conjecture that if $q(x)$ is ergodic and $\sigma(L(q))$ is entirely absolutely continuous spectrum, then in fact $q(x)$ is almost periodic. The inverse spectral problem for $L(q)$ will give us strong partial results toward this appealing statement.

Our first result is a general formula for the trace of $L(q) - L_D(q$ which is related to (2.4).

Lemma 2.2: *For $q(x) \in C(\mathbb{R})$, then $L(q) - L_D(q)$ has a trace, and in fact*

$$\text{(2.5)} \qquad \operatorname{tr}(L(q) - L_D(q)) = \frac{1}{2} q(0) \ .$$

In terms of quantities more clearly related to the spectrum,

$$\text{(2.6)} \qquad \operatorname{tr}(L(q) - L_D(g)) =$$

$$= \lim_{\varepsilon \to 0} \frac{1}{2\pi} \int_{-\infty}^{\infty} \lambda \frac{d}{d\lambda} \left(\operatorname{im} \log g(0,0;\lambda + i\varepsilon) - \frac{\pi}{2} \right) d\lambda$$

From Lemma 2.1 the Greens function has values in $\mathbb{C}^+$ for $\lambda \in \mathbb{C}^+$, thus the boundary values $g(0,0;\lambda + i0)$ exist almost every where, and $\operatorname{im} \log g \in [0,\pi]$. The proofs of formulae (2.5) and (2.6) appear in [3]. We are interested in the cases in which the right hand side of (2.6) simplifies. For every $\lambda \in \mathbb{R} - \sigma(L(q))$, the Greens function is analytic and real, that is

$$\text{(2.7)} \qquad \lim_{\varepsilon \to 0} \operatorname{im} g(0,0;\lambda + i\varepsilon) = 0 \ .$$

Consider those $q(x) \in C(\mathbb{R})$ which satisfy an additional condition, that for almost all $\lambda \in \sigma(L(q))$

$$\text{(2.8)} \qquad \lim_{\varepsilon \to 0} \operatorname{re} g(0,0;\lambda + i\varepsilon) = 0 \ .$$

Every periodic $q(x)$ will satisfy (2.8). The potentials of Dubrovin, Matveev and Novikov [5], of Levitan [8], of Moser [12] and of Avron and Simon [1] also satisfy (2.8). Futhermore, the scattering potentials discovered by Bargmann [2] satisfy (2.8); these are potentials whose reflection coefficient vanishes, which are the solitons for the KdV equation. With this analogy we call the class of potentials whose Greens function satisfies (2.8) the 'reflectionless' potentials, or the potentials 'of Bargmann type'. For this class, the formula (2.6) simplifies, recovering the equation (2.4).

Theorem 2.3: *Let $q(x) \in C(\mathbb{R})$ be 'reflectionless', and assume that* $\sum_{j=1}^{\infty} |\lambda_{2j} - \lambda_{2j-1}| < \infty$. *Then the trace formula yields*

$$(2.9) \qquad q(0) = \lambda_0 + \sum_{j=1}^{\infty} \lambda_{2j} + \lambda_{2j-1} - 2\mu_j .$$

Conversely if (2.9) holds for arbitrary x, that is, for $q(x)$ with $\mu_j = \mu_j(x)$, then $q(x)$ is 'reflectionless'.

Ergodic potentials are brought into this picture by a beautiful result of Kotani [6] . In our setting it can be stated:

Lemma 2.4: *Those ergodic potentials $q(x) \in C(\mathbb{R})$ such that $\sigma(L(q))$ is entirely absolutely continuous spectrum, satisfy (2.7), and thus are 'reflectionless'.*

§3. Analysis on a torus

Formula (2.9) suggests very natural spectral data from which one may recover a 'reflectionless' potential. Consider the potential after translation by ξ ; the spectrum $\sigma(L(q(\cdot + \xi))) = \sigma(L(q))$, with the gap endpoints $\{\lambda_j\}_{j=0}^{\infty}$ invariant. The Greens function at the new origin, expressed in terms of the original one, is $g(\xi,\xi;\lambda)$. If $\mu_j(\xi)$ are the zeros of $g(\xi,\xi;\lambda)$ in the spectral gap $(\lambda_{2j-1},\lambda_{2j})$, then the trace formula recovers $q(\xi) = \lambda_0 + \sum_{j=r}^{\infty} \lambda_{2j} + \lambda_{2j-r} - 2\mu_j(\xi)$

Thus the auxilliary spectrum $\{\mu_j(\xi)\}_{j=r}^{\infty}$ determine $q(\xi)$. It remains to discuss the dynamical behavior of $\mu_j(\xi)$:

$$0 = \frac{d}{d\xi} g(\xi,\xi;\mu_j(\xi)) \tag{3.1}$$

$$= \partial_\xi g(\xi,\xi;\mu_j(\xi)) + \partial_\lambda g(\xi,\xi;\mu_j(\xi)) \frac{d\mu_j}{d\xi} .$$

A computation shows that $\partial_\xi g(\xi,\xi,\mu_j) = \pm 1$, where the p occurs if μ_j is an eigenvalue of the left $1/2$ line problem of (2.2), and the minus occurs if μ_j is a right $1/2$ line eigenvalue (it must be one or the other if $\mu_j \in (\lambda_{2j-1}, \lambda_{2j})$). Setting $\sigma_j = \pm 1$ appropriately,

$$\frac{d\mu_j}{d\xi} = \frac{\sigma_j}{\partial_\lambda g(\xi,\xi;\mu_j)} \equiv \sigma_j V_j(\mu,\xi) . \tag{3.2}$$

The V_j are interpreted as components of a vector field, the analog of the Dubrovin equation in the periodic case. Explicit dependence of V_j on ξ is not desirable; a simple application of a result in harmonic analysis proves the following.

<u>*Lemma 3.1*</u>. *The Dubrovin vector fields $\vec{V}(\vec{\mu})$ is autonomous, depending only upon $\{\lambda_j\}_{j=0}^{\infty}$ as parameters, and $\{\mu_j\}_{j=1}^{\infty}$ as independent variables.*

With a change of variables, the vector field (3.2) is seen to live on a torus. Letting $\xi_j = (\lambda_{2j} + \lambda_{2j-1})/2$ a gap midpoint, and $\varsigma_j = (\lambda_{2j} - \lambda_{2j-1})/2$ the radius, define an angle φ_j by

$$\varsigma_j \cos \varphi_j = \mu_j - \xi_j \ , \tag{3.3}$$

where $0 \le \varphi_j < \pi$ if $\sigma_j = +1$, and $\pi \le \varphi_j < 2\pi$ if $\sigma_j = -1$. The angles $\{\varphi_j \in [0, 2\pi)\ ,\ j = 1, 2, \ldots\} \equiv \varphi$ define global variables on a torus T. Define a topology on T with the distance function

$$d(\varphi, \psi) = \sup_j \varsigma_j^{1/2} \left| |\varphi_j - \psi_j|_{\mathrm{mod}\ 2\pi} \right| . \tag{3.4}$$

We assume that $\sum_{j=1}^{\infty} \varsigma_j < \infty$; then T is compact with respect to this topology. Our first result on the structure of a 'reflectionless' potential is the following.

<u>Theorem 3.2:</u> *The translation $q(x) \to q(x + \xi)$ gives a continuous path $\{\varphi_j(\xi)\ ;\ j = 1, 2, \ldots\}$ on T. Thus a 'reflectionless' potential is recovered by the trace formula from a continuous path on T.*

The change of variables (3.3) transforms the vector field (3.2) into the equation

$$\frac{d\varphi_j}{d\xi} = W_j(\varphi) \quad , \tag{3.5}$$

with $W_j(\varphi) = -\sigma_j V_j(\varsigma_j \cos \varphi_j + \xi_j) / \varsigma_j \sin \varphi_j$. The results of global existence of orbits of (3.5), and continuous dependence upon initial position depend upon the regularity of the vector field, that is, whether $W(\varphi) \in C^1(T)$. Our estimate to this end depends upon certain geometrical properties of the spectrum as a closed set in $\mathbb{R}$. Define the two-point function

$$Z_2 = \sup_{j\geq 0} \sum_{k\neq j} \frac{(\xi_j \xi_k)^{1/2}}{\rho_{jk}} \|W_j\|_{L^\infty} .$$

where $\rho_{jk} = \min\{|\lambda_{2k-1} - \lambda_{2j}|, |\lambda_{2j-1} - \lambda_{2k}|\}$ is the distance between the jth and k^{th} spectral gaps. For convenience define the one-point function $Z_1 = \sum_{j=1}^{\infty} \xi_j$. If Z_2 is infinite the set $\sigma(L(q))$ is quite 'wild' ; by contrast standard Cantor sets with relative Lebesque measure $> 6/7$ are tame [3].

Theorem 3.3. *Let $\sigma \subseteq \mathbb{R}$ be a closed set whose compliment*

$$(-\infty, \lambda_0) \cup \bigcup_{j=1}^{\infty} (\lambda_{2j-1}, \lambda_{2j})$$ *is associated with a torus T.*

If both Z_1, $Z_2 < \infty$, then $W(\varphi) \in C^1(T)$.

Corollary 3.4: *If $q(x)$ is 'reflectionless', and $\sigma(L(q))$ has both Z_1 and Z_2 finite, then the hull*

$$\mathcal{H}(q) = \overline{\{q(x+\xi) \,;\, \xi \in \mathbb{R}\}}$$

is compact. The closure is taken with respect to uniform convergence on compact sets.

The proofs of the therems 3.2 and 3.3 appear in [3].

This corollary has immediate bearing on the case of an ergodic potential, for the hull $\mathcal{H}(q) = \text{supp}\, P \subseteq C(\mathbb{R})$, for P the probability measure determing the statistics of $q(x)$. Suppose that $q(x)$ is ergodic, and yet $\sigma_{sing}(L(q)) = \sigma_{pp}(L(q)) \cup \sigma_{sc}(L(q)) = \emptyset$. The corollary and Lemma 2.4 imply that either the support of the probability measure P is very small, or the spectrum is a very complicated set.

Theorem 3.5: *If $q(x)$ is ergodic, and $\sigma(L(q)) = \sigma_{ac}(L(q))$, then either (i) supp P is compact, or else (ii) the set $\sigma(L(q))$ is 'wild', with Z_1 or Z_2 infinite.*

§4. The heirarchy of the KdV equation

The Dubrovin vector field $W(\varphi)$ on the torus T gives rise to a flow; this is the evolution of a Hamiltonian system corresponding to the conservation of momentum for the Korteweg deVries equation (KdV). For reflectionless potentials $q(x)$, the vector fields for the other Hamiltonian systems of the KdV heirarchy can also be represented as tangent vector fields to T. These are obtained from higher traces of $L(q)$ and $L_D(q)$.

Lemma 4.1: *Consider $q(x) \in C(\mathbb{R})$, which is r-times continuously differentiable at zero. Then the β-traces exist for $2(\beta - 1) \leq r$;*

$$(4.1) \qquad P_\beta(q)(0) = \left(-\frac{1}{2}\right)^\beta \operatorname{tr}\left(L^\beta(q) - L^\beta_D(q)\right) .$$

where P_β is a universal polynomial in $q(0), \ldots, \partial_x^r q(0)$ with rational coefficients. If $q(x)$ is reflectionless, and the moment

$$\sum_{j=1}^{\infty} \xi_j^{\beta-1} \varsigma_j < \infty , \quad \text{then}$$

$$(4.2) \qquad P_\beta(q)(0) = L_\beta \equiv \lambda_0^\beta + \sum_{j=1}^{\infty} \lambda_{2j}^\beta + \lambda_{2j-1}^\beta - 2\mu_j^\beta .$$

There is a well known relationship between the higher traces (4.2) and gradients of the KdV Hamiltonians [9], arising from the relationship between two formal expansions of the Greens function $g(0,0;\lambda)$:

$$(4.3) \qquad 2\sqrt{-\lambda}\, g(0,0;\lambda) \sim \exp\left(-\frac{1}{2}\sum_{\gamma=1}^{\infty}\frac{1}{\gamma\lambda^{\gamma}} L_{\gamma}\right)$$

$$\sim \sum_{\beta=0}^{\infty}\left(\frac{1}{2\lambda}\right)^{\beta} \operatorname{grad} H_{\beta} \ .$$

Equating terms in the two series in (4.3), the relationship is described as

$$(4.4) \qquad \operatorname{grad} H_{\beta} = R_{\beta}(L_1, \ldots, L_{\beta}) \ ,$$

where R_{β} is another universal polynomial with rational coefficients. For example; $\operatorname{grad} H_1(q) = L_1$, giving rise to translation flow, and $\operatorname{grad} H_2(q) = L_2 - \frac{1}{2}(L_1)^2$ gives the KdV equation. Through (4.4) we obtain tangent vector fields on T .

<u>Lemma 4.2:</u> *The Hamiltonian vector field in the KdV heirarchy*

$$(4.5) \qquad \partial_x \operatorname{grad} H_{\beta}(q)$$

corresponds to the vector field on T *given by*

$$(4.6) \qquad W_j^{\beta}(\varphi) = \partial_{\mu_j} R_{\beta}(L_1, \ldots, L_{\beta})\, W_j(\varphi) \ .$$

If both $Z_1, Z_2 < \infty$, *and the moment* $\sum_{j=1}^{\infty} \xi_j^{\beta-1} \zeta_j < \infty$,

then $W^{\beta}(\varphi) \in C^1(T)$. *Furthermore,* $W(\varphi)$ *and* $W^{\beta}(\varphi)$ *commute.*

The boundedness and regularity of $W^{\beta}(\varphi)$ in lemma 4.2 gives the immediate consequence of global existence and continuous dependence of orbits $\varphi(t)$ of

$$\frac{d}{dt}\varphi_j = W_j^\beta(\beta) \quad . \tag{4.7}$$

Multiply each component by $-\xi_j \sin\varphi_j$, sum over j, and use (4.6) we obtain a solution of the βth evolution equation of the KdV heirarchy,

$$\frac{d}{dt}q = \partial_x \operatorname{grad} H_\beta(q) \quad . \tag{4.8}$$

The orbit $\{\varphi(t)\,;\, -\infty < t < \infty\}$ gives the value of $q(x,t)$ at $x = 0$. Since $[W, W^\beta] = 0$ the value of $q(x,t)$ for all x is recovered by the translation flow.

Theorem 4.3: *Among the 'reflectionless' potentials* $q(x) \in C^{2(\beta-1)}(\mathbb{R})$ *with spectra* $\sigma(L(q))$ *satisfying* $Z_1, Z_2 < \infty$ *and the moment condition* $\sum_{j=1}^{\infty} \xi_j^{\beta-1} \zeta_j > \infty$, *the evolution equations of the KdV heirarchy have global solutions.*

This extends the results of McKean, Van Moerbeke and Trubowitz [9] [10], Dubrovin, Matveev and Novikov [5] and others. The results on the convergence of the quantities (4.1) and (4.2), and the smoothness of the vector field (4.6) appear in [4].

§5. The reflectionless spectral class. The last question that we discuss in this article is the characterization of the reflectionless spectral class of a closed set $\sigma \subseteq \mathbf{R}$. In previous sections, we have found that reflectionless potentials $q(x) \in C(\mathbf{R})$ with spectrum $\sigma = \sigma(L(q)) \subseteq \mathbf{R}$ can be recovered via the trace formula from an orbit of the Dubrovin vector field $\vec{W}(\varphi)$ (3.5) on a torus T. From Lemma 3.1 this torus and vector field are defined uniquely given the set σ. A natural question is whether every orbit of $\vec{W}(\varphi)$ gives rise to a reflectionless potential with spectrum σ; that is, whether T represents the full reflectionless spectral class of σ.

To rephrase this problem, given a set $\sigma \in \mathbf{R}$ define the torus $T = \{\varphi; \mu_j(\varphi) = \zeta_j \cos \varphi_j + \xi_j \in (\lambda_{2j-1}, \lambda_{2j})\}$ and two functionals

$$Q(\varphi) = \lambda_0 + \sum_{j=1}^{\infty} \lambda_{2j} + \lambda_{2j-1} - 2\mu_j(\varphi) \tag{5.1}$$

the trace formula, and

$$G(\varphi;\lambda) = \frac{1}{2\sqrt{\lambda_0 - \lambda}} \prod_{j=1}^{\infty} \frac{\mu_j(\varphi) - \lambda}{\sqrt{(\lambda_{2j} - \lambda)(\lambda_{2j-1} - \lambda)}}\,. \tag{5.2}$$

If $Z_1 < \infty$ then $Q \in C(T)$, and if additionally $Z_2 < \infty$, then $G(\cdot;\lambda) \in C^1(T)$. Furthermore, the vector field (3.5) is well defined. Given a complete orbit $\{\varphi(x); x \in \mathbf{R}\}$, we may define $q(x) = Q(\varphi(x)) \in C(\mathbf{R})$. The question is (i) whether $q(x)$ is reflectionless, and (ii) if the diagonal of the Green's function for $L(q)$ coincides with the function $G(\varphi(x);\lambda)$. If so, the torus T is identified as the reflectionless spectral class of σ.

As a corollary, we obtain partial results on another question; which closed sets $\sigma \subset \mathbf{R}$ are spectra for some Schrödinger operator with potential $q(x) \in C(\mathbf{R})$? For a large class of sets σ, we construct reflectionless potentials with $\sigma(L(q)) = \sigma$; in fact, we construct the full torus T, the reflectionless spectral class. Certain other potentials may also have spectrum σ. These, of course, will violate the reflectionless condition (2.8), and will not be parameterized by T.

Theorem 5.1 *Let $\{\varphi(x); -\infty < x < \infty\} \subseteq T$ be a complete orbit of the Dubrovin vector field (3.5) on T. Form a potential from the trace formula*

$$q(x) = Q(\varphi(x)) .$$

In order that the Green's function $g(x,x;\lambda)$ for $L(q)$ coincide with $G(\varphi(x);\lambda)$, it suffices (i) that $w(\varphi) \in C^2(T)$, and (ii) that the following identity is satisfied:

$$\text{(5.3)} \qquad \frac{1}{4G^2(\varphi;\lambda)} + \lambda - Q(\varphi) = \frac{1}{4}\sum_{j=1}^{\infty}\left(\frac{1}{(\lambda-\mu_j)^2} + \frac{\omega_j}{\lambda-\mu_j}\right)V_j^2(\mu) .$$

The $V_j(\mu)$ are components of the vector field in (3.2), and

$$\omega_j(\mu) = \frac{1}{\lambda_0-\mu_j} + \frac{1}{\lambda_{2j}-\mu_j} + \frac{1}{\lambda_{2j-1}-\mu_j} + \sum_{k\neq j}\frac{1}{\lambda_{2k}-\mu_j} + \frac{1}{\lambda_{2k-1}-\mu_j} - \frac{2}{\mu_k-\mu_j} .$$

The idea of Theorem 5.1 is straightforward, for the diagonal of the Green's function $g(x,x;\lambda)$ satisfies the differential equation

$$\text{(5.4)} \qquad \partial_x^2 g/g - (\partial_x g/g)^2 = 2(q(x)-\lambda) - 1/2g^2 ,$$

and vice versa any function satisfying (5.4) will be the diagonal of the Green's function for $L(q)$. The left hand side of (5.4) yields the right hand side of (5.3) by direct computation, as long as the vector field $\vec{W} \in C^2(T)$. These are analogous formulae to ones appearing in McKean and Van Moerbeke [9].

It remains only to discuss geometrical conditions on the set σ under which Theorem 5.1 (i) and (ii) will hold. The one and two point functions Z_1, Z_2 were defined in §3; define analogously

$$Z_3 = \sup_j \left\{ \sum_{l,m \neq j} \frac{(\zeta_j \zeta_l \zeta_m)^{1/2}}{\rho_{jl}\rho_{jm}} C_j, \quad \sum_{m \neq j} \frac{\zeta_j \zeta_m^{1/2}}{\rho_{jm}^2 C_j} \right\},$$

with $C_j = \| W_j(\varphi) \|_{L^\infty(T)}$.

Lemma 5.2 *If all Z_1, Z_2, and $Z_3 < \infty$, then $W(\varphi) \in C^2(T)$.*

This satisfies point (i) of Theorem 5.1. The proof appears in [4], and is similar to the direct estimates of Theorem 3.3.

The approach of Jacobi via the residue calculus is used to address point (ii) of the theorem. Use the Cauchy integral formula for $1/4G^2(\varphi;\lambda) + \lambda - Q(\varphi)$ over the contour $\Gamma = \Gamma_{outer} \cup \Gamma_{inner}$, with $\Gamma_{outer} = \{Re^{i\theta},\ R \gg |\lambda|, \arctan(\varepsilon/R) < \theta < 2\pi - \arctan(\varepsilon/R)\}$ and $\Gamma_{inner} = \{\xi \pm i\varepsilon,\ \lambda_0 - \varepsilon \leq \xi \leq R\} \cup \{\lambda_0 - \varepsilon + i\eta; -\varepsilon \leq \eta \leq \varepsilon\}$. As $R \to \infty$ the outer integral vanishes. For fixed R as $\varepsilon \to 0$ the inner integral should pick up only the residues of the double poles at $\{\mu_j; j \in \mathbf{Z}\}$. Since however these poles typically have accumulation points, and in fact the entire set σ may be entirely accumulation points, the integral over Γ_{inner} must be controlled as $\varepsilon \to 0$. Denote $R(\varphi;\lambda)$ the right hand side of identity (5.3).

Lemma 5.3 *If $Z_1, Z_2 < \infty$, then $R(\varphi;\lambda) \in C(T)$.*

Thus it suffices to obtain identity (5.3) in the case that all but finitely many $\mu_j \equiv \xi_j$; that is, they are centered in their gap $(\lambda_{2j-1}, \lambda_{2j})$.

Lemma 5.4 *If $Z_1, Z_2 < \infty$, and additionally the moment $M_2 = \sum_{l=1}^{\infty} l^2 \zeta_l$ is finite, then (5.3) is satisfied.*

The proof involves control of a sequence of contour integrals $\Gamma_{inner} = \Gamma_{inner}^{\varepsilon}$ as $\varepsilon \to 0$, and appears in [4].

Thus under relatively simple geometrical conditions on the closed set σ; that Z_1, Z_2, Z_3 and $M_2 < \infty$, the associated torus represents the reflectionless spectral class of σ, and the analog of the inverse spectral procedure of [5][9][10] carries through.

§6. References

[1] J. Avron and B. Simon, Almost periodic Schrödinger operators. I. limit periodic potentials. Commun Math. Phys. 82 (1981) p.101-120.

[2] V. Bargmann, Remarks on the determination of a central field of force from the elastic scattering phase shift. Phys. Rev. 75 (1949) p. 301-303

[3] W. Craig, The trace formula for Schrödinger operators on the line. preprint

[4] W. Craig, Trace formulae and the higher KdV flows. in preparation.

[5] A. Dubrovin, V. Matveev and S. Novikov, Nonlinear equations of Korteweg deVries type, finite zone linear operators and Abelian varieties. Russian Math. Surveys 31 p. 51-146.

[6] S. Kotani, One dimensional random Schrödinger operators and Herglotz functions. Proc. Taniguchi Symp. on Prob. Methods in Math. Physics, Katata. to appear.

[7] B. Levitan, Almost periodicity of infinite zone potentials. Math USSR Isv. 18 (1982) p. 249-273.

[8] B. Levitan, Approximation of infinite zone potentials by finite zone potentials. Math USSR Isv. 20 (1983) p. 55-87

[9] H. McKean and P. van Moerbeke, The spectrum of Hills' equation. Invent. Math. 30 (1975) p. 217-274.

[10] H. McKean and E. Trubowitz, Hills' operator and hyperelliptic function theory in the presence of infinitely many branch points. Commun. Pure Appl. Math. 29 (1976) p. 143-226.

[11] J. Moser, Integrable Hamiltonian Systems and Spectral Theory Lezioni Fermiane. Academia Nazionale dei Lincei-Scuola Normale Superiore-Pisa 1981.

[12] J. Moser, An example of a Schrödinger operator with almost periodic potential and nowhere dense spectrum. Comment Math. Helv. 56 (1981) p. 198-224.

HUNTING OF THE QUARTON

Anthony P. Csizmazia
Mathematical Sciences Research Institute
1000 Centennial Dr., Berkeley, CA 94720, U.S.A.

Abstract

A summary is given of some of the author's results concerning the "CS equations" (C: Congruent S: Schroedinger operators) and the "ACS eqs." (A: Altered). The CS eqs., respectively ACS eqs. are strongly 2+1-D (two spatial dimensions in addition to time) generalizations of the cornerstone (higher order) Korteweg-de Vries (KdV) eqs.,respectively IMKdV eqs. (I: Integrated, M: Modified) of soliton theory. The operator theoretic context which engenders the CS eqs. and the ACS eqs. is developed. The key to the relation between the CS eqs. and the ACS eqs. is either of the presented semifactorizations of the 2-D Schroedinger operator with potential function. Semifactorizations of the 4-D and 8-D Schroedinger operator are also achieved. Topics for further research are delineated.

Let x, t, (t:time) be real variables. Say u is a real valued function of x, t, It is required that on the line, u and each of its spatial derivatives vanishes rapidly at infinity. The case with u periodic in x is also of interest. The KdV$_{\ell+1}$ eq. has the form $\dot{u} = -2\partial[T_\ell(u)]$, $T_\ell = P\partial_x^{-1}$ with P the differential operator part of $(\partial_x - \partial_x^{-1} \circ u)^{2\ell+1}$.

We make assumptions about the function m which parallel those on u. The IMKdV$_{\ell+1}$ eq. is $\dot{m} = -2E_\ell(m_x)$ with E_ℓ the differential operator part of $(\partial - m_x\partial^{-1} \circ m_x)^{2\ell}$. The MKdV$_{\ell+1}$ eq. is obtained by applying ∂_x to both sides of the IMKdV$_{\ell+1}$ eq.

Variant of the Miura transformation: $u = m_{xx} + (m_x)^2 = J(m)$.

Linearization of J: $(-\partial_x^2 + u)(\phi) = 0$, $\phi = e^m$.

u is a solution of the KdV$_{\ell+1}$ eq. provided $u = J(m)$ and m is a solution of the IMKdV$_{\ell+1}$ eq.

Let M be the plane or the torus with standard coordinates x, y Define $d = \partial_x + i\partial_y$. Then $d^* = -\partial_x + i\partial_y$, $\Delta = \partial_x^2 + \partial_y^2 = -d^*d = -dd^*$. Take $b_{\underline{k}} = (d^*)^k(b)$. At each instant in time let u be a real valued function defined on M. When M is the plane require that u and each of its spatial derivatives vanish rapidly at infinity. Let $s = \omega(u)$, with $\omega = d^{-1}d^*$. When M is the plane s is required to vanish at infinity. On the torus s is taken so as to have 0 as its mean value. The CS eq. is $\dot{u} = 2\mathrm{Re}\, d^*(u_{\underline{2}} + 3su)$. The higher order CS eqs. are specified next. Set $L' = (d^*)^{-1} \circ (-\Delta + u) = d + (d^*)^{-1} \circ u$. Say $K = Td^*$ with T a pseudodifferential operator (ψdo) of the form

$$1 + (d^*)^{-1} \circ a_1 (d^*)^{-1} + (d^*)^{-2} \circ a_2 (d^*)^{-2} + \cdots$$

Note that K is a d^* operator, i.e. K involves only d^* derivatives or their inverses. Assume: *K commutes with L'.* Let $r = 2\ell + 1$. Take $H\langle r\rangle$ to be the total part $H(r)^+$ with order ≥ 0 of $H(r) = 2\mathrm{Re}(K^r)$. The CS$_{\ell+1}$ eq. is $\dot{u} = H\langle r\rangle^*(u)$. The CS eq. is CS$_2$.

Set $(L, H) = LH + H^*L$. The CS eqs. arise [1], [2] from the C eq. (C: Congruence) $\dot{L}_t = -(L_t, H_t)$. Indeterminacy principle: *Assume $L^* = L$. Then $(L, H) = (L, H')$, if*

$H = H' + SL$ *with* $S^* = -S$. The "special" C eq. has L_t equal to the 2-D Schroedinger operator $-\Delta + u$. H_t is required to be a differential operator with real valued coefficients. Each solution H_t of the special C eq. can be reduced, via a judicious choice of S, to the canonical form

$$H = Td^* - T^*d, \quad T = T_\ell = \sum_n (d^*)^{\ell-n} \circ a_n (d^*)^{\ell-n}, \; a_0 = 1$$

Theorem 1 *The C eq. with H_t in canonical form holds iff for some ℓ: $H_t = H\langle 2\ell+1\rangle$ and u evolves according to the $CS_{\ell+1}$ eq.*

The CS_j eq. is a *strongly* 2+1-D generalization of the KdV_j eq. in the sense of Theorem 2 below. Say M is the plane. Let x', y' be obtained from x, y by rotation through the angle θ. Assume $f(x', t_1)$ [respectively $g(y', t_2)$] is a solution of the KdV_j eq. in the coordinates x', t_1 (respectively y', t_2). Set $t_1 = (-1)^j t \cos(r\theta)$, $t_2 = (-1)^{j-1} t \sin(r\theta)$, $r = 2j - 1$.

Theorem 2 *The $CS_{\ell+1}$ eq. in x, y, t coordinates has, for each θ, the solution*

$$u(x, y, t) = f(x', t_1) + g(y', t_2).$$

Note: This u is not required to vanish at infinity.

> *Research Topic.* Find a strongly 2+1-D generalization (see Theorem 2) of the sine-Gordon eq.

Strong evidence is presented in [2] that the CS_j eq. has the sequence of independent invariants $\int_M T_k(u)$, $k = 0, 1, \ldots$

Next the ACS eq. is given. Say m is a function. The assumptions on m parallel those on u. Set $w = d(m)$. Let $\beta = \omega(|w_{\underline{1}}|^2)$. The ACS eq. is $\dot{m} = 2\text{Re}\{m_{\underline{3}} + [3\beta + (m_{\underline{1}})^2]m_{\underline{1}}\}$. We now present the $\text{ACS}_{\ell+1}$ eq. It is a strongly 2+1-D generalization of the $\text{IMKdV}_{\ell+1}$ eq. Take A_ℓ to be the differential operator part of $(1-N)^{-1}K^r(1+N)^{-1}$, with $r = 2\ell+1$ and $N = (d^*)^{-1} \circ \overline{w}$. The $\text{ACS}_{\ell+1}$ eq. is $\dot{m} = 2\text{Re}A_\ell(m)$. The ACS eq. is ACS_2.

The equation $(-\Delta + u)(\phi) = 0$, $\phi = e^m$ leads to the 2-D generalization of the variant of the Miura transformation $u = \Delta(m) + (m_x)^2 + (m_y)^2 = J(m)$.

Theorem 3 *u is a solution of the $CS_{\ell+1}$ eq., if $u = J(m)$ and m is a solution of the $ACS_{\ell+1}$ eq.* (See Lemma 1 below.)

Each time invariant of the CS_j eq. yields an invariant of the ACS_j eq. on replacing u with $J(m)$. In particular, the prenorm $[\int (m_x)^2 + (m_y)^2]^{1/2}$ is an invariant of ACS_j. If m is a solution of the ACS_j eq. then so is $-m$. This gives rise to a Bäcklund transformation of CS_j.

> *Research Topics.* What interpretations does the ACS eq. have in physics? On the plane does the ACS eq. have soliton-like solutions vanishing rapidly at infinity? Make computer studies of this equation. Express the ACS_j eq. as an infinite dimensional Hamiltonian system, if possible.

The relation between the CS eqs. and the ACS eqs. of Theorem 3 is assured by Lemma 1 (or Lemma 2) below. In two or more dimensions $-\Delta + u$ has no nontrivial factorization into a product of differential operators. Let M be the 2-D plane or torus. Set $u = J(m)$ and $L = -\Delta + u$. Let $D = d - w$ with $w = d(m)$. The first "semifactorization" of L is $L = \operatorname{Re} D^*D$. Let τ denote the operator transpose without conjugation.

Lemma 1 *The C eq. $-\dot{L} = (L, H)$ holds, if the MC eq. (M: Modified)*

$$-\dot{D} = DH + \gamma_0 D + \gamma_1 \overline{D}$$

holds, H, γ_0, γ_1 being differential operators with $\overline{H} = H$, $\gamma_0^* = -\gamma_0$ and $\gamma_1^\tau = -\gamma_1$.

The $CS_{\ell+1}$ eq. is the progeny of the C eq. with $H = H\langle r\rangle$, $r = 2\ell + 1$. Define the $MCS_{\ell+1}$ eq. to be the equation obtained by applying d to both sides of the $ACS_{\ell+1}$ eq. $MCS_{\ell+1}$ is a strongly 2+1-D generalization of the $MKdV_{\ell+1}$ eq. The $MCS_{\ell+1}$ eq. arises from the MC eq. with $H = H\langle r\rangle$, $\gamma_0 = Q - Q^*$. Q is taken to be a d^* operator: $Q = -(d^*)^r +$ lower order terms. Then $Q = -(B^{-1}K^r B)^+$ with $B = 1 - N$. $\gamma_1 = \gamma^+$. Here $\gamma = [w + (d^*)^{-1} d \circ \overline{w}, Q](d^* - \overline{w}N)^{-1}$. In the case of dependence only on x, t: $\gamma_1 = 0$.

The identification $\phi \mapsto \begin{pmatrix} \phi \\ \phi \end{pmatrix}$ induces an identification of operators $L(\phi) \mapsto V \begin{pmatrix} \phi \\ \phi \end{pmatrix}$, where $V = \begin{pmatrix} -\Delta + |w|^2 & \Delta(m) \\ \Delta(m) & -\Delta + |w|^2 \end{pmatrix}$. V has the factorization, $V = (D')^* D$, where $D = \begin{pmatrix} w & -d \\ d^* & \overline{w} \end{pmatrix}$. This constitutes the second semifactorization of L.

Let $\mathcal{O}$ be the complex algebra of differential operators on M. Say A, B are in $\mathcal{O}$. Set $\langle A, B\rangle = \begin{pmatrix} A & B \\ \overline{B} & \overline{A} \end{pmatrix}$. Such $\langle A, B\rangle$ constitute a real $*$-algebra $\mathcal{A}$. Allow ϕ to be complex-valued. Each element of $\mathcal{A}$ maps the real vector space of all $\begin{pmatrix} \phi \\ \overline{\phi} \end{pmatrix}$ into itself. $\mathcal{O}$ over the reals embeds in $\mathcal{A}$ via $A \mapsto \langle A, 0\rangle$. Let $\alpha = 2^{-1}\langle 1, 1\rangle$, $\alpha' = 2^{-1}\langle 1, -1\rangle$.

Lemma 2 *The C eq. $-\dot{L} = (L, H)$ holds , if the "factor eq."*

$$-\dot{D}' = D'(H + G\alpha') + SD', \quad S^* = -S$$

holds with $H = \overline{H}$, $G = \overline{G}$ in $\mathcal{O}$ and S in $\mathcal{A}$.

Say $S = \langle \gamma_0, \gamma_1\rangle$. The MC eq. is the image of the factor eq. under $\langle A, B\rangle \mapsto A + B$.

> *Research Topic.* Consider the role of the ACS eq. in Lemma 1. Is there a 2+1-D version of the sine-Gordon eq. which plays an analogous role in a variant of Lemma 1? Hope is offered by certain results on the sine-Gordon eq. in 1+1-D . See pages 133, 143 and page 159, exercise 6, of [3].

The most important problem in soliton theory is for $n \geq 3$ to:

(*)*Find strongly $n + 1 - D$ analogues of the 1+1-D soliton equations.*

We have solved (*) for $n = 2$ using novel semifactorizations of the 2-D Schroedinger operator. Semifactorizations of the 4-D and 8-D Schroedinger operators are obtained below. The algebra of quaternions and the (nonassociative) Cayley algebra of octonions are used.

> *Research Topics.* Produce 4+1-D "soliton" equations using either of the semifactorizations of the 4-D Schroedinger operator described below. "Quarton": Soliton roaming about in four spatial dimensions. Is it a mythical beast? Embed the CS eq. (or the ACS eq.) into a higher dimensional soliton eq. in the strong sense suggested by Theorem 2.

The first semifactorization of the 2-D Schroedinger operator is generalized to 4-D in Lemma 3 below. Let M_v be the plane or torus of dimension $v = 2a$. Say $x_1, y_1, \ldots, x_a, y_a$ are the standard coordinates for M. Set $\Delta_v = \sum_h \partial_\alpha^2 + \partial_\beta^2$ with $\alpha = x_h$, $\beta = y_h$. Assume u, m are real valued functions on M. Take $L\langle v\rangle$ to be the Schroedinger operator $-\Delta_v + u$. Suppose $L\langle v\rangle(e^m) = 0$. This amounts to $u = \Delta(m) + \sum(m_\alpha)^2 + (m_\beta)^2$. Δ_4 factors via quaternions: $\Delta_4 = \bar{d}d = d\bar{d}$. $d = d_1 + d_2 j$, $d_h = \partial_\alpha + i\partial_\beta$, $j^2 = -1$ and $ji = -ij$. This generalizes to a semifactorization of $L\langle 4\rangle$. Take $w = d(m)$ and $D = d - w$. The differential operators on M with quaternion (respectively complex)valued coefficients constitute a real associative algebra $\mathcal{Q}$ (respectively $\mathcal{C}$). Say A_1, A_2 are in $\mathcal{C}$. Allow $c = \pm 1$. Define $\langle A_1, A_2\rangle = \begin{pmatrix} A_1 & A_2 \\ c\overline{A_2} & \overline{A_1} \end{pmatrix}$. $A_1 + A_2 j$ is identified with $\langle A_1, A_2\rangle_{-1}$. Thus $A^* = A_1^* - A_2^\tau j$.

Lemma 3 $L\langle 4\rangle = \mathit{Re}(D^*D)$.

Proof. Say $A = A_1 + A_2 j$, $B = B_1 + B_2 j$ with A_h, B_h in $\mathcal{C}$. We define

$$A \diamond B = A_1 B_1 - \overline{A}_2 B_2 + (A_2 B_1 + \overline{A}_1 B_2) j\,.$$

This "twisted product" is obtained from BA by replacing the subproducts $B_h A_h$, $B_h \overline{A}_h$ with $A_h B_h$, $\overline{A}_h B_h$ respectively. If a, b are quaternion valued functions on M, then $a \diamond b = ba$. $\overline{AB} = \overline{A} \diamond \overline{B}$. So $L\langle 4\rangle = -\Delta + |w|^2 + \mathrm{Re}A$ with $A = \bar{d} \circ w - w \diamond \bar{d} = \bar{d}(w)$.

Note that $(A \diamond B)^* = B^* \diamond (A^*)$.

The second semifactorization of $L\langle 2\rangle$ is generalized to $L\langle 4\rangle$ by Eq. (1) together with Theorem 4 below. Let ϕ range over the real valued infinitely differentiable functions defined on M_4. Define $V = \langle -\Delta + |w|^2, \Delta(m)\rangle_1$. The identification,

$$L\langle 4\rangle(\phi) \mapsto V \begin{pmatrix} \phi \\ \phi \end{pmatrix} \tag{1}$$

holds. Later we factor V in an 8-D real algebra described next. Allow $c = \pm 1$. Say A, A', B, B' are in $\mathcal{Q}$. Let $\underline{A} = \langle A, A'\rangle_c$, $\underline{B} = \langle B, B'\rangle_c$. Define

$$\underline{AB} = \langle AB + cA' \diamond B', A \diamond B' + A'B\rangle_c\,. \tag{2}$$

When A, A', B, B' are quaternion valued functions, Eq. (2) defines the usual product for octonions. $\mathcal{Q}^2$ endowed with the product defined by Eq. (2) is an algebra. Take $\ell = \langle 0, 1\rangle_c$, $\underline{A} = A + A'\ell$, $\underline{A}^* = A^* + c(\overline{A}')^*\ell$. $(\underline{AB})^* = \underline{B}^*\underline{A}^*$. Say $c = 1$. Set $\underline{D}' = w - d\ell$.

Theorem 4 $V = \underline{D}'^{*}\underline{D}'$.

We pass to 8-D by considering operators defined on M_8. Take $c = -1$. Set $\underline{d} = d + d'\ell$, with $d' = d_3 + d_4 j$. $-\Delta_8 = \underline{d}^{*}\underline{d} = \underline{d}\underline{d}^{*}$. Let $\underline{w} = \underline{d}(m)$ and $\underline{D} = \underline{d} - \underline{w}$.

Theorem 5 $L\langle 8\rangle = Re(\underline{D}^{*}\underline{D})$.

> *Research Topics.* Factorizations of the Laplace operator on a plane M of arbitrary (finite) dimension are achieved by employing the Cayley numbers [4]. Generalize this to a semifactorization of the Schroedinger operator with potential function defined on M.

References

[1] A. Csizmazia, *Invariants under Congruence for Infinite Dimensional Operators*, Thesis, University of California, Berkeley, (1984).

[2] A. Csizmazia, *The Heisenberg-Lax Equation Generalized*, in "Some Topics on Inverse Problems", Proceedings of the XVIth Workshop on the Interdisciplinary Study of Inverse Problems, (Montpellier, France, Nov.30–Dec. 4, 1987), World Scientific Publishing Co., Singapore, (1988).

[3] G. L. Lamb, Jr., *Elements of Soliton Theory*, John Wiley & Sons, Inc., New York, (1980).

[4] B. L. van der Waerden, *Group Theory and Quantum Mechanics*, Vol. 214, Springer-Verlag, Berlin, (1974).

NEKHOROSHEV'S THEOREM AND PARTICLE CHANNELING IN CRYSTALS

H. Scott Dumas
Departments of Mathematics and Physics
The State University of New York at Albany
Albany, NY 12222
USA

I would like to report on recent work[1] I have done in developing a mathematical theory of particle channeling in crystals. This theory is rudimentary in that it is entirely classical and deterministic and applies only to positively charged particles moving in a highly idealized crystal. Nevertheless, it is the first time the existence of channeling motions has been rigorously deduced for a full three-degree-of-freedom model, and it gives an interesting mathematical unity to a range of high-energy particle motions in crystals, including the separate phenomena of "axial" and "planar" particle channeling. The theory makes use of results in Hamiltonian perturbation theory due to Nekhoroshev[2,3], and more recently to Benettin, Gallavotti, Galgani, and Giorgilli [4,5]. In such short space I cannot give a reasonable introduction to the relevant purturbation theory, but I am aided somewhat by A. Giorgilli's presentation in these proceedings of a different application of Nekhoroshev's results; in fact, my indebtedness to Professor Giorgilli goes beyond this conference, as it was primarily from his and his colleagues' work that I learned the perturbation techniques used here.

1. Particle Channeling

Channeling is the process by which fast charged particles are steered by the rows or planes of atoms in a perfect crystal. This can occur when a crystal is immersed in a particle beam in such a way that the beam direction is nearly aligned with a major axis or plane of the crystal; it can also occur when particles are created inside a crystal. Channeling has led to a better understanding of the properties of solids, and has had numerous technological applications. For example, it has been used as a material analysis tool to study crystal defects, surfaces and interfaces, and to measure the the location of crystal impurities. It has been used to measure nuclear lifetimes, to study the strain in "strained-layer superlattices," and to deflect high energy particle beams. In fact, a bent crystal can deflect beams of much higher energy than a traditional magnetic septum, though of course transmission is drastically reduced. Channeled electrons and positrons also emit electromagnetic radiation, an effect that has received increasing attention as a possible source of coherent, monoenergetic gamma radiation. The reader interested in these and other applications is referred to Refs. 6,7, and to the references therein.

2. Channeling Theories

The problem of understanding channeling theoretically was first taken up by J. Lindhard in the 1960's.[8,9] Lindhard viewed the problem of high-energy particle motion near a particular crystal axis or plane as a series of correlated grazing collisions with individual nuclei, and from this view he argued that, in the limit of small collision angles, the resulting particle motion is well-approximated by the motion of the same particles in a potential obtained by averaging over the axis or plane in question. Models using these directionally averaged potentials are known as "axial continuum" or "planar continuum" models, and Lindhard's arguments for them are especially convincing in the axial case. The same arguments are not so convincing in the planar case however, since a planar direction in a crystal is distinguished from an axial direction precisely by the fact that particles "see no periodicity" along that direction, so that the notion of "correlated grazing collisions with nuclei" is lost. Although the arguments leading to the continuum models were not mathematically complete, Lindhard's insight was vindicated through experiment, and his continuum models remain the starting point for theoretical discussions of channeling.

The mathematical situation was clarified somewhat when T. Burns and J. Ellison pointed out[10] that ODE averaging methods could be used to rigorously derive the continuum model in the axial case, providing not only estimates of the error between exact and continuum model motions on long time intervals, but also a systematic expansion of the exact motion in terms of the continuum model motion on the same interval. I was able to collaborate with Ellison and others[11,12] in refining the axial-averaging calculations, and the ideas presented here are an outgrowth of that collaboration. I am also indebted to P. Lochak for pointing out to me how averaging methods in the Hamiltonian case lead naturally to the work of Nekhoroshev.

3. The Perfect Crystal Model

Nekhoroshev-type results can be brought to bear on the channeling problem by first reducing it to the "cubic perfect crystal model"; that is, to a classical Hamiltonian system

$$\mathcal{H}(p,q) = \frac{1}{2m}p^2 + V(q), \tag{1}$$

where $p, q \in \mathbf{R}^3$ and the crystal potential V is real analytic and has period d in each component of q. The requirement that V be analytic is not as severe as it might seem, as perfect crystal potentials are typically constructed by summing screened, thermally averaged Coulombic atomic potentials centered on the lattice sites in $\mathbf{R}^3$. The thermal averaging is acheived by convolving the screened atomic potentials with the spatial probability distribution of the nuclei around the lattice sites, and this convolution renders the potential analytic. It should be pointed out that while this is the periodic potential which best models the thermal lattice vibrations of the crystal, since it has no "hard core" it cannot faithfully represent the process of close encounter (Rutherford scattering) between charged particles and nuclei. This objection is removed, at least mathematically, by introducing what I call the *channeling criterion*. If we assume that the potential V has been adjusted by means of an additive constant so that its minimum value is zero and its maximum value over $\mathbf{R}^3$ is $\mathcal{E}_M$, then for energies $\mathcal{E}_\perp$ $(0 \le \mathcal{E}_\perp \le \mathcal{E}_M)$, we may consider the subsets of configuration space

$$\mathcal{B}(\mathcal{E}_\perp) = \{q \in \mathbf{R}^3 \mid V(q) \ge \mathcal{E}_\perp\}. \tag{2}$$

If, as is assumed here, the potential governs the motion of positively charged particles, then clearly for sufficiently large $\mathcal{E}_\perp < \mathcal{E}_M$, the set $\mathcal{E}_\perp$ is the disjoint union of (slightly deformed) balls centered on the lattice sites. By choosing a physically suitable value for $\mathcal{E}_\perp$, we may distinguish particle trajectories which come too close to nuclei to be governed by the thermally averaged potential as those which enter the region $\mathcal{B}(\mathcal{E}_\perp)$. More precisely, fix $\mathcal{E}_\perp$, and consider a solution $(p(\tau), q(\tau))$ of the equations of motion corresponding to (1). Such a solution is a *channeling solution on the time interval $\mathcal{I}$* provided

$$q(\tau) \notin \mathcal{B}(\mathcal{E}_\perp), \quad \text{or equivalently} \quad V(q(\tau)) < \mathcal{E}_\perp \quad \forall \tau \in \mathcal{I}. \tag{3}$$

This is the channeling criterion, and it is assumed that the perfect crystal model is a good approximation for particle trajectories that satisfy it. A trajectory which fails to satisfy this criterion at time t_1 is assumed to suffer a "close encounter" with a nucleus, and is not viewed as a good approximation to an actual particle trajectory for subsequent times $t \geq t_1$.

The Hamiltonian (1) may be transformed to nondimensional, nearly integrable form as follows. Restricting attention to particles of fixed energy $\mathcal{E}$, i.e. to trajectories $(p(\tau), q(\tau))$ satisfying $\mathcal{H}(p(\tau), q(\tau)) = \mathcal{E}$, we define the scaled momentum (actions) $I \in \mathbf{R}^3$, the scaled position (angles) $\theta \in T^3$, the scaled potential W and scaled time t by

$$I = (m\mathcal{E})^{-1/2} p, \quad \theta = \frac{1}{d} q, \quad W(\theta) = \frac{1}{\mathcal{E}_\perp} V(\theta d), \quad t = \frac{1}{\tau_0} \tau. \tag{4}$$

Here $\tau_0 = d/\sqrt{m\mathcal{E}}$ is the time required for a particle to travel a distance $\sqrt{2}d$ in the potential-free case. The choice of scaling is motivated by the assumption $\mathcal{E} \gg \mathcal{E}_\perp$, so that trajectories satisfying the channeling criterion maintain a kinetic energy of approximately $\mathcal{E}$. The transformed Hamiltonian H now reads $H(I, \theta) = \frac{1}{2} I^2 + (\mathcal{E}_\perp/\mathcal{E}) W(\theta)$, or, writing $\epsilon = \mathcal{E}_\perp/\mathcal{E} \ll 1$, this becomes

$$H(I, \theta) = \frac{1}{2} I^2 + \epsilon W(\theta), \quad \text{where} \quad W \in C^\omega(T^3), \tag{5a}$$

$$\text{and} \quad 0 \leq W(\theta) \leq \mathcal{E}_M/\mathcal{E}_\perp. \tag{5b}$$

It is no surprise that the scaled Hamiltonian appears in action-angle form, since this scaling views the potential as a perturbation of rectilinear motion in the lattice T^3. System (5) is called the "nearly integrable form of the perfect crystal model" or the "scaled perfect crystal model."

The region of close encounter $\mathcal{B}(\mathcal{E}_\perp)$ described in equation (2) is immediately reformulated in terms of the scaled variables as

$$\mathcal{C}(1) = \{\theta \in T^3 \mid W(\theta) \geq 1\}, \tag{6}$$

and so the criterion (3) for a solution $(I(t), \theta(t))$ of Hamilton's equations corresponding to (5) to be a channeling trajectory on the time interval $\mathcal{I}$ becomes

$$\theta(t) \notin \mathcal{C}(1), \quad \text{or equivalently} \quad W(\theta(t)) < 1 \quad \forall\, t \in \mathcal{I}. \tag{7}$$

4. Application of Perturbation Theory

As it stands, system (5) is ripe for applications of peturbation techniques, but these applications must be interpreted with care. First of all, we note that solutions of the unperturbed system are simply straight windings of T^3, or rectilinear motions in the crystal with fixed direction $I(t) = I(0)$. The KAM theorem applies to this model and says that for sufficiently small perturbations, solutions with "highly nonresonant" initial directions $I(0)$ will remain uniformly close to the unperturbed rectilinear trajectories $I(t) = I(0)$ for all time. It can be shown that highly nonresonant initial directions are a subset of the nonchanneling directions, in the sense that particles with these initial directions quickly enter the neighborhoods $\mathcal{C}(1)$ surrounding the lattice sites. The conclusion of the KAM theorem may then be interpreted physically by saying that particles impinging on the crystal in highly nonresonant directions continue through the crystal in almost unperturbed straight lines, passing through the softened nuclei like high speed projectiles. This is the most spectacular example of how the thermally averaged potential fails to model close encounter processes adequately. Nevertheless, an important result may be derived which says that, for a set of initial directions consisting of a "nice" set in I-space which contains and closely approximates the highly nonresonant directions, trajectories are nearly rectilinear until they encounter nuclei. Morever, these encounters occur quickly, within a time which may be estimated in terms of ϵ and the Diophantine conditions defining the highly nonresonant directions. Together, I call these results the *spatial continuum model* by analogy with the other continuum models; lack of space prohibits a further discussion here, but the interested reader may consult Ref. 1 for details.

The next natural question is to ask what can be learned from Nekhoroshev's theorem, and it is here, I believe, that we acheive the most useful perturbation approach to the channeling problem. To get an idea of how this works, let us follow the Italian authors in seeking a near-identity canonical change of variables $(I, \theta) \mapsto (J, \phi)$ which will bring (5) into "normal form" by (almost) replacing the perturbing potential with a suitable average. The Lie method is the conceptually simplest way of introducing canonical transformations in this context; for this purpose we consider a function $\chi : D \to \mathbf{R}$ (the Lie generating function) analytic on a certain domain in phase space $D = K \times T^3$, where K is a "nice" subset of action space. Formally, the ϵ-near-identity transformation T generated by such a function acts on smooth functions of phase space by way of the operator $e^{\epsilon L_\chi}$, where $L_\chi = \{\ ,\chi\}$ is the Poisson bracket of χ with operands. To first order, the transformed Hamiltonian H' obtained in this way from H in (5) is

$$\begin{aligned} H'(J,\phi) &= (e^{\epsilon L_\chi} H)(J,\phi) = H(J,\phi) + \epsilon\{H,\chi\} + O(\epsilon^2) \\ &= \frac{1}{2}J^2 + \epsilon(W(\phi) - J \cdot \frac{\partial \chi}{\partial \phi}) + O(\epsilon^2). \end{aligned} \tag{8}$$

Attempting to eliminate the $O(\epsilon)$ term leads to what Arnol'd calls the homological equation. Expressed in terms of the complex Fourier coefficients of χ and W, the homological equation reads

$$2\pi i(k \cdot J)\chi_k(J) = W_k \quad \forall k \in \mathbf{Z}^3. \tag{9}$$

Since (9) has no solution for $k = 0$, we see immediately that the best that can be done is to reduce the $O(\epsilon)$ term to W_0, the space average of W. Even so, because of the small denominators $k \cdot J$, the convergence of the Fourier series for χ defined by solving (9) for the remaining k is problematic, and the nicest domain of convergence is obtained by disregarding Fourier coefficients with indices of order higher than some "ultraviolet cutoff" $N(\epsilon)$. Using the exponential decrease of the Fourier coefficients with increasing $|k|$, these high-order terms may be lumped with the $O(\epsilon^2)$ remainder, and the resulting truncated series comprising the low-order harmonics then has well-behaved small denominators on the subset of action space defined by

$$\{J \in \mathbf{R}^3 \mid |k \cdot J| \geq C\epsilon^{\alpha}|k|^{-p} \ \forall\, 0 < |k| \leq N\}, \tag{10}$$

where $|k| = |k_1| + |k_2| + |k_3|$, and $C, \alpha, p > 0$ are appropriate constants. It is this domain which converges with increasing N to a highly nonresonant subset of action space as occurs typically in the KAM theorem.

Although the domain (10) has nonempty interior and is "large" (i.e., its complement vanishes with ϵ), it does not include the low-order channeling directions; that is, directions J near to where the small denominators $k \cdot J$ vanish for nonzero $k \in \mathbf{Z}^3$ with small norm. It turns out that these directions are contained in the domains of transformations which reduce W not to its spatial average, but instead to its average over angles parallel to these directions. To see how this comes about, note that if a denominator $k \cdot J$ vanishes, then so does any denominator $j \cdot J$, where j is codirectional with k; i.e., lies in the maximal one-dimensional submodule $\mathcal{M}_k$ spanned by k. (The modular structure of $\mathbf{Z}^3$ is recalled in the next section.) If one deletes from the Fourier series of the transformation all Fourier coefficients with small denominators generated by a submodule $\mathcal{M}_k$, this modified transformation will be defined in and around the J-plane $k \cdot J = 0$, and it will reduce W to the "resonant subseries" $\sum_{j \in \mathcal{M}_k} W_j e^{2\pi i j \cdot \phi}$, which is precisely the average of W over angles parallel to the J-plane. In this case the transformed Hamiltonian is said to be in resonant normal form to first order. What Nekhoroshev calls an "analytic lemma" is then obtained by carrying the transformation to optimal order in ϵ for any suitable submodule of $\mathbf{Z}^3$; this results, roughly speaking, in a near-identity canonical transformation

$$T : K \times T^3 \to T(K \times T^3), \tag{11}$$

where K is a compact subset of action space which is *nonresonant with respect to* $\mathcal{M}$ *to order* N:

$$I \in K \ \Rightarrow\ |k \cdot I| \geq \frac{3}{2} A\epsilon^{\alpha}|k|^{-p} \quad \text{for all } k \notin \mathcal{M} \quad \text{with } |k| \leq N \tag{12}$$

for suitable positive constants A, α, and p. On $K \times T^3$, T transforms the Hamiltonian (5) to the resonant normal form $H' = H \circ T$, which may be written

$$H'(J,\phi) = \frac{1}{2}J^2 + \epsilon G(J,\phi) + R(J,\phi), \quad \text{where} \tag{13a}$$

$$G(J,\phi) = \sum_{k \in \mathcal{M}} G_k(J) e^{2\pi i k \cdot \phi}, \quad \text{and} \tag{13b}$$

$$R(J,\phi) = \sum_{k \notin \mathcal{M}} R_k(J) e^{2\pi i k \cdot \phi}. \tag{13c}$$

Furthermore, the remainder R is exponentially small in the sup norm over the relevant domain:

$$\|R\|_{K\times T^3} \leq 2E\epsilon e^{-\epsilon^{-\tau/4}}, \tag{14}$$

where τ, E are again suitable prespecified positive numbers.

5. Preliminary Definitions and the Channeling Theorem

In order to state an abbreviated form of the channeling theorem, a few definitions and notational conventions must be spelled out. First, in order to discuss resonances, we recall that a finite subset $\{k^{(i)}\}_{i=1}^m$ of $\mathbf{Z}^3$ generates a submodule of $\mathbf{Z}^3$ defined as $\{z \in \mathbf{Z}^3 | z = \sum n_i k^{(i)}, n_i \in \mathbf{Z}\}$. Each submodule of $\mathbf{Z}^3$ has dimension 0, 1, 2, or 3, in the sense that that the smallest vector subspace of $\mathbf{R}^3$ containing the submodule has that dimension. If $\{k^{(i)}\}_{i=1}^m$ generates a submodule of dimension n, then the maximal submodule generated by $\{k^{(i)}\}_{i=1}^m$ is the largest submodule of dimension n containing $\{k^{(i)}\}_{i=1}^m$. In the remainder of this article, the symbol $\mathcal{M}$ will refer to a 1- or 2-dimensional maximal submodule of $\mathbf{Z}^3$. Not surprisingly, every $\mathcal{M}$ is generated respectively by integer combinations of a 1- or 2-element basis in $\mathbf{Z}^3$; the *order* $|\mathcal{M}|$ of a maximal submodule is the smallest integer r such that $\mathcal{M}$ admits a basis of vectors with norm less than or equal to r.

To each $\mathcal{M}$ corresponds a resonance in action space, namely the orthogonal complement of $\mathcal{M}$. Actions in the neighborhood of a resonance (and sufficiently far from other low-order resonances) are channeling directions; this is formalized by first defining the *resonant zone corresponding to* $\mathcal{M}$ by

$$\mathcal{Z}_{\mathcal{M}}(C,\alpha) \;=\; \{I \in \mathbf{R}^3 \mid |k\cdot I| \leq C\epsilon^{\alpha}|k|^{-p}\ \forall k \in \mathcal{M}, 0 < |k| \leq |\mathcal{M}|\} \tag{15}$$

for appropriate positive values of C, α and p. In the case where $\mathcal{M}$ is one dimensional, the zone $\mathcal{Z}_{\mathcal{M}}(C,\alpha)$ will also be denoted $\mathcal{Z}_k(C,\alpha)$, where k is the unique generator (up to inversion) of $\mathcal{M}$. Finally, given $\mathcal{M}$, its associated *resonant block of order* $N \geq |\mathcal{M}|$ is defined as the subset of action space

$$\widehat{\mathcal{Z}}_{\mathcal{M}}^{C,\alpha}(N, C_1, \alpha_1) = \mathcal{Z}_{\mathcal{M}}(C_1,\alpha_1) \setminus \bigcup_{\substack{k\notin\mathcal{M}\\ |k|\leq N}} Int\, \mathcal{Z}_k(C,\alpha), \tag{16}$$

where *Int* denotes interior. These actions correspond to channeling directions; physically, it is interesting to note that when $dim\mathcal{M} = 1$ (corresponding to a planar direction), low-order "axial" zones are removed, as they should be.

We turn now to the relevant assumptions and definitions concerning the scaled potential W. Given $\mathcal{M}$ and given $\theta \in T^3 \approx \mathbf{R}^3/\mathbf{Z}^3$, write $\theta = \theta^* + \widehat{\theta}$, where θ^* and $\widehat{\theta}$ are the projections onto $span\mathcal{M}$ and $(span\mathcal{M})^\perp$ in $\mathbf{R}^3$. The resonant subseries $\Pi_{\mathcal{M}}W(\theta^*)$ corresponding to $\mathcal{M}$ is then just the average over the variables $\widehat{\theta}$ determined by $\mathcal{M}$; that is, $\Pi_{\mathcal{M}}W(\theta^*)$ is a continuum potential. We next consider regions in configuration space bounded by equipotential surfaces of the continuum potentials. Given $\mathcal{M}$ and any number $Q \geq 0$, define the closed subset $\mathcal{A}_{\mathcal{M}}(Q)$ as

$$\mathcal{A}_{\mathcal{M}}(Q) \;=\; \{\theta \in T^3 \,|\, \Pi_{\mathcal{M}}W(\theta^*) \leq Q\}. \tag{17}$$

Our assumption about W is then the following:

Assumption A. *There exists a critical order $M^* \geq 1$ such that for any $\mathcal{M}$ with order $|\mathcal{M}| \leq M^*$, there are numbers Q' and Q'', $0 \leq Q' < Q'' < 1$ with the property that given any $Q \in [Q', Q'']$, $\mathcal{A}_{\mathcal{M}}(Q) \neq \emptyset$ and $\mathcal{A}_{\mathcal{M}}(Q) \cap \mathcal{C}(1) = \emptyset$, where $\mathcal{C}(1) = \{\theta \in T^3 \,|\, W(\theta) \geq 1\}$.*

This physical assumption says that at sufficiently low order, there are equipotential surfaces of the continuum potential $\Pi_{\mathcal{M}} W$ which do not intersect the restricted region $\mathcal{C}(1)$. If $dim\mathcal{M} = 1$, these surfaces are planes; if $dim\mathcal{M} = 2$, the surfaces are cylindrical sheets or tubes. The assumption therefore says that at sufficiently low order, there are clear planar or axial pathways through the crystal which do not meet the close encounter region $\mathcal{C}(1)$. We will not examine this assumption further, since a calculation of M^* requires specific knowledge about the location of lattice sites in the crystal which we do not assume here. We simply remark that the assumption is satisfied by any physical cubic crystal for reasonable values of $\mathcal{E}_\perp$, which ultimately defines the size of $\mathcal{C}(1)$.

Finally, we define the transverse energy of a trajectory with respect to a particular continuum potential by

$$E_\perp(I, \theta) = \frac{1}{2}(I^*)^2 + \epsilon \Pi_{\mathcal{M}} W(\theta^*), \tag{18}$$

where again I^* and θ^* denote the projections of I and θ onto $span\mathcal{M}$. We now state an abbreviated form of the

Channeling Theorem. *Let $\mathcal{M}$ be a 1- or 2-dimensional maximal submodule of $\mathbf{Z}^3$ with order $|\mathcal{M}| \leq M^*$. Fix suitable positive values for E (related to $\mathcal{E}$, $\mathcal{E}_\perp$) and A (which determines the width of resonant zones), and assume certain consistency relations among the positive numbers α, τ, d, c, p, a which allow for a transformation (11) of the Hamiltonian (5) to the resonant normal form (13). (An example of such numbers is $\alpha = 1/8, \tau = 1/72, d = 1/2, c = 5/8, p = 5, a = 5/24$.) Fix the maximum initial (scaled) transverse energy $Q \geq Q'$ and the maximum change in (scaled) transverse energy $\delta > 0$ such that $Q + \delta \leq Q''$, where $Q' < Q'' < 1$ are defined in Assumption A ; set $C = |\mathcal{M}|^{p+1}(2Q/3)^{1/2}$. Then there exists a sufficiently small $\epsilon > 0$ such that any initial condition (I_0, θ_0) for (5) with initial transverse energy*

$$E_\perp(I_0, \theta_0) \leq \epsilon Q \tag{19}$$

and with suitable initial direction

$$I_0 \in \widehat{\mathcal{Z}}_{\mathcal{M}}^{\frac{5}{2}A,\alpha}(N, C, 1/2) \tag{20}$$

gives rise to a solution (I, θ) of (5) which satisfies the channeling criterion (7) on the exponentially long time interval $[0, T_0]$, where

$$T_0 = \frac{\sigma}{48E}\epsilon^\tau e^{\epsilon^{-\tau/4}} - 1. \tag{21}$$

This solution is approximated by a "generalized continuum model" solution in the sense that $(I, \theta) = T(J, \phi)$, where T is the near-identity transformation (11) bringing (5) into the normal form (13) and (J, ϕ) is the solution to (13) with initial condition $(J_0, \phi_0) =$

$T^{-1}(I_0, \theta_0)$. *Furthermore, on the interval* $[0, T_0]$, *the longitudinal momentum* $\widehat{I}(t)$ *is nearly constant:*

$$\|\widehat{I}(t) - \widehat{I}_0\|_\infty \leq \frac{3}{4} A\epsilon^c, \tag{22}$$

as is the transverse energy:

$$\frac{1}{\epsilon}|E_\perp(I(t), \theta(t)) - E_\perp(I_0, \theta_0)| \leq \delta. \tag{23}$$

A complete statement and a proof of this theorem can be found in Ref. 1. It should be stressed that the normal form (13) cannot be immediately applied to obtain the result; the theorem relies on the near-conservation of transverse energy and longitudinal momentum to ensure that trajectories beginning well inside the domain of a particular resonant normal form remain inside that domain. This phenomenon of "trapping into resonance" is a general characteristic of quasi-convex Hamiltonians (Hamiltonians whose unperturbed energy surfaces are convex) as was originally pointed out by Nekhoroshev and described in detail in Ref. 5.

As it stands, the channeling theorem corroborates the picture of channeling which has emerged experimentally over the last two decades: solutions are approximated by continuum models on very long time intervals, and transverse energy and longitudinal momentum are nearly conserved. It remains to be seen whether this deterministic, Hamiltonian approach can serve as the foundation for more comprehensive particle channeling theories, perhaps incorporating the statistical effects of lattice vibrations and of scattering by individual electrons.

References

1. H.S. Dumas, Ph.D. dissertation, University of New Mexico, 1988
2. N.N. Nekhoroshev, *Russian Math. Surveys* **32** (6): 1-65 (1977)
3. N.N. Nekhoroshev, in *Topics in Modern Mathematics, Petrovskii Seminar No. 5*: 1-58, O.A. Olcinik, ed., Consultants Bureau, London, 1980
4. G. Benettin, L. Galgani, and A. Giorgilli, *Celestial Mech.* **37**: 1-25 (1985)
5. G. Benettin and G. Gallavotti, *J. Stat. Phys.* **44**: 293-338 (1986)
6. J.U. Andersen, *Notes on Channeling*, University of Århus, 1986 (unpublished)
7. D.S. Gemmell, *Rev. Mod. Phys.* **46**: 129-227 (1974)
8. J. Lindhard, *Mat. Fys. Medd. Dan. Vid. Selsk.* **34**, no. 14 (1965)
9. Ph. Lervig, J. Lindhard, and V. Nielsen, *Nucl. Phys.* **A 96**: 481-504 (1967)
10. T.J. Burns and J.A. Ellison, *Phys. Rev.* **B 29**: 2790-2792 (1984)
11. H.S. Dumas and J.A. Ellison, in *Lecture Notes in Physics* **252**: 200-230, A.W. Sáenz, W.W. Zachary, and R. Cawley, eds., Springer, New York, 1986
12. H.S. Dumas, J.A. Ellison and A.W. Sáenz (in preparation)

Linearized Maps for the Davey-Stewartson I Equation

A.S. Fokas
Department of Mathematics and Computer Science
and
Institute for Nonlinear Studies
Clarkson University
Potsdam, New York 13676, U.S.A.

January 1989

INS #116[1]

Abstract

Simple equations are derived which express the deformation of the scattering data as a function of the deformation of the potential and vise versa. Applications include the characterization of the higher flows, and computation of the Poisson bracket of scattering data.

1 Introduction

We consider the system of equations

$$iq_{1_t} + \frac{1}{2}(q_{1xx} + q_{1yy}) - (\varphi_x - q_1 q_2)\, q_1 = 0,$$

$$-iq_{2_t} + \frac{1}{2}(q_{2xx} + q_{2yy}) - (\varphi_x - q_1 q_2)\, q_2 = 0, \tag{1.1}$$

$$\varphi_{xx} - \varphi_{yy} - 2(q_1 q_2)_x = 0.$$

The reduction $q_2 = \pm q_1^*$, leads to the Davey-Stewartson (DS) I system of equations [1]. This system is the shallow water limit of the Benney-Roskes equation [2], where q is the amplitude of a surface wave packet and φ characterizes the mean motion generated by this surface wave. One assumes small amplitude, nearly monochromatic, nearly one-dimensional waves with dominant surface tension. The DS equation provides a two dimensional generalization of the celebrated nonlinear Schrödinger equation and furthermore can be derived from rather general asymptotic considerations [3].

Introducing characteristic coordinates $\xi = x + y, \eta = x - y$, defining

$$U_1 \doteqdot -\varphi_\eta + \frac{1}{2} q_1 q_2, \quad U_2 \doteqdot -\varphi_\xi + \frac{1}{2} q_1 q_2, \tag{1.2}$$

and integrating equation (1.1.c), equations (1.1) reduce to

---

[1]Presented at the Workshop on Integrable Systems and Applications at Ile d'Oleron, France, June 20-24, 1988.

$$iq_{1_t} + q_{1\xi\xi} + q_{1\eta\eta} + (U_1 + U_2)q_1 = 0,$$

$$-iq_{2_t} + q_{2\xi\xi} + q_{2\eta\eta} + (U_1 + U_2)q_2 = 0, \tag{1.3}$$

$$U_1 = -\tfrac{1}{2}\textstyle\int_{-\infty}^{\xi} d\xi'(q_1q_2)_\eta + u_1, \qquad U_2 = -\tfrac{1}{2}\textstyle\int_{-\infty}^{\eta} d\eta'(q_1q_2)_\xi + u_2,$$

where $u_1(\eta,t) \doteq U_1(-\infty,\eta,t), u_2(\xi,t) \doteq U_2(\xi,-\infty,t)$.

An initial value problem associated with equations (1.1), where $q_1(x,y,0)$, $q_2(x,y,0)$ are given and decaying for large x, y, was solved in [4]. However, recently there has been a renewed interest for equations (1.1): (a) Schultz and Ablowitz [5] have emphasized the importance of the boundary conditions for φ. In particular they have shown that the case solved in [4] corresponds to $u_1 = u_2 = 0$.

(b) It has recently been shown [6] that DSI with $q_2 = -q_1^*$ admits a special type of a localized solution, a two dimensional breather solution, exponentially decaying in both special coordinates. This is a remarkable development since a disappointing feature of all dispersive multidimensional equations studied so far, has been the lack of two dimensional exponentially decaying solitons.

In this note we first review the direct and inverse problem associated with DSI. In §3 we explain the ideas used recently by Santini and the author in order to establish that DSI, for generic nonzero boundary values of u_1 adn u_2, possesses two dimensional solitons. In §4 we study the linearized maps between scattering data and q_1, q_2. In §5 we use the results of §4 to: (i) Derive the higher flows in terms of asymptotics of appropriate eigenfunctions. (ii) Show that the evolution of the scattering data is simple if $u_1 = u_2 = 0$. This is consistent with the results of §3. (iii) Show how the Poisson brackets of scattering data can be computed.

2 The Direct and Inverse Problem

Equations (1.1) are associated with the Lax equation

$$(\partial_x + J\partial_y)\Psi + Q\Psi = 0, \quad J \doteq \begin{pmatrix} 1 & 0 \\ 0 & -1 \end{pmatrix}, \quad Q \doteq \begin{pmatrix} 0 & q_1 \\ q_2 & 0 \end{pmatrix}. \tag{2.1}$$

Using characteristic coordinates and letting $\Psi = exp[ik(Jx - y)]M$, the Lax equation for $M(\xi,\eta,k)$ is solved by the following linear integral equations

$$\begin{aligned} M_{11}^+ &= 1 - \tfrac{1}{2}\textstyle\int_{-\infty}^{\xi} d\xi' q_1 M_{21}^+ & M_{12}^+ &= -\tfrac{1}{2}\textstyle\int_{-\infty}^{\xi} d\xi' q_1 M_{22}^+ e^{ik(\xi-\xi')} \\ M_{21}^+ &= \tfrac{1}{2}\textstyle\int_{\eta}^{\infty} d\eta' q_2 M_{11}^+ e^{ik(\eta'-\eta)} & M_{22}^+ &= 1 - \tfrac{1}{2}\textstyle\int_{-\infty}^{\eta} d\eta' q_2 M_{12}^+. \end{aligned} \tag{2.2}$$

Equations (2.2) are Volterra integral equations, thus M^+ is analytic in the upper half k-complex plane. Similarly if M^- satisfies equations similar to those of (2.2), with the integrals in M_{21}^+, M_{12}^+ replaced by $\int_{-\infty}^{\eta}$ and $\int_{\xi}^{\infty}$ respectively, it follows that M^- is analytic in the lower half k-complex plane.

The eigenfunctions M^+, M^- are related via the scattering equations

$$\begin{aligned} \begin{pmatrix} M_{11}^+(k) \\ M_{21}^+(k) \end{pmatrix} - \begin{pmatrix} M_{11}^-(k) \\ M_{21}^-(k) \end{pmatrix} &= \textstyle\int_R d\ell S_{21}(k,\ell) e^{-i\ell\xi - ik\eta} \begin{pmatrix} M_{12}^+(\ell) \\ M_{22}^+(\ell) \end{pmatrix}, \\ \begin{pmatrix} M_{12}^+(k) \\ M_{22}^+(k) \end{pmatrix} - \begin{pmatrix} M_{12}^-(k) \\ M_{22}^-(k) \end{pmatrix} &= \textstyle\int_R d\ell S_{12}(k,\ell) e^{i\ell\eta + ik\xi} \begin{pmatrix} M_{11}^-(\ell) \\ M_{21}^-(\ell) \end{pmatrix}, \end{aligned} \tag{2.3}$$

where the scattering data are given by

$$S_{12}(k,\ell) = -\frac{1}{4\pi}\int_{R^2} d\xi d\eta q_1 M_{22}^-(k) e^{-ik\xi - i\ell\eta}, \quad S_{21}(k,\ell) = \frac{1}{4\pi}\int_{R^2} d\xi d\eta q_2 M_{11}^+(k) e^{ik\eta + i\ell\xi}. \tag{2.4}$$

Equations (2.3) define a nonlocal Riemann-Hilbert problem the solutions of which yields M^+, M^- in terms of S_{12}, S_{21}. Then q_1, q_2 follow from

$$q_1 = -\frac{1}{\pi}\int_{R^2} dk d\ell S_{12}(k,\ell) e^{i\ell\eta + ik\xi} M_{11}^-(\ell), \quad q_2 = \frac{1}{\pi}\int_{R^2} dk d\ell S_{21}(k,\ell) e^{-i\ell\xi - ik\eta} M_{22}^+(\ell). \tag{2.5}$$

Equations (2.2) were studied by Kaup [7] and are also precisely the equations studied in [4] in the special case of $N = 2$ where one uses characteristic coordinates. The scattering equations considered in [4] are more complicated than the above, because in [4] $M^+ - M^-$ was related to M^-. The scattering equations (2.3) are those used in [8] and for the case of $q_2 = \pm q_1^*$ were studied rigorously by Kaup [7] using a Gel'fand-Levitan-Marchenko formulation (see also [9]).

3 Soliton Solutions

We have recently shown that DSI supports, for generic non-zero values of u_1, u_2, two dimensional solitons. We recall that in the case of $u_1 = u_2 = 0$, which is typical of what has occurred so far with dispersive multidimensional problems, arbitrary initial data disperse away as $t \to \infty$. The situation changes dramatically if the boundary conditions are non-zero: (i) For arbitrary time-independent boundary conditions, any arbitrary initial disturbance will decompose into a number of two-dimensional breathers as $t \to \infty$. (ii) For arbitrary time-dependent boundary conditions, any arbitrary initial disturbance will decompose into a number of two-dimensional solitons as $t \to \infty$. A simple such soliton solution is

$$q = \frac{4\rho\sqrt{\mu_R\lambda_R} exp\left\{-\mu_R\hat{\xi} - \lambda_R\hat{\eta} + i\left[\mu_I(\hat{\xi}+\bar{\xi}) + \lambda_I(\hat{\eta}+\bar{\eta}) + (|\mu|^2+|\lambda|^2)t + arg(c\bar{c})\right]\right\}}{\left[1+exp\left(-2\mu_R\hat{\xi}\right)\right]\left[1+exp(-2\lambda_R\hat{\eta})\right] + |\rho|^2} \tag{3.1}$$

where

$$\mu_R, \lambda_R \epsilon R^+, \quad \bar{\xi}, \bar{\eta} \epsilon R, \quad \mu, \lambda, \rho, c, \bar{c} \epsilon \mathbf{C}, \hat{\xi} \doteqdot \xi - 2\mu_I t - \bar{\xi}, \quad \hat{\eta} \doteqdot \eta - 2\lambda_I t - \bar{\eta}.$$

The solution of the initial-boundary value problem (1.3) with $q_2 = -q_1^* = -q^*$, and with $q(\xi,\eta,0), u_1(\eta,t), u_2(\xi,t)$ given, involves the following novel aspects: (i) In the presence of nontrivial boundary conditions the scattering data (2.4) satisfy nontrivial time-evolution equations. These equations can be derived using an algorithmic approach introduced in [11]. We find it convenient to work with the Fourier transform of S_{12}, S_{21}. The evolution of S_{12} is given by

$$i\hat{S}_{12_t} + \hat{S}_{12_{\xi\xi}} + \hat{S}_{12_{\eta\eta}} + (u_2(\xi,t) + u_1(\eta,t))\hat{S}_{12} = 0, \quad \hat{S}_{12}(\xi,\eta) \doteqdot \int_{R^2} dk d\ell e^{ik\xi + i\ell\eta} S(k,\ell). \tag{3.2}$$

The evolution for $\hat{S}_{21}$ is similar where i is replaced by $-i$. Thus, given u_1, u_2 and $q(\xi,\eta,0)$ one computes $\hat{S}_{12}(\xi,\eta,0)$, then (3.2) implies $\hat{S}_{12}(\xi,\eta,t)$ and hence $S_{12}(k,\ell,t)$. Then the inverse problem of §2 yields $q(\xi,\eta,t)$.

(ii) To solve equation (3.2) we use separaton of variables and utilize the spectral theory of $\Psi_{xx} + (u(x)+k^2)\Psi = 0$ and of $i\Psi_t + \Psi_{xx} + (u(x,t)+k^2)\Psi = 0$. (iii) As $t \to \infty$, $S_{12}(k,\ell)$ becomes degenerate, i.e. $S_{12}(k,\ell) = \sum_j a_j(k)b_j(\ell)$. Then the inverse problem can be solved in closed form. We recall that several authors have used degenerate kernels to obtain exact solutions, see for example [12], [13].

4 Linearized Maps

Let

$$S \doteqdot \begin{pmatrix} 0 & S_{12} \\ S_{21} & 0 \end{pmatrix}, \; M \doteqdot \begin{pmatrix} M_{11}^- & M_{12}^+ \\ M_{21}^- & M_{22}^+ \end{pmatrix}, \; m \doteqdot \begin{pmatrix} M_{11}^+ & M_{12}^- \\ M_{21}^+ & M_{22}^- \end{pmatrix}, \; E(k,\ell) \doteqdot \begin{pmatrix} e^{i\ell\eta+ik\xi} & 0 \\ 0 & e^{-ik\eta-i\ell\xi} \end{pmatrix}. \tag{4.1}$$

Then the equations (2.4), (2.5) can be written as

$$S(k,\ell) = -j\frac{1}{8\pi}\int_{R^2} d\xi d\eta E^{-1}(k,\ell)Qm(k), \quad Q = -j\frac{1}{2\pi}\int_{R^2} d\ell dk M(\ell)E(k,\ell)S(k,\ell), \tag{4.2}$$

where $ja = [J, a]$. Equations (4.2) motivate the introduction of the following matrix-valued scalar products (see [14]).

$$(f, g) \doteqdot \frac{1}{8\pi}\int_{R^2} d\xi d\eta (f^+ g)^{\text{off}}, \langle f(k,\ell), g(k,\ell)\rangle \doteqdot \frac{1}{2\pi}\int_{R^2} dk d\ell \left(f(k,\ell)g^+(-k,-\ell)\right)^{\text{off}}, \tag{4.3}$$

where f^+ is the transpose of f and superscript off denotes the off-diagonal part.

Let $\psi = m diag(exp(ik\eta), exp(-ik\xi))$, then we shall show that

$$S = -j(e_{-\ell}, Q\psi(k)), \dot{S} = -j(\tilde{\psi}(-\ell), \dot{Q}\psi(k)), \tag{4.4}$$

where $e_\ell = diag\left(e^{i\ell\eta}, e^{-i\ell\xi}\right)$ and $\psi(k), \tilde{\psi}(k)$ are defined by

$$\psi(k) = e_k + \begin{pmatrix} -\frac{1}{2}\int_{-\infty}^{\xi} d\xi' q_1\psi_{21} & \frac{1}{2}\int_{\xi}^{\infty} d\xi' q_1\psi_{22} \\ \frac{1}{2}\int_{\eta}^{\infty} d\eta' q_2\psi_{11} & -\frac{1}{2}\int_{-\infty}^{\eta} d\eta' q_2\psi_{12} \end{pmatrix} \doteqdot e_k + gQ\psi \tag{4.5}$$

$$\tilde{\psi} = e_k - gQ^+\tilde{\psi}. \tag{4.6}$$

Equation (4.4a) is equation (4.2a) where one uses the fact that only the off-diagonal part is relevant. To derive equation (4.4b) we note that the direct problem implies that ψ satisfies equation (4.5). Hence

$$\psi = (I - gQ)^{-1}e_k, \quad \text{thus } \dot{\psi} = (I - gQ)^{-1}g\dot{Q}\psi, \quad \text{hence}$$

$$\dot{Q}\psi + Q\dot{\psi} = \dot{Q}\psi + Q\{g\dot{Q}\psi + gQg\dot{Q}\psi + \cdots\} = (I - Qg)^{-1}\dot{Q}\psi.$$

Hence,

$$\dot{S} = -j(e_{-\ell}, (Q\dot{\psi})) = -j\left(e_{-\ell}, (I - Qg)^{-1}\dot{Q}\psi(k)\right) = -j\left((I - \tilde{g}Q^+)^{-1}e_{-\ell}, \dot{Q}\psi(k)\right),$$

where $\tilde{g}$ is the adjoint of g. But

$$\int_{R^2} d\xi d\eta b(\xi,\eta)\int_{-\infty}^{\xi} d\xi' a(\xi',\eta) = \int_{R^2} d\xi d\eta a(\xi,\eta)\int_{\xi}^{\infty} d\xi' b(\xi',\eta),$$

and hence if g is defined by (4.5), then $\tilde{g}$ is $-g$.

Equation (4.6) implies that the adjoint problem is obtained by replacing Q by $-Q^+$, thus if $\tilde{S}(k,\ell)$ is the scattering data of the adjoint problem, then

$$\tilde{S}(k,\ell) = j\left(e_{-\ell}, Q^{+}\tilde{\psi}(k)\right). \tag{4.7}$$

It turns out that there exists a simple relationship between S and $\tilde{S}$:

$$\tilde{S}(k,\ell) = S^{+}(-\ell,-k). \tag{4.8}$$

To derive (4.8) we note that

$$Q\psi = Q(I-gQ)^{-1}e_k = Q(e_k + gQe_k + gQgQe_k + \cdots) = (I-Qg)^{-1}Qe_k.$$

Thus

$$S(k,\ell) = -j\left(e_{-\ell},(I-Qg)^{-1}Qe_k\right) = -j\left(Q^{+}(I+gQ^{+})e_{-\ell},e_k\right) = -j(Q^{+}\tilde{\psi}(-\ell),e_k).$$

Hence, $S^{+}(k,\ell) = j(e_k, Q^{+}\tilde{\psi}(-\ell))$, where we have used that $-j(A,B) = (j(B,A))^{+}$. Comparing the above with (4.7) we find (4.8).

We now compute the deformation of Q induced by a deformation of S. We shall show that

$$Q = -j\langle M(\ell)E(k,\ell)S(k,\ell), I\rangle \qquad \dot{Q} = -j\langle M(\ell)E(k,\ell)\dot{S}(k,\ell), \tilde{M}(k)\rangle \tag{4.9}$$

where $M, \tilde{M}$ are defined by

$$M = I + \begin{pmatrix} -\frac{1}{2}\int_{-\infty}^{\xi} d\xi' q_1 M_{21} & -\frac{1}{2}\int_{-\infty}^{\xi} d\xi' q_1 M_{22} e^{-ik(\xi'-\xi)} \\ -\frac{1}{2}\int_{-\infty}^{\eta} d\eta' q_2 M_{11} e^{ik(\eta'-\eta)} & -\frac{1}{2}\int_{-\infty}^{\eta} d\eta' q_2 M_{12} \end{pmatrix}, \tag{4.10}$$

and $\tilde{M}$ satisfies an equation similar to (4.10) with $Q \to -Q^{+}$.

Equation (4.9a) is equation (4.2b). To derive equation (4.9b) we note that the inverse problem implies that M satisfies

$$M(k) = I + G(TM), TM \doteqdot M(\ell)E(k,\ell)S(k,\ell) \;\; G(M) = \left(P_k^{-}\int_R d\ell (M)_1, P_k^{+}\int_R d\ell (M)_2\right), \tag{4.11}$$

where $P^{\pm}$ denote the usual projection operators, and M_1, M_2 denote the first and second column of the matrix M. Thus as before $(T\dot{M}) = (I-TG)^{-1}\dot{T}M$, and $\dot{Q} = -j\langle \dot{T}M, \tilde{M}(k)\rangle$, where $\tilde{M}$ is defined by $\tilde{M} = I + \tilde{G}\tilde{T}\tilde{M}$ and $\tilde{G}, \tilde{T}$ denote the adjoint of G, T w.r.t. $<,>$. We first compute the adjoint of T:

$$\langle TM(k), f(k,\ell)\rangle = \int_{R^2} dk d\ell M(\ell)E(k,\ell)S(k,\ell)f^{+}(-k,-\ell) = \int_{R^2} dk d\ell M(k)E(\ell,k)S(\ell,k)f^{+}(-\ell,-k)$$

$$= \int_{R^2} dk d\ell M(k)\left(f(-\ell,-k)S^{+}(\ell,k)E(\ell,k)\right)^{+} = \langle M(k), (\tilde{T}f)(k,\ell)\rangle.$$

Thus

$$(\tilde{T}f)(-k,-\ell) = f(-\ell,-k)S^{+}(\ell,k)E(\ell,k) = f(-\ell,-k)E(-k,-\ell)S^{+}(\ell,k),$$

hence

$$(\tilde{T}f)(k,\ell) = f(\ell,k)E(k,\ell)S^{+}(-\ell,-k).$$

To compute the adjoint of G we note that the adjoint of the operator $P_k^{\pm}\int_R d\ell$ is $P_k^{\pm}\int_R d\ell$. Thus if M satisfies (4.11), then $\tilde{M}$ satisfies

$$\tilde{M} = I + \left(P^- \int_R d\ell(\tilde{T}\tilde{M})_1, P^+ \int_R d\ell(\tilde{T}\tilde{M})_2\right),$$

where $(\tilde{T}\tilde{M})_1$ and $(\tilde{T}\tilde{M})_2$ are the first and second columns of $\tilde{T}\tilde{M}$ and $\tilde{T}\tilde{M} = \tilde{M}(\ell)E(k,\ell)\tilde{S}(k,\ell)$. Hence, $\tilde{M}$ is the adjoint of M.

Equations similar to (4.4) and (4.9) for the case of DSII are given in [14].

5 Some Applications

Equations (4.4) and (4.9) as well as equation (4.8) and $S(k,\ell) = -j(Q^+\tilde{\psi}(-\ell), e_k)$ will be used here to derive several interesting results.

5.1 The Higher Flows

We shall show that if $\dot{S} = k^n SJ - \ell^n JS$ then

$$\dot{Q} = -j\left\{\begin{pmatrix} m_{11} & M_{12} \\ m_{21} & M_{22} \end{pmatrix}(k)J\begin{pmatrix} \tilde{M}_{11} & \tilde{m}_{12} \\ \tilde{M}_{21} & \tilde{m}_{22} \end{pmatrix}(-k) + \begin{pmatrix} M_{11} & m_{12} \\ M_{21} & m_{22} \end{pmatrix}J\begin{pmatrix} \tilde{m}_{11} & \tilde{M}_{12} \\ \tilde{m}_{21} & \tilde{M}_{22} \end{pmatrix}(-k)\right\}_{n+1}, \tag{5.1}$$

where $\{a(k)\}_n$ denotes the coefficient of the $1/(k)^n$ in the expansion of $a(k)$. The eigenfunctions $M, \tilde{M}, m, \tilde{m}$ are defined in (4.5), (4.6), (4.10), with $m = \psi diag(exp(-ik\eta), exp(ik\xi)), \tilde{m} = \tilde{\psi} diag(exp(ik\eta), exp(-ik\xi))$.

To derive the above we substitute (5.1a) in (4.9b) to obtain

$$\dot{Q} = -j\langle (k^n + \ell^n)M(\ell)ESJ, \ \ \tilde{M}(k)\rangle.$$

Thus

$$q_{1_t} = -\frac{1}{\pi}\int_{R^2} d\ell dk(k^n + \ell^n)\left(\tilde{M}_{21}(-k)e_2 S_{21} M_{12}(\ell) - \tilde{M}_{22}(-k)e_1 S_{12} M_{11}(\ell)\right), \tag{5.2}$$

where $E(k,\ell) \doteqdot diag(e_1, e_2)$. Recall that m, M satisfy

$$\begin{pmatrix} m_{11} - M_{11} \\ \\ m_{21} - M_{21} \end{pmatrix}(k) = \int_R d\ell e_2 S_{21}\begin{pmatrix} M_{12}(\ell) \\ \\ M_{22}(\ell) \end{pmatrix}, \quad \begin{pmatrix} M_{12} - m_{12} \\ \\ M_{22} - m_{22} \end{pmatrix}(k) = \int_R d\ell e_1 S_{12}\begin{pmatrix} M_{11}(\ell) \\ \\ M_{21}(\ell) \end{pmatrix}.$$

Similarly

$$\begin{pmatrix} \tilde{m}_{11} - \tilde{M}_{11} \\ \\ \tilde{m}_{21} - \tilde{M}_{21} \end{pmatrix}(-k) = -\int_R d\ell e_1(\ell,k)S_{12}(\ell,k)\begin{pmatrix} \tilde{M}_{12}(-\ell) \\ \\ \tilde{M}_{22}(-\ell) \end{pmatrix},$$

$$\begin{pmatrix} \tilde{M}_{12} - \tilde{m}_{12} \\ \\ \tilde{M}_{22} - \tilde{m}_{22} \end{pmatrix}(-k) = -\int_R d\ell e_2(\ell,k)S_{21}(\ell,k)\begin{pmatrix} \tilde{M}_{11}(-\ell) \\ \\ \tilde{M}_{21}(-\ell) \end{pmatrix}.$$

The above equations can be used to simplify (5.2). For example,

$$\int_{R^2} dkd\ell k^n \tilde{M}_{21}(-k)M_{12}(\ell)e_2 S_{21} = \int_R dkk^n \tilde{M}_{21}(-k)(-M_{11}(k)+m_{11}(k)),$$

$$\int_{R^2} dkd\ell \ell^n \tilde{M}_{21}(-k)M_{12}(\ell)e_2 S_{21} = \int_{R^2} dkd\ell k^n \tilde{M}_{21}(-\ell)M_{12}(k)e_2(\ell,k)S_{21}(\ell,k) =$$

$$= \int_R dkk^n M_{12}(k)(-\tilde{m}_{22}(-k)+\tilde{M}_{22}(-k)),$$

where we have used that $e_2(\ell,k)=e_1(-k,-\ell)$. Similar simplifications are valid for the second term of (5.2) and hence

$$q_t = -\frac{1}{\pi}\int_R dkk^n \tilde{M}_{21}(-k)m_{11}(k) - M_{12}(k)\tilde{m}_{22}(-k) + \tilde{M}_{22}(-k)m_{12}(k) - M_{11}(k)\tilde{m}_{21}(-k).$$

But $\tilde{M}_{21}, m_{11}, M_{12}, \tilde{m}_{22}$ are (+) functions, while $\tilde{M}_{22}, m_{12}, M_{11}, \tilde{m}_{21}$ are (-) functions, hence the above yields (5.1b).

The above analysis shows that if one fixes the evolution of the scattering data then there exists a unique evolution for q. This evolution will be in general nonlocal, since the asymptotic expansion of the eigenfunctions involves nonlocal terms.

5.2 Evolution of the Scattering Data

We now show that if boundary terms are allowed in the evolution of q then in general the evolution of the scattering data is not given by $\dot{S} = k^n SJ - \ell^n JS$. To fix ideas we consider $n=2$. Recall that

$$S = -j(e_{-\ell}, Q\psi(k)) = -j(Q^+\tilde{\psi}(\ell), e_k).$$

Also if $D \doteqdot diag(\partial_\eta^2, \partial_\xi^2)$, then $De_{-\ell} = -\ell^2 e_{-\ell}$, $De_k = -k^2 e_k$. Hence

$$\ell^2 S = \left(\ell^2 e_{-\ell}, -j(Q\psi(k))\right) = (De_{-\ell}, j(Q\psi)) = (e_{-\ell}, jDQ\psi) =$$

$$= \left((I-\tilde{g}Q^+)(I-\tilde{g}Q^+)^{-1}e_{-\ell}, jDQ\psi\right) = \left(\tilde{\psi}(-\ell), (I-Qg)jDQ\psi(k)\right).$$

Similarly

$$k^2 S = \left((I-Q^+\tilde{g})jDQ^+\tilde{\psi}(-\ell), \psi(k)\right).$$

Thus

$$(\ell^2+k^2)S = \frac{1}{4\pi}\begin{pmatrix} 0 & \sigma_{12} \\ \sigma_{21} & 0 \end{pmatrix}, \sigma_{21} = \int_{R^2} d\xi d\eta \left(\tilde{\psi}_{12}q_1\partial_\xi^2\psi_{21} - \tilde{\psi}_{22}\partial_\xi^2 q_2\psi_{11} - \psi_{11}\partial_\eta^2 q_2\tilde{\psi}_{22} + \psi_{21}q_1\partial_\eta^2\tilde{\psi}_{12}\right).$$

However, it can be shown (see the end of this subsection) that

$$\sigma_{21} = \int_{R^2} d\xi d\eta \left[(q_{1\xi\xi}+q_{1\eta\eta}+Aq_1)\tilde{\psi}_{12}\psi_{21} - (q_{2\xi\xi}+q_{2\eta\eta}+Aq_2)\tilde{\psi}_{22}\psi_{11}\right],$$

where in the remaining of §5.2, $\tilde{\psi}_{ij} = \tilde{\psi}_{ij}(-k)$ and

$$A \doteqdot -\frac{1}{2}\left[\int_{-\infty}^{\eta} d\eta'(q_1q_2)_\xi + \int_{-\infty}^{\xi} d\xi'(q_1q_2)_\eta\right].$$

Equation (4.4) can be written as

$$-J\dot{S} = \frac{1}{4\pi}\begin{pmatrix} 0 & \Sigma_{12} \\ \Sigma_{21} & 0 \end{pmatrix}, \Sigma_{21} = \int_{R^2} d\xi d\eta \left[\dot{q}_1\tilde{\psi}_{12}\psi_{21} + \dot{q}_{22}\psi_{11}\right].$$

Thus

$$\dot{S} = i(\ell^2 + k^2)SJ \;\text{ iff }\; i\dot{q}_1 = q_{1\xi\xi} + q_{1\eta\eta} + Aq_1, \quad -i\dot{q}_2 = q_{2\xi\xi} + q_{2\eta\eta} + Aq_2.$$

These equations correspond to $u_1 = u_2 = 0$. If u_1, u_2 are different than zero, then the evolution of the scattering data will involve an extra term. This term for $\dot{S}_{12}$ is proportional to

$$\int_{R^2} d\xi d\eta \left(\psi_{21}\tilde{\psi}_{12}q_1 + \tilde{\psi}_{22}\psi_{21}q_2\right)(u_1 + u_2).$$

In order to simplify σ_{21} one needs to simplify $B \doteq \psi_{21}\tilde{\psi}_{12_\xi}q_{1_\xi} - \tilde{\psi}_{22}\psi_{11_\xi}q_{2_\xi}$ as well as $\psi_{11}\tilde{\psi}_{22_\eta}q_{2_\eta} - \tilde{\psi}_{12}\psi_{21_\eta}q_{1_\eta}$. The term B can be written as $\frac{1}{4}(q_1q_2)_\xi\tilde{\psi}_{22}\int_\eta^\infty d\eta'(q_2\psi_{11})$ plus a total derivative. Furthermore

$$\int_\eta^\infty d\eta'\tilde{\psi}_{22}(q_2\psi_{11}) = \tilde{\psi}_{22}\int_\eta^\infty q_2\psi_{11} + \int_\eta^\infty(\tilde{\psi}_{22})_\eta\int_\eta^\infty d\eta''(q_2\psi_{11}) = \tilde{\psi}_{22}\int_\eta^\infty q_2\psi_{11} + \int_\eta^\infty d\eta' q_1\tilde{\psi}_{12}\psi_{21}.$$

Thus

$$\tilde{\psi}_{22}\int_\eta^\infty q_2\psi_{11} = \int_\eta^\infty d\eta'\left(q_2\tilde{\psi}_{22}\psi_{11} - q_1\tilde{\psi}_{12}\psi_{21}\right),$$

and

$$\int_R d\eta(q_1q_2)_\xi\tilde{\psi}_{22}\int_\eta^\infty q_2\psi_{11} = \int_R d\eta\left(q_2\tilde{\psi}_{22}\psi_{11} - q_1\tilde{\psi}_{12}\psi_{21}\right)\int_{-\infty}^\eta d\eta'(q_1q_2)_\xi.$$

5.3 Poisson Brackets of the Scattering Data

Equation (4.4b) implies

$$\frac{\delta S(k,\ell)}{\delta q_1} = -\frac{\pi}{2}j\tilde{\psi}(-\ell)^+\begin{pmatrix} 0 & 1 \\ 0 & 0 \end{pmatrix}\psi(k), \quad \frac{\delta S(k,\ell)}{\delta q_2} = -\frac{\pi}{2}j\tilde{\psi}(-\ell)\begin{pmatrix} 0 & 0 \\ 1 & 0 \end{pmatrix}\psi(k).$$

Thus

$$\frac{\delta S_{12}}{\delta q_1} = -\pi\psi_{22}(k)\tilde{\psi}_{11}(-\ell), \; \frac{\delta S_{12}}{\delta q_2} = -\pi\psi_{12}(k)\tilde{\psi}_{21}(-\ell), \; \frac{\delta S_{21}}{\delta q_1} = \pi\psi_{21}(k)\tilde{\psi}_{12}(-\ell), \; \frac{\delta S_{21}}{\delta q_2} = \pi\psi_{11}(k)\tilde{\psi}_{22}(-\ell). \tag{5.3}$$

Using equations (5.3) it is straightforward to compute the Poisson brackets of any scattering data. Consider for example,

$$\{S_{12}(k,\ell), S_{21}(p,q)\} \doteq \frac{1}{\pi^2}\int_{R^2} d\xi d\eta\left(\frac{\delta S_{12}}{\delta q_1}\frac{\delta S_{21}}{\delta q_2} - \frac{\delta S_{12}}{\delta q_2}\frac{\delta S_{21}}{\delta q_1}\right) =$$

$$= \int_{R^2} d\xi d\eta\left[-\psi_{22}(k)\tilde{\psi}_{11}(-\ell)\psi_{11}(p)\tilde{\psi}_{22}(-q) + \psi_{12}(k)\tilde{\psi}_{21}(-\ell)\psi_{21}(p)\tilde{\psi}_{12}(-q)\right].$$

But,

$$\left[\psi_{22}(k)\tilde{\psi}_{22}(-q)\right]_\eta = -\frac{1}{2}q_2\psi_{12}(k)\tilde{\psi}_{22}(-q) + \frac{1}{2}q_1\psi_{22}(k)\tilde{\psi}_{12}(-q) =$$

$$= -\psi_{12}(k)\tilde{\psi}_{12_\xi}(-q) - \tilde{\psi}_{12}(-q)\psi_{12}(k)_\xi = -\left[\psi_{12}(k)\tilde{\psi}_{12}(-q)\right]_\xi.$$

Thus

$$\left[\psi_{22}(k)\tilde{\psi}_{22}(-q)\right]_\eta = -\left[\psi_{12}(k)\tilde{\psi}_{12}(-q)\right]_\xi, \quad \text{hence} \quad \begin{array}{l} F_{1_\xi} = \psi_{22}(k)\tilde{\psi}_{22}(-q) \\ F_{1_\eta} = -\psi_{12}(k)\tilde{\psi}_{12}(-q) \end{array}. \tag{5.4}$$

Similarly

$$\left[\psi_{11}(p)\tilde{\psi}_{11}(-\ell)\right]_\xi = -\left[\psi_{21}(p)\tilde{\psi}_{21}(-\ell)\right]_\eta, \quad \text{hence} \quad \begin{array}{l} F_{2_\xi} = -\psi_{21}(p)\tilde{\psi}_{21}(-\ell) \\ F_{2_\eta} = -\psi_{11}(p)\tilde{\psi}_{11}(-\ell) \end{array}. \tag{5.5}$$

Substituting equations (5.4), (5.5) in $\{S_{12}, S_{21}\}$ we find

$$\{S_{12}(k,\ell), S_{21}(p,q)\} = \int_{R^2} d\xi d\eta \left(F_{1_\eta}F_{2_\xi} - F_{1_\xi}F_{2_\eta}\right) = \int_{R^2} d\xi d\eta \left[\left(F_1 F_{2_\xi}\right)_\eta - \left(F_1 F_{2_\eta}\right)_\xi\right].$$

Equations (4.5), (4.6) imply,

$$\lim_{\eta\to\infty} F_{2_\xi} = 0, \lim_{\xi\to\infty} F_{1_\eta} = 0, \lim_{\xi\to-\infty} F_{2_\eta} = e^{i(p+\ell)\eta}, \lim_{\eta\to-\infty} F_{1_\xi} = e^{-i(q+k)\xi}. \tag{5.6}$$

Also, since $\lim_{\eta\to-\infty} F_1 = -\int_\xi^\infty d\xi' F_{1_{\xi'}} = -\int_\xi^\infty d\xi' exp\left[i(-q-k+i0)\xi'\right]$,

$$\int_R d\xi F_1 F_{2_\xi}|_{\eta=-\infty} = \frac{1}{4}\int_R d\xi \frac{e^{-i(q+k)\xi}}{(-q-k+i0)} \left(\int_R d\eta q_2 \psi_{11}(p)\right)\left(\int_R d\eta q_1 \tilde{\psi}_{11}(-\ell)\right).$$

But

$$\int_R d\eta q_2 \psi_{11}(p) = 2\int_R dp' e^{-ip'\xi} S_{21}(p,p'), \int_R d\eta q_1 \tilde{\psi}_{11}(\ell) = -2\int_R d\ell' e^{i\ell'\xi} \tilde{S}_{21}(\ell,\ell').$$

Thus

$$\int_R d\xi F_1 F_{2_\xi}|_{\eta=-\infty} = i\int_{R^3} d\xi dp' d\ell' \frac{e^{-i(q+k+p'+\ell')\xi}}{(-q-k+i0)} S_{21}(p,p') S_{12}(\ell',\ell). \tag{5.7}$$

Similarly

$$\lim_{\xi\to-\infty} F_{1_\eta} = \frac{1}{4}\left(\int_R d\xi q_1 \psi_{22}(k)\right)\left(\int_R d\xi q_2 \tilde{\psi}_{22}(q)\right) = -\left(\int_R dk' S_{12}(k,k') e^{ik'\eta}\right)\int_R dq' \tilde{S}_{12}(q,q') e^{-iq'\eta},$$

and since

$$\lim_{\xi\to-\infty} F_2 = -\int_\eta^\infty d\eta' F_{2_{\eta'}} = -\int_\eta^\infty d\eta' exp\left[i(p+\ell+i0)\eta'\right],$$

$$\int_R d\eta F_1 F_{2_\eta}|_{\xi=-\infty} = -\int_R d\eta F_{1_\eta} F_2|_{\xi=-\infty} = -i\int_{R^3} d\eta dk' dq' \frac{e^{i(p+\ell+k'+q')\eta}}{p+\ell+i0} S_{12}(k,k') S_{21}(q',q). \tag{5.8}$$

Substituting (5.7), (5.8) in $\{S_{12}, S_{21}\}$ we find

$$\{S_{12}(k,\ell), S_{21}(p,q)\} = -2i\pi\int_{R^2} dk' dq' \frac{\delta(p+\ell+k'+q')}{p+\ell+i0} S_{12}(k,k') S_{21}(q',q)$$

$$+2i\pi \int_{R^2} dp' d\ell' \frac{\delta(q+k+p'+\ell')}{q+k-i0} S_{21}(p,p') S_{12}(\ell',\ell). \tag{5.9}$$

Similar results are given in [15].

Acknowledgements

This work was partially supported by the Air Force Office of Scientific Research under Grant Number 87-0310, Office of Naval Research under Grant Number N00014-88K-0447, and National Science Foundation under Grant Number DMS-8803471. I am grateful to Vas Papageorgiou for several suggestions.

References

[1] A. Davey and K. Stewartson, Proc. R. Soc. London Ser. A **338**, 101 (1974).

[2] D.J. Benney and G.J. Roskes, Stud. Appl. Math. **48**, 377 (1969).

[3] F. Calogero and W. Eckhaus, Inverse Problems **3** (1987); F. Calogero and A. Maccari, in Inverse Problems, ed. by P.C. Sabatier, V.E. Zakharov and E.A. Kuznetsov, Physica **18D**, 455 (1986).

[4] A.S. Fokas, Phys. Rev. Lett., **51**, No. 1, 3-6 (1983).

[5] C. Schultz and M.J. Ablowitz, Trace Formual for DSII in $\bar{\partial}$ Limit Case, preprint, Clarkson University, (1988).

[6] M. Boiti, J. Leon, L. Martina, and F. Pempinelli, Scattering of Localized Solitons in the Plane, Physics Lett. A, **132**, 432 (1988).

[7] D.J. Kaup, Physica **1D**, 45-67 (1980).

[8] S.V. Manakov, private communication.

[9] A.S. Fokas and L.-Y. Sung, Inverse Problems on the Plane and the Davey-Stewartson Equations, preprint, Clarkson University (1989).

[10] A.S. Fokas and P.M. Santini, Solitons in Multidimensions, INS #106, Clarkson University, (1988).

[11] A.S. Fokas and C. Schultz, preprint, Clarkson University, (1989).

[12] V.E. Zakharov and A.B. Shabat, Func. Anal. Appl. **8**, 43 (1974).

[13] D.J. Kaup, J. Math. Phys., **22**, 1176 (1981).

[14] R. Beals and R.R. Coifman, The Spectral Problem for the Davey-Stewartson and Ishimori Hierarchies, preprint, Yale University (1988).

[15] P.P. Kulish and V.D. Lipovsky, Phys. Lett. A, **127**, 413 (1988).

On the Role of Nonlinearities in Classical Electrodynamics

Luigi GALGANI
Dipartimento di Matematica dell'Università di Milano, Via Saldini 50 – 20133 MILANO (Italia)

Abstract:

A report is given of some preliminary investigations on the dynamical properties of the classical Hamiltonian model for the interaction of electromagnetic radiation with a nonrelativistic charged point particle.

1. Introduction.

It is well known that after the sixties the scientific community became fully aware of the role of nonlinearities in classical dynamical systems, up to the point that classical dynamical systems started to become even more fashionable than many standard subjects of theoretical physics. Many studies in this framework were devoted to internal mathematical aspects of the problem, although one of the very first studies, namely that of Fermi Pasta and Ulam[1], was devoted to a foundational problem in theoretical physics, namely the validity of the ergodic hypothesis in dynamical systems of physical interest, with its consequences on equipartition of energy for systems of weakly coupled harmonic oscillators. The first physicist who made a connection between such a work and the mathematical literature (KAM theorem[2]) was certainly Chirikov, and following him then came Joseph Ford. As is well known, these scientists interpreted the first numerical results as indicating that the contribution of Fermi Pasta Ulam would not be significant for statistical physics; on the other hand the italian school, which entered the game in the year 1970, maintained the contrary. A very interesting and fair dispute thus started, and the conclusion is not yet completely clear. My personal opinion, which I don't pretend should be shared by my dear opponents, is that within 85.3 percent we are right. From the conceptual point of view, a very relevant element for the discussion is the choice of considering as relevant very long (as opposed to infinite) times, in the spirit of Boltzmann and Jeans; in this connection one makes use of the technical results of the mathematical russian school of Arnold (Nekhoroshev theorem[3]), to which some nontrivial contributions were also given by the italian school. For the philosophy of Boltzmann and Jeans, one can see for example my paper at the school of Noto[3], and for the Nekhoroshev–like technique one can see, in addition to the original russian works, the paper of Benettin at the same school, or the paper of Giorgilli at the present conference, or the recent book of Lochak and Meunier[3].

So it is clear that dynamical systems are relevant for mechanics, and possibly for statistical mechanics. But what about classical electrodynamics? The first scientists that raised this problem in the present framework were Bocchieri and Loinger[4], who just asked whether one could not find some results of the Fermi Pasta Ulam type in classical electrodynamics. To such an end they introduced a very simple model that was repeatedly investigated[5], namely a charged plate which can move in a given direction and is acted upon by a nonlinear mechanical force. One writes down the equations of motion for the Fourier coefficients of the electromanetic field, and finds a system of infinite harmonic oscillators with some nonlinear coupling, so that the analogy with the Fermi Pasta Ulam problem is quite clear. Again the numerical computations appear to indicate nonequipartition of energy, although the problem is still undecided.

The reason why Bocchieri and Loinger chose their model was that they did not want to enter any discussion about the structure of the electron, and so the simplest way out was just to take a rigid charged body, as their charged plate is. In the way it was formulated, their model turned out to be essentially linear, and the nonlinearity was introduced, as recalled above, through a mechanical force on the plate. Obviouly, those authors knew very well that one can have internal nonlinearities of pure electromagnetic origin, but it so happened that no one worked this out explicitly.

I personally was very fascinated by the problem raised by Bocchieri and Loinger, and to become familiar with it I gave a small contribution to the study of their model, in a paper with my friend Benettin[5]. After some time, I decided to start a new study, and there were many discussions with many people, such as Cercignani, Scotti, Loinger, Prosperi, Benettin, Giorgilli and Valz–Gris from around Milan, and Guerra from Rome; E. Nelson too discussed with us for some days of a beautiful week passed together, and by the way the choice of the boundary conditions in the model described below is due to his suggestion; also J.Kijowski from Warsaw started collaborating with us on a possible generalization to an elastic electron; finally there was the contribution of three students, Stefano Cicale[6], Carlo Angaroni[7] and Lia Forti[8].

The conclusion of all these discussions was that there was no need at all to invent a model of classical nonlinear electrodynamics, because the prototype of such a model was there since always, and known to everybody. In a sense, the circumstance that such a model was not really considered as a classical nonlinear dynamical system in the modern sense is a very interesting fact itself, especially from the point of view of history of science. Indeed, the Hamiltonian of such a model is written in all books in quantum electrodynamics; but the corresponding equations of motion, being nonlinear, are not studied; in a paper of Fermi[9], just half of them are explicitly written down, and in some papers[10] they are explicitly written down in a linear approximation, namely the dipole approximation. But, in such a way, many interesting features are lost. Instead, such a model is usually studied after quantization, just because the mathematics of quantum mechanics is much simpler than that of classical mechanics.

2. The model.

The model is the following one. One considers a charged point particle interacting with the electromagnetic field in the nonrelativistic approximation, neglecting the Coulomb selfinteraction of the particle, as usual in electrostatics. Namely, one considers the Maxwell equations with the current due to a point particle (a delta function centered on the particle's position), and the Newton equation for the particle with the Lorentz force. This model was first introduced by Fermi[11] and then discussed for example by Heitler[12]; people working in quantum electrodynamics are familiar with it from the work of Bloch and Nordsieck[13]. Following Fermi, we consider such a model "*relevant for all problems in radiation theory which do not involve the structure of the electron*". The pragmatic attitude is then the following: this model was studied since the thirties in its quantum version, so let us study it, as it stands, in its classical version.

From a technical point of view, following Fermi, I prefer to choose the Coulomb gauge, so that one has just transversal modes in the field, while the scalar potential disappears as a dynamical variable, being reduced to the static Coulomb potential, which acts formally as an external potential when there are several particles, and is just neglected in the case of one particle. To be definite, let us consider the field in a cubic box of side L with periodic boundary conditions. Then one starts describing the vector potential $\mathbf{A}$ through a set of Lagrangian coordinates q_s with s in a suitable range, by

$$\mathbf{A}(\mathbf{x},t) = \sum_s q_s(t)\mathbf{a}_s(\mathbf{x}) \ , \tag{1}$$

where $\mathbf{a}_s$ are the eigenfunctions of the Laplacian in the box

$$\nabla^2 \mathbf{a}_s + k_s^2 \mathbf{a}_s = 0 \tag{2}$$

satisfying the transversality condition

$$\nabla \cdot \mathbf{a}_s = 0 \ . \tag{3}$$

Thus, a vector basis of eigenfunctions for the Laplacian satisfying the transversality conditions (3) is given by the system of *even* and *odd* functions

$$\begin{aligned} \mathbf{a}^e_{nj}(\mathbf{x}) &= c\sqrt{8\pi}\ \mathbf{u}_{nj} \cos(\mathbf{k}_n \cdot \mathbf{x}) \\ \mathbf{a}^o_{nj}(\mathbf{x}) &= c\sqrt{8\pi}\ \mathbf{u}_{nj} \sin(\mathbf{k}_n \cdot \mathbf{x}) \end{aligned} \tag{4}$$

with $n = (n_1, n_2, n_3)$ and nonnegative integers n_i, c being the velocity of light; the wave vectors $\mathbf{k}_n$ are given by

$$\mathbf{k}_n = (\frac{2\pi}{L} n_1, \frac{2\pi}{L} n_2, \frac{2\pi}{L} n_3) \ , \tag{5}$$

while $\{\mathbf{u}_{nj}\}$, $j = 1, 2$, is a basis of unit polarization vectors orthogonal to $\mathbf{k}_n$. Then the Hamiltonian of the system is

$$\begin{aligned} H(\mathbf{x}, \mathbf{p}, \{q_s\}, \{p_s\}) = &\frac{1}{2m}\left(\mathbf{p} - \frac{e}{c}\sum_{n,j} q^e_{n,j}\mathbf{a}^e_{n,j}(\mathbf{x}) - \frac{e}{c}\sum_{n,j} q^o_{n,j}\mathbf{a}^o_{n,j}(\mathbf{x})\right)^2 + \\ &+ \frac{1}{2}\sum_{n,j}\left(\frac{(p^e_{n,j})^2}{L^3} + L^3\omega_n^2 (q^e_{n,j})^2 + \frac{(p^o_{n,j})^2}{L^3} + L^3\omega_n^2 (q^o_{n,j})^2\right) \\ &+ V(\mathbf{x}) \ , \end{aligned} \tag{6}$$

where $\omega_n = c|\mathbf{k}_n|$ are the angular frequencies of the modes, and $p_{n,j}$ are the momenta conjugated to $q_{n,j}$, while V is a possible *external* potential. So much for the case of one particle. In the case of N particles one just substitutes the first term at the right hand side of (6) by

$$\sum_{i=1}^{N} \frac{1}{2m_i}\left(\mathbf{p}_i - \frac{e_i}{c}\sum_s q_s \mathbf{a}_s(\mathbf{x}_i)\right)^2 \ ,$$

and $V(\mathbf{x})$ by $V(\mathbf{x}_1, \ldots, \mathbf{x}_n)$; in particular V contains the Coulomb interaction between *different* particles as if it were a given external potential, while the Coulomb self–interaction of each of the particles is still neglected. The unknowns of the problem are *both* the coordinates of the particles *and* the coordinates $\{q_s\}$ of the vector potential of the field.

It is of interest to look at the explicit form of the equations of motion. From the Hamiltonian (6), considering for example the case of a single particle with no external potential, they are easily found to be

$$
\begin{aligned}
\ddot{q}^e_{nj} + \omega_n^2 q^e_{nj} &= \frac{e\sqrt{8\pi}}{L^3}\ \dot{\mathbf{x}}\cdot\mathbf{u}_{nj}\cos(\mathbf{k}_n\cdot\mathbf{x})\\
\ddot{q}^o_{nj} + \omega_n^2 q^o_{nj} &= \frac{e\sqrt{8\pi}}{L^3}\ \dot{\mathbf{x}}\cdot\mathbf{u}_{nj}\sin(\mathbf{k}_n\cdot\mathbf{x})\\
m\ddot{\mathbf{x}} &= -e\sqrt{8\pi}\sum_{nj}\mathbf{u}_{nj}\left[\dot{q}^e_{nj}\cos(\mathbf{k}_n\cdot\mathbf{x}) + \dot{q}^o_{nj}\sin(\mathbf{k}_n\cdot\mathbf{x})\right]\\
&\quad + e\sqrt{8\pi}\ \dot{\mathbf{x}}\wedge\sum_{nj}(\mathbf{k}_n\wedge\mathbf{u}_{nj})\left[-q^e_{nj}\sin(\mathbf{k}_n\cdot\mathbf{x}) + q^o_{nj}\cos(\mathbf{k}_n\cdot\mathbf{x})\right]\ .
\end{aligned}
\tag{7}
$$

Notice the nonlinear structure of the right hand sides of equations (7); in the equation for the charged particle this is clearly exhibited also in the particular case where the particle is constrained to move on a straight line, for example the x-axis with periodicity L, so that it has only one degree of freedom. Indeed in such a case the magnetic part of the Lorentz force is not effective and the second term at the right hand side of the particle equation vanishes, so that the equation of motion for the particle takes the simpler form

$$
m\ddot{x} = -e\sqrt{8\pi}\sum_{nj}\alpha_{nj}\left[\dot{q}^o_{nj}\cos(\kappa_n x) + \dot{q}^e_{nj}\sin(\kappa_n x)\right]\ , \tag{8}
$$

where $\alpha_{nj} = \mathbf{u}_{nj}\cdot\mathbf{i}$, $\kappa_n = \mathbf{k}_n\cdot\mathbf{i}$, and $\mathbf{i}$ is the unit vector of the x–axis.

It is instructive to check the conservation of energy by multiplying both sides of equations (7) by $L^3\dot{q}^e_{nj}, L^3\dot{q}^o_{nj}, \dot{\mathbf{x}}$ respectively, and adding. One has then $\frac{dE}{dt} = 0$, with

$$
E = \frac{m}{2}\left|\dot{\mathbf{x}}\right|^2 + \frac{L^3}{2}\sum_{nj}\left[(\dot{q}^{o2}_{nj} + \omega_n^2 q^{o2}_{nj}) + (\dot{q}^{e2}_{nj} + \omega_n^2 q^{e2}_{nj})\right]\ ,
$$

namely, conservation of the sum of kinetic energy of the particle and of the field oscillators' energies, in agreement with the form (6) of the Hamiltonian.

Consider now in particular what happens in the dipole approximation. Such an approximation is defined by $\mathbf{k}\cdot\mathbf{x} = 0$, so that in the equations of motion given above all terms containing $\sin\mathbf{k}\cdot\mathbf{x}$ just disappear, while $\cos\mathbf{k}\cdot\mathbf{x}$ is replaced by 1; thus the equations for the odd modes are decoupled, and the system takes the form

$$
\begin{aligned}
\ddot{q}^e_{nj} + \omega_n^2 q^e_{nj} &= \frac{e\sqrt{8\pi}}{L^3}\ \dot{\mathbf{x}}\cdot\mathbf{u}_{nj}\\
\ddot{q}^o_{nj} + \omega_n^2 q^o_{nj} &= 0\\
m\ddot{\mathbf{x}} &= -e\sqrt{8\pi}\sum_{nj}\mathbf{u}_{nj}\ \dot{q}^e_{nj}\\
&\quad + e\sqrt{8\pi}\ \dot{\mathbf{x}}\wedge\sum_{nj}(\mathbf{k}_n\wedge\mathbf{u}_{nj})\ q^o_{nj}\ .
\end{aligned}
\tag{9}
$$

In such a way, all internal nonlinearities of the dynamical system are lost, and the nonlinearities are just the trivial ones due to a possible external potential.

Coming back to the general nonlinear model it is important to remark that the model obviously presents the property of time reversal symmetry, corresponding to the transformation

$$
(\mathbf{x}(t), \mathbf{p}(t), q_s(t), p_s(t)) \to (\mathbf{x}(-t), -\mathbf{p}(-t), -q_s(-t), p_s(-t))\ .
$$

This is well known, and exploited in the quantum version of the model, but often forgotten in dealing with the classical problem. So, clearly it is not possible that the particle emits energy in all motions which are solutions of the considered system of equations; indeed, if it occurs that there exists one such motion, then for any such motion there corresponds another one with suitable inverted initial data (for the complete system) in which the charged particle always emits energy.

Another obvious remark is that in this complete model one finds nothing as as the term with the third derivative of $\mathbf{x}$, characteristic of the Dirac[14] approach, because such a term just comes from suitable approximations[15], while the attempt here is to take fully into consideration the field, without eliminating it.

3. Some preliminary results.

I report now on some preliminary results, published in a joint paper with Giorgilli, Angaroni, Forti and Guerra[16]. The first notrivial result, due to Lia Forti, is that, in the case of only one particle with no external force, the particle can perform uniform rectilinear motion only with certain initial data for the field, which are uniquely determined by the particle's velocity. This fact might appear as trivial, because one can think of the analogy with the relativistic particle, where for the particle in uniform rectilinear motion one has a unique field, which is the Lorentz transform of the static Coulomb field for the particle at rest. The nontrivial point here is that such a field appears as having a dynamical role, allowing a particle to perform uniform motion notwithstanding the nonlinear character of system of equations considered. In a sense one can say that one is here in presence of a kind of solitonic solution, because one has something moving rigidly with arbitrary velocity $\mathbf{v}$ with $|\mathbf{v}| \neq c$ in a nonlinear system. Moreover, such a solution turns out to be present for any arbitrary truncation on the field modes.

The solution is the following one:

$$\begin{aligned} \mathbf{x}(t) &= \mathbf{x}_0 + \mathbf{v}_0 t \\ q^e_{nj}(t) &= A_{nj} \cos(\mathbf{k}_n \cdot \mathbf{v}_0)t \\ q^o_{nj}(t) &= A_{nj} \sin(\mathbf{k}_n \cdot \mathbf{v}_0)t \ , \end{aligned} \tag{10}$$

with

$$A_{nj} = \frac{e}{cL^3} \frac{\mathbf{v}_0 \cdot \mathbf{u}_{nj}}{\omega_n^2 - (\mathbf{k}_n \cdot \mathbf{v}_0)^2} \ ,$$

and arbitrary $\mathbf{x}_0$, $\mathbf{v}_0$ with $|\mathbf{v}_0| \neq c$. Notice that, with $\omega_n = c|\mathbf{k}_n|$, the denominator in A_{nj} never vanishes for $|\mathbf{v}_0| \neq c$, while it might vanish for $|\mathbf{v}_0| = c$. By the way, this fact indicates that the coupling of the Newton–Lorentz equation for the charged particle with Maxwell's equations for the field requires, for pure reasons of internal consistency, the condition $|\mathbf{v}_0| \neq c$.

To see that the motion described by (10) is in fact a solution, try a solution whose projection on $\mathbf{x}$ is $\mathbf{x}(t) = \mathbf{x}_0 + \mathbf{v}_0 t$, and substitute it at the right hand side of the first two equations (7). Then, omitting for the moment for notational simplicity the indices nj, the equations for the even and odd modes are just of the elementary familiar type

$$\begin{aligned} \ddot{q}^e + \omega^2 q^e &= \gamma \cos \Omega t \\ \ddot{q}^o + \omega^2 q^o &= \gamma \sin \Omega t \end{aligned} \tag{11}$$

with a suitable γ and with Ω proportional to ω, precisely $\Omega = \mathbf{k} \cdot \mathbf{v}_0$, and the well known general solution is

$$q^e(t) = a\cos\omega t + b\sin\omega t + \frac{\gamma}{\omega^2 - \Omega^2}\cos\Omega t$$
$$q^o(t) = \tilde{a}\cos\omega t + \tilde{b}\sin\omega t + \frac{\gamma}{\omega^2 - \Omega^2}\sin\Omega t\ , \tag{12}$$

with arbitrary constants $a, b, \tilde{a}, \tilde{b}$. Notice that, by $|\mathbf{k}| = \frac{\omega}{c}$, for $|\mathbf{v}_0| < c$ it is always $\Omega < \omega$, so that one is dealing here with a *nonresonant forcing.* In correspondence with the choice $a = b =, \tilde{a} = \tilde{b} = 0$ one easily checks that the net force on the particle, expressed by the right hand side of the third equation (7), identically vanishes, so that (10) is a solution. This occurs even for any truncation of the frequencies, due to the fact that the cancellation at the right hand side of the third equation (7) occurs independently for every pair of even and odd terms in the sum, corresponding to any n, j.

For some preliminary numerical results, see ref. 16.

4. Some comments.

I add now some free comments, inspired by the particular solution discussed above. First, one remarks that the energy of each mode of frequency ω turns out to be proportional to $1/\omega^2$. Thus, for what concerns the energy of the whole field corresponding to such a solution, considering the case of no truncation and of L so large to ensure the validity of the asymptotic formula for the density of modes as proportional to ω^2, as a consequence one clearly has an infinite energy, and with a spectrum asymptotically flat as in the *white noise*; and this is quite analogous to the type of divergence of the Coulomb potential energy at the position of the particle. Here, however, such an infinite energy appears to have a dynamical role, allowing the particle to perform uniform motion. On the other hand, for any other generic set of initial data the modes will instead disturb the particle which will continously emit and absorb energy.

So, it is quite natural to consider as particularly relevant, for example in scattering problems, those initial data with the field adapted to the initial velocity $\mathbf{v}$ of the particle, in the sense that it allows for its uniform rectilinear motion according to the formulae given above. But the situation is completely different in the case of more that one particle. Indeed, if $\mathbf{v}_1, \cdots, \mathbf{v}_N$ are the initial velocities of the N particles and if one takes as initial data for the field the superposition of the N data adapted to $\mathbf{v}_1, \cdots, \mathbf{v}_N$ respectively, then, due to the nonlinearity of the equations of motion, there is no reason to expect for special cancellations of the forces at the positions of the particles. So one is reasonably led to conceive that the motions will have instead qualitative features somehow reminiscent of motions under the influence of white noise. This is the first qualitative expectation that one can infer from an elementary study of the equations of motion of our model.

Because of this, one should expect that the particles will in general have motions of a higly fluctuating character. Thus regular motions for the particles can only come about through a smoothing procedure which might be relevant for macroscopic bodies, as stressed for example in the works of De Groot[17], but are unjustified a priori for microscopic objects; for the latter objects, the presence of a kind of background field somehow of a stochastic character, as advocated by Nelson[18], seems to turn out to be a dynamical consequence of the nonlinearity of the equations, when at least two particles are considered. Moreover, one might remark that the fluctuations so induced on the particle motions, althoug being somehow nonlocal because such is the character of the field modes, should however manifest a kind of local character, being small at the position of each of the particles when they are far away. In this sense one might then hope to have the possibility of recovering some weak form of the superposition principle.

I don't have time to enter in many intersting problems that would be naturally raised, in discussing the present model, such as: connection with the Larmor formula and role of the initial data for the field (and so also connection with the so–called stochastic electrodynamics of De la Pena, Cetto, Boyer and so on); role of the width of the box defining the model, and many others (see for example ref. 19). Neither I have time to enter into details concerning an interesting generalization to the model of a nonpoinlike electron, on which good results are being found in Milan by D.Bambusi[20]. I would like however to mention at least a very interesting case that I studied with Giorgilli, namely the case of a material oscillator with one field mode of the same frequency, already mentioned in ref. 21. In such a case perturbation theory was developed formally at first order, and an interesting structure was found, that we liked to call *ravioli–like*. From such a structure it turns out that a situation with zero energy for the material oscillator is unstable, and that in the presence of sufficiently high energy for the field, the material oscillator immediately takes up a fluctuating energy which is very near to mc^2, independently of the frequency of the oscillator and of the width of the box. We are not yet certain of the physical significance of this result. However, in consideration of the fact that our interest in the present model was mainly motivated by the possibility of understanding something on Planck's constant, in the light of the so–called second theory of Planck[19], this possible result was anyhow commented by us with some humour by saying that, as Colombo was looking for India and found America, so we were looking for $h\nu$ and might have found mc^2.

[1] E.Fermi, J.Pasta, S.Ulam, in E.Fermi, *Collected Works*, (Roma, 1965), p. 978.

[2] A.N.Kolmogorov, Dokl. Akad. Nauk SSSR, 98, 527 (1954);
V.I.Arnold, Usp. Mat. Nauk, 18, 13 (1963);
J.Moser, Nachr. Akad. Wiss. Gottingen, Math.-Phys. Kl., 2, 1 (1962).

[3] N.N. Nekhoroshev, Russ. Math. Surveys **32** N.6,1 (1977);
N.N. Nekhoroshev, Trudy Sem. Petrovs. N.5, 5 (1979).
G.Benettin, L.Galgani and A.Giorgilli, Comm. Math. Phys. 113,87 (1987);
G.Benettin, L.Galgani and A.Giorgilli, Comm. Math. Phys. in print;
G.Benettin, L.Galgani and A.Giorgilli, Phys. Lett. A 120,23 (1987);
G.Benettin, in G.Gallavotti and P.Zweifel eds., *Nonlinear evolution and chaotic phenomena*, Plenum P.C. (New York, 1988);
A.Giorgilli, in G.Turchetti ed. *Nonlinear dynamics*, World Scientific (Singapore, 1989);
L.Galgani, in G.Gallavotti and P.Zweifel eds., *Nonlinear evolution and chaotic phenomena*, Plenum Publ. Corp. (New York, 1988);
P.Lochak and C.Meunier, *Multiphase averaging for classical dynamical systems*, Springer (Berlin, 1988).

[4] P.Bocchieri, A.Loinger, Lett. Nuovo Cimento, 4,310 (1970);
P.Bocchieri, A. Crotti, A.Loinger, Lett. Nuovo Cimento, 4,341 (1972);
P.Bocchieri, A.Loinger, F.Valz–Gris, Nuovo Cim. 19B,1 (1974).

[5] G.Casati, I.Guarneri, F. Valz–Gris, Phys.Rev A16, 1273 (1977);
I.Guarneri, G.Toscani, Lett. Nuovo Cim. 14, 101 (1975);
I.Guarneri, G.Toscani, Bollettino U.M.I. 14B, 31 (1977);
G.Benettin, L.Galgani. J.Stat.Phys. 27, 153 (1982);
G.Casati. I.Guarneri. F.Valz–Gris, J.Stat.Phys. 30,195 (1983);
R.Livi, M.Pettini, S.Ruffo and A.Vulpiani, J.Phys. 20A,577 (1987).
C.Alabiso, M.Casartelli and S.Sello, J.Stat.Phys., in print.

[6] S.Cicale, Thesis, University of Milan (1985).

[7] C.Angaroni, Thesis, University of Milan (1986).

[8] L.Forti, Thesis, University of Milan (1987).
[9] E.Fermi, Ann.Inst.H.Poincaré, 1, 53 (1931).
[10] G.Morpurgo, Nuovo Cim. 9, 809 (1952).
[11] E.Fermi, Rev.Mod.Phys., 4, 131 (1932).
[12] W.Heitler, *The quantum theory of radiation*, Clarendon Press,(Oxford, 1950).
[13] F.Bloch and A.Nordsieck, Phys.Rev. 52, 54 (1937).
[14] P.A.M.Dirac, Proc.Roy.Soc.London A167, 148 (1938).
[15] A.M.Cetto and L.de la Pena–Auerbach, Revista Mexicana de Fisica, 29, 537 (1983).
[16] L.Galgani, A.Giorgilli, C.Angaroni, L.Forti and F.Guerra, preprint.
[17] S.R.De Groot, *The Maxwell equations. Nonrelativistic and relativistic derivation from electron theory,* North-Holland P.C. (Amsterdam,1969).
[18] E.Nelson, *Quantum fluctuations*, Princeton U. P. (Princeton, 1985).
[19] A.Einstein, Physik. Zeits. 10, 185 (1909);
M.Planck, Ann.d.Phys. 37,642-656 (1912);
J.A.Wheeler, R.P.Feynman, Rev. Mod. Phys., 17, 2 (1945).
[20] D.Bambusi, in preparation.
[21] L.Galgani and A.Giorgilli, in *Nonlinear phenomena in Vlasov plasmas*, F.Doveil ed., Editions de Physique (Orsay, 1989).

This article was processed by the author using the TEX Macropackage from Springer-Verlag.

UPPER BOUNDS ON THE LYAPUNOV EXPONENTS FOR DISSIPATIVE PERTURBATIONS OF INFINITE DIMENSIONAL HAMILTONIAN SYSTEMS

Jean-Michel Ghidaglia
Laboratoire d'Analyse Numérique
C.N.R.S. et Université Paris-Sud
91405 Orsay (France)

ABSTRACT. We try to present in a rather unified framework some recent results on the estimate of the dimension of attractors for nonlinear partial differential equations which are zero order dissipative perturbations of hamiltonian systems. This includes local and nonlocal dispersive wave equations (like e.g. the K-dV equation, the Benjamin-Ono equation,...) and nonlinear Schrödinger equations. Some of these applications are new.

RESUME. Nous donnons une présentation relativement unifiée de quelques résultats récents sur l'estimation de la dimension des attracteurs pour des équations aux dérivées partielles qui sont des perturbations dissipatives d'ordre zéro de systèmes hamiltoniens. Ceci comprend des équations non linéaires, locales et non locales, d'ondes dispersives (équation de K-dV, de Benjamin-Ono,...) ainsi que des équations de Schrödinger non linéaires. Certaines de ces applications sont nouvelles.

1. INTRODUCTION.

A very large variety of evolution partial differential equations arising from Physics or Mechanics are dissipative perturbations of conservative ones. Most of the later can be written as infinite dimensional hamiltonian systems :

$$\frac{du}{dt} + JH'(u) = 0. \tag{1.1}$$

The function $u : t \longrightarrow u(t)$ assumes its values into an infinite dimensional Hilbert space $\mathcal{H}$. The operator J is linear and possibly unbounded on $\mathcal{H}$ and the nonlinear functional H is the hamiltonian of the system. The notation H' stands for the gradient of H i.e.

$$(H'(u), v) = \lim_{\epsilon \to 0} \frac{1}{\epsilon}(H(u+\epsilon v) - H(u)) \ ;$$

and $(\cdot,\cdot)$ denotes the scalar product on $\mathcal{H}$.

In general, (1.1) describes an ideal situation and in some cases one must consider the influence of perturbations. This leads to

$$\frac{du}{dt} + JH'(u) + D(u) = f, \tag{1.2}$$

where the external force f is either time-independent or time-periodic and $D(u)$ is a dissipative term. When (1.2) is written in an adimensional form, the perturbative terms are "small". However the order of the dissipative term (roughly speaking the number of derivatives appearing in $D(u)$) is quite important ; for instance : $D(u) = -\epsilon\Delta u$, $\Delta = \partial^2/\partial x_i \partial x_i$ (an order two dissipation) can change drastically the mathematical (and physical) properties of the original system.

In this paper, our aim is to study some properties of the long time behavior of the solutions to (1.2) when the dissipative term is a zero order damping : $D(u) = \gamma u$, $\gamma > 0$. But before that let us make a few remarks on the unperturbed case (1.1). As it is easily seen, the trajectories of this dynamical system lay on a level set of the function H, determined by the initial datum. At this level of generality, not much seems to be known on the long time behavior of the solutions, apart from the cases where (1.1) can be linearized through a functional transform like I.S.T. (see e.g. Ablowitz and Segur [2]). For example, Friedlander [5] has shown that the nonlinear wave equation (which is of the form (1.1)) where u depends periodically on $x \in \mathcal{R}$,

$$\frac{\partial^2 u}{\partial t^2} - \frac{\partial^2 u}{\partial x^2} + u^3 = 0,$$

preserves a finite measure on a suitable functional space. Hence by Poincaré's Theorem, the flow of this equation has the returning property modulo a set of zero measure. Although other particular equations have been considered (see e.g. Rudnicki [12]), no general theory seems to be known. We also refer to Chueshov [4] who discuss this problem in relation with statistical quantum mechanics. It seems plausible that for a wide class of hamiltonian H, most of the trajectories have an infinite dimensional omega limit set included in their energy level, $H =$ constant. When this occurs, the long time behavior of the solutions to (1.1) can be seen as truely infinite dimensional.

Here we shall deal with the perturbed equation

$$\frac{du}{dt} + JH'(u) + \gamma u = f. \tag{1.3}$$

As a matter of fact, this equation and (1.1) have almost the same mathematical properties with regards to the Cauchy problem on a finite time interval $[-\tau, \tau]$, $0 < \tau < \infty$. As follows from the main result of this paper, Theorem 5.1, the long time behavior of solutions to (1.3) is definitely different from that of (1.1). Indeed we shall prove, in particular, that the omega limit sets of the solutions to (1.3) are **finite dimensional**, contrasting with the unperturbed equation.

This paper is organized as follows. In the next Section we introduce the two classes of applications that we have in view, namely nonlinear dispersive waves equations and nonlinear Schrödinger equations. The third Section deals with existence of bounded absorbing sets and global attractors. In the fourth Section we define the uniform Lyapunov exponents in relation with the linearized equation. The main result, i.e. an upper bound on these exponents is given in the fifth Section. Finally the sixth Section contains the applications, including some new ones for nonlocal equations.

2. EXAMPLES OF APPLICATIONS.

Let us describe the two class of applications we have in mind.

Example 1 : Nonlinear dispersive waves equations. In this first order, real valued equation, the unknown function $u = u(x,t)$ satisfies

$$\frac{\partial u}{\partial t} + u\frac{\partial u}{\partial x} - \mathcal{L}\frac{\partial u}{\partial x} + \gamma u = f, \tag{2.1}$$

where $\mathcal{L}$ denotes a linear self-adjoint (pseudo)-differential operator acting on the space variable. This kind of equations occurs in models of unidirectional propagation of small amplitude, nonlinear dispersive long waves. The most popular example is the usual Korteweg-de Vries equation in which $\mathcal{L} = -\partial^2/\partial x^2$. But $\mathcal{L}$ can also be a nonlocal operator, as it is the case for the Benjamin-Ono equation or the Intermediate Long-Wave Equation, for which we refer to Section 6.2.

Example 2 : Nonlinear Schrödinger equations. In this first order, complex valued equation, the unknown function $v = v(x,t)$ satisfies

$$\frac{\partial v}{\partial t} + i\frac{\partial^2 v}{\partial x^2} + ig(|\, v\,|^2)v + \gamma v = f, \tag{2.2}$$

and the nonlinear potential g is real valued.

In both examples we consider the case where the unknown function is space-periodic with a given period $L > 0$. These two examples can be written as (1.1) with the following notations. Concerning Example 1, we take $\mathcal{H} = L^2(0,L)$ that we endow with its usual scalar product and norm

$$(u,v) = \int_0^L uv\,dx, \|\, v\,\| = (v,v)^{1/2};$$

and set

$$H(u) = \frac{1}{2}\int_0^L \{u\mathcal{L}u - u^3/3\}dx, \quad J = \partial/\partial x. \tag{2.3}$$

In Example 2, we set $u = (Rev,\ Imv) \in \mathcal{R}^2$, $\mathcal{H} = L^2(0,L)^2$ and

$$H(v) = \frac{1}{2}\int_0^L \{|\, v_x\,|^2 - 2G(|\, v\,|^2)\}dx, \quad J = i, \tag{2.4}$$

where $G(\sigma) = \int_0^\sigma g(s)ds$.

3. BOUNDED ABSORBING SETS AND ATTRACTORS.

We assume that the Cauchy problem associated with (1.3) is well posed on a Hilbert space $E \subset H$. That is for every $s \in \mathcal{R}$ and $u_0 \in E$, equation (1.3) admits a unique solution $u \in \mathcal{C}(\mathcal{R}, E)$ such that $u(s) = u_0$. Moreover denoting by $S(s,t)$ the following mapping on E

$$S(t,s)u_0 = u(t) \text{ solution of (1.3) with } u(s) = u_0,$$

we ask $S(t,s)$ to be a homeomorphism on E. Since the force is time-periodic i.e. $\exists\, T > 0$ such that

$$f(t+T) = f(t), \ \forall\, t \in \mathcal{R}, \tag{3.1}$$

one sees easily that

$$S(t+T, s+T) = S(t,s), \ \forall\, t, s \in \mathcal{R}. \tag{3.2}$$

Hence for each $t \in \mathcal{R}$, the family $\{S(t+mT, t\}_{m \in Z}$ forms a discrete group. In the special case where f is time-independent, $\{S(t,0)\}_{t \in \mathcal{R}}$ forms a continuous group and we have $S(t,s) = S(t-s, 0)$. We also note that in the general case we have

$$S(t+mT, t) = S(t,0)S(T,0)^m S(0,t), \ \forall\, t \in \mathcal{R}, \ \forall\, m \in Z. \tag{3.3}$$

As shown by (2.3) and (2.4), the hamiltonians H we consider can be written as

$$H(u) = \frac{1}{2}a(u,u) + r(u). \tag{3.4}$$

We assume that a is a continuous bilinear form on a suitable subspace V of $\mathcal{H}$ such that the imbedding of V into $\mathcal{H}$ is dense and compact. An important property of a is its coerciveness i.e. there exist $\alpha > 0$ and $\lambda > 0$ such that

$$a(v,v) \geq \alpha \parallel v \parallel_V^2 - \lambda \parallel v \parallel^2, \ \forall\, v \in V. \tag{3.5}$$

Let us briefly describe this setting in Example 1 (with $\mathcal{L} = -\partial^2/\partial x^2$) and Example 2.

Example 1 (continued). We set

$$V = H_L^1 = \{v, v(x+L) = v(x) \text{ and } v, \frac{dv}{dx} \in L^2(0,L)\}, \tag{3.6}$$

$$\parallel v \parallel_V^2 = \int_0^L \{v^2 + L^2 v_x^2\} dx, \tag{3.7}$$

$$a(u,v) = \int_0^L u_x v_x dx. \tag{3.8}$$

Then (3.5) holds true with $\alpha = \lambda = L^{-2}$. As it is well known, K-dV is well posed on $H_L^2 = \left\{ v \in H_L^1, \frac{d^2 v}{dx^2} \in L^2(0,L) \right\}$ and we can take $E = H_L^2$.

Example 2 (continued). Here we take

$$V = \left(H_L^1\right)^2, \ \parallel v \parallel_V^2 = \int_0^L \left(\mid v \mid^2 + L^2 \mid v_x \mid^2\right) dx, \tag{3.9}$$

$$a(u,v) = \int_0^L \left(\frac{du_1}{dx}\frac{dv_1}{dx} + \frac{du_2}{dx}\frac{dv_2}{dx} \right) dx, \ u = u_1 + iu_2 ... \tag{3.10}$$

Here again (3.5) holds true with the same values of α and λ. Moreover, provided

$$\lim_{s \to +\infty} \frac{G_+(s)}{s^3} = 0; \ \exists\, \omega > 0, \ \limsup_{s \to +\infty} \frac{h(s) - \omega G(s)}{s^3} \leq 0, \tag{3.11}$$

where $G_+(s) = \mathrm{Max}(G(s), 0)$ and $h(s) = sg(s)$; the nonlinear Schrödinger equation (2.2) is well posed on either $E = H^1_L$ or $E = H^2_L$ (see [7] and the references therein).

We recall now the definition of an absorbing set.

DEFINITION 3.1. We are given a norm N on E (not necessarily equivalent to that of E). We say that a set $B_a \subset E$ is a bounded absorbing set for N when
(i) *B_a is bounded with respect to N,*
(ii) *for every subset $B \subset E$, bounded w.r. to N, there exists $T(B) \in \mathcal{R}$, such that*

$$t - s \geq T(B) \implies S(t,s)B \subset B_a.$$

A straightforward use of this definition is as follows

PROPOSITION 3.1. Assume that $f \in C(\mathcal{R}, \mathcal{H})$ and

$$(JH'(v), v) = 0, \ \forall\, v \in E. \tag{3.12}$$

Then (1.3) possesses a bounded absorbing set in the $\mathcal{H}$-norm.

Proof. Thanks to (3.12), we deduce from (1.3) that

$$\frac{1}{2}\frac{d}{dt} \| u \|^2 + \gamma \| u \|^2 = (f, u).$$

Hence by the Cauchy-Schwarz inequality

$$\frac{d}{dt} \| u \| + \gamma \| u \| \leq \| f \|_\infty, \tag{3.13}$$

where we have set $\| f \|_\infty = \sup_{t \in \mathcal{R}} \| f(t) \|$, which is finite due to (3.1). Therefore using Gronwall's lemma, we deduce from (3.13) that

$$\| u(t) \| \leq \| u(s) \| e^{-\gamma(t-s)} + \| f \|_\infty (1 - e^{-\gamma(t-s)})/\gamma. \tag{3.14}$$

It follows from this last inequality that the point (ii) of Definition 3.1 holds true when we take e.g.

$$B_a = \{v \in E, \ \| v \| \leq 2 \| f \|_\infty \gamma^{-1}\}.$$

Remark 3.1. It is clear, in the hamiltonian case i.e. for equation (1.1) that one cannot expect existence of bounded absorbing sets. Existence of such sets can be taken as a versatile definition of dissipativity.

When Proposition 3.1 applies the positive orbits $\mathcal{O}^+(u_0) = \{S(mT,0)u_0,\ m \in \mathcal{N}\}$ are bounded and the omega-limit sets

$$\omega(u_0) = \bigcap_{n \geq 0} \text{cl}\ (\mathcal{O}^+(S(nT,0)u_0)),$$

where the closures (cl) are understood with respect to the $\mathcal{H}-$norm, are included in B_a and therefore bounded independently of $u_0 \in E$. Since we are in an infinite dimensional space, it is not obvious whether or not $\omega(u_0)$ is non empty and a compactness property of $\mathcal{O}^+(u_0)$ is necessary at this point. It is worthwhile to note that such a property is subtle here. Indeed, since the $S(nT,0)$ are homeomorphisms on E, these mappings are not compact. We shall return latter to this question (see Remark 3.2).

The existence of a bounded absorbing set shows that all the trajectories of (1.3) are confined into a bounded region of the phase space (E). A natural question is then whether or not these trajectories are attracted by an invariant set. More precisely, one can ask for the existence of a global attractor in the sense of the following definition.

DEFINITION 3.2. A subset $\mathcal{A} \subset E$ is termed as the global attractor for (1.3) if

(i) *$\mathcal{A}$ is non empty and compact in E,*
(ii) *$\mathcal{A}$ is invariant : $S(mT,0)\mathcal{A} = \mathcal{A}$, $\forall\ m \in Z$,*
(iii) *$\mathcal{A}$ is attracting : for every bounded set B in E,*

$$\lim_{m \to +\infty} d(S(mT,0)B, \mathcal{A}) = 0,$$

where d denotes the distance induced by the norm of E.

A very large class of nonlinear p.d.e.'s do have global attractors like e.g. dissipative parabolic equations, damped wave equations,... Existence of global attractors in Examples 1 and 2 is an open problem. This leads us now to introduce a slightly different definition.

DEFINITION 3.3. A subset $\mathcal{A} \subset E$ is termed as the global weak-attractor for (1.3) if the conditions (i) *to* (iii) *in Definition 3.2 hold true when E is endowed with its weak-topology.*

Concerning Example 1 (with $\mathcal{L} = -\partial^2/\partial x^2$) and Example 2, we have the following results

THEOREM 3.1. ([8]). We assume that $f \in C(\mathcal{R}; H^2_L)$. The perturbed K-dV equation (2.1) possesses a global weak-attractor in H^2_L.

THEOREM 3.2. ([7]). We assume that $f \in C^1(\mathcal{R}; L^2(0,L)^2)$. The perturbed nonlinear Schrödinge equation (2.2) possesses global weak-attractors in $\left(H^1_L\right)^2$ and $\left(H^2_L\right)^2$.

Remark 3.2. We have mentionned previously that a compactness property of cl $\left(\mathcal{O}^+(u_0)\right)$ was necessary. Using the weak topology instead of the strong topology allows one to deduce compactness from boundedness. But a technical problem arises : we have to show that the $S(t,s)$ are continuous w.r. to the weak topology. It is indeed the case in the previous examples.

4. UNIFORM LYAPUNOV EXPONENTS.

Let us consider a non empty subset $X \subset E$, which is invariant under the evolution of (1.3)

$$S(T,0)X = X. \tag{4.1}$$

This set X could be a $T-$periodic orbit or a more complicated omega-limit set or the universal weak attractor (when it exists). In the hamiltonian case, X could be a level set of H and in that case X is infinite dimensional. In the dissipative case, we are going to derive sufficient conditions ensuring that X is finite dimensional. These conditions are fulfilled by the examples of Section 2. This result is based on an estimate of the global Lyapunov exponents along the trajectories. And therefore it is natural to linearize (1.3) as follows :

$$v_t + JH''(u)v + \gamma v = 0, \tag{4.2}$$

where $u_0 \in X$ and $u = u(t) = S(t,0)v_0$. Since we take $u_0 \in X \subset E$, the function $u = u(t)$ is continuous from $\mathcal{R}$ into E and (4.2) is a linear evolution equation with mild time-dependent coefficients. We assume that given $v_0 \in v$, (4.2) is well-posed on V i.e. that there exists $v \in \mathcal{C}(\mathcal{R},V)$ solution to (4.2), satisfying $v(0) = v_0$ and the linear operator $L(t,u_0)$ defined as follows

$$L(t,u_0)v_0 = v(t), \ \forall\, u_0 \in X, \ \forall\, v_0 \in V \tag{4.3}$$

is linear and continuous on V. As it is well known, we expect $L(t,u_0)$ to be the differential of the flow $S(t,0)$ at the point u_0. We will make the following hypotheses.

For every $m \in Z$,

$$\lim \frac{| S(mT,0)v_0 - S(mT,0)u_0 - L(mT,u_0)(v_0-u_0) |_V}{| v_0 - u_0 |_V} = 0\,, \tag{4.4}$$

where the limit is taken for $u_0, v_0 \in X, | u_0 - v_0 |_V \to 0$;

$$\sup_{u_0 \in X} \ | L(mT,u_0) |_{\mathcal{L}(V)} < \infty. \tag{4.5}$$

Then we introduce the numbers

$$\bar{\omega}_p(m) = \sup_{w \in \mathcal{O}^+(u_0)} \ \omega_p(L(mT;w)) \tag{4.6}$$

where

$$\omega_p(L) = \| \Lambda^p L \|_{\mathcal{L}(\Lambda^p V)} \tag{4.7}$$

is the norm of the pth exterior product of $L \in \mathcal{L}(V)$. Thanks to the differentiation chain rule and the fact that $\omega_p(L_1 \circ L_2) \le \omega_p(L_1)\omega_p(L_2)$, we have

$$\bar{\omega}_p(m_1+m_2) \le \bar{\omega}_p(m_1)\bar{\omega}_p(m_2).$$

Hence the following limit exists

$$\Pi_p = \Pi_p(u_0) = \lim_{m \to +\infty} \bar{\omega}_p(m)^{1/(mT)} < \infty. \tag{4.8}$$

The uniform Lyapunov exponents on the positive orbit $\mathcal{O}^+(u_0)$ are then defined recursively from the Π_p by

$$\mu_1 = \text{ Log } \Pi_1 \text{ and } \mu_j = \text{ Log } \Pi_j - \text{ Log } \Pi_{j-1} \text{ for } j \geq 2. \tag{4.9}$$

These numbers are related to the divergence of trajectories of the dynamical system generated by (1.3). As a result, if for some $p \geq 1$ we have $\mu_1 + \mu_2 + ... + \mu_p < 0$, then the omega-limit set of the positive orbit $\mathcal{O}^+(u_0)$ is finite dimensional in the sense of the Hausdorff and fractal dimensions ([3],[9]). Indeed under this condition, the differential of the group $\{S(mT, 0), m \in \mathcal{Z}\}$, i.e. (4.2), shrinks the $p-$dimensional volumes as $m \longrightarrow \infty$. And this produces the result on the dimensions.

5. AN UPPER BOUND ON THE UNIFORM LYAPUNOV EXPONENTS.

When (3.12) holds true, the hamiltonian equation (1.1) has two invariants, namely $\| u \|^2$ and $H(u)$. On the other hand, in order to estimate the $\bar{\omega}_p(m)$ given in (4.6), we have to study the evolution of the volume build on m solutions $v_1, ..., v_m$ to the linearized equation (4.2) ; the volume being understood in the sense of the $V-$norm. The usual method (Constantin-Foias and Temam [3]) is based on the evolution of $a(v, v)$, when v is solution to (4.2). This method fails here and one must consider an other quantity (see q_μ in (5.3) below), that can be motivated as follows. The unperturbed equation (1.1) conserves the hamiltonian $H(u)$ and this follows by taking the scalar product in $\mathcal{H}$ of (1.1) with $H'(u)$. By analogy we take the scalar product of the linearized equation (4.2) with the linearized multiplier : $H''(u)v$. We find

$$(v_t, H''(u)v) + \gamma(v, H''(u)v) = 0, \tag{5.1}$$

which can also be written as

$$\frac{1}{2}\frac{d}{dt}(v, H''(u)v) + \gamma(v, H''(u)v) = \frac{1}{2}(H^{(3)}(u)u_t v, v). \tag{5.2}$$

Using the decomposition (3.4), we see that

$$(H''(u)v, v) = a(v, v) + (r''(u)v, v),$$

and with regards to (3.5) it is natural to introduce $(u = u(t) = S(t, 0)u_0)$

$$q_\mu(t, v) \equiv (H''(u), v, v) + \mu \| v \|^2, \ \forall\, t \in \mathcal{R}, \ \forall\, v \in V, \tag{5.3}$$

where $\mu > 0$ and $u_0 \in E$ are arbitrary for the moment. Now, provided we set

$$r_\mu(t, v) \equiv \frac{1}{2}(H^{(3)}(u)u_t v, v) + \mu(JH''(u)v, v), \tag{5.4}$$

we deduce from (4.2) and (5.2) the following identity :

$$\frac{1}{2}\frac{d}{dt}\{q_\mu(t, v)\} + \gamma q_\mu(t, v) = r_\mu(t, v). \tag{5.5}$$

In what follows, we strengthen (3.12) by asking

$$a(w, Jw) = 0, \ (Jr'(w), w) = 0, \ \forall \ w \in E. \tag{5.6}$$

Hence (5.4) reads

$$r_\mu(t, v) = \frac{1}{2}(r^{(3)}(u)u_t v, v) + \mu(Jr''(u)v, v) \tag{5.7}$$

i.e. r_μ is independent of a.

Let us now return to the examples.

Example 1 (continued) : We consider the K-dV case i.e. $\mathcal{L} = -\partial^2/\partial x^2$. We have

$$r(u) = -\frac{1}{6}\int_0^L u^3 dx, \ \ J = \partial/\partial x. \tag{5.8}$$

The identities (5.6) hold true and

$$q_\mu(t, v) = \int_0^L \{\mu v^2 + v_x^2 - uv^2\} dx, \tag{5.9}$$

$$r_\mu(t, v) = -\frac{1}{2}\int_0^L (\mu u_x + u_t) v^2 dx. \tag{5.10}$$

Example 2 (continued) : We have $u = u_1 + iu_2$,

$$r(u_1, u_2) = -\int_0^L G(u_1^2 + u_2^2) dx, \ J = i. \tag{5.11}$$

Here again, (5.6) holds true and, $v = v_1 + iv_2$,

$$q_\mu(t, v_1, v_2) = \int_0^L \{\mu \mid v \mid^2 + \mid v_x \mid^2\} dx - \tag{5.12}$$

$$- \int_0^L \{g \mid v \mid^2) \mid v \mid^2 + 2g'(\mid u \mid^2)[Re(u\bar{v})]^2\} dx,$$

$$r_\mu(t, v_1, v_2) = -2\mu \int_0^L g'(\mid u \mid^2) Re(u\bar{v}) Im(u\bar{v}) dx - \tag{5.13}$$

$$- \int_0^L \{\mid v \mid^2 (g(\mid u \mid^2)/2)_t + [Re(u\bar{v})]^2 (g'(\mid u \mid^2)/2)_t +$$

$$+ 2g'(\mid u \mid^2) Re(u\bar{v}) Re(u_t \bar{v})\} dx.$$

In both cases, q_μ is a "lower order" perturbation of the norm $\int_0^L \{\mu \mid v \mid^2 + \mid v_x \mid^2\} dx$. More precisely, returning to the abstract formulation, we have the following property.

There exists $\theta \in]0,1]$ such that for every $R > 0$, and $u \in E$, $\mid u \mid_E \leq R$ one can find a constant $C_0 = C_0(R)$ such that for every $w \in V$,

$$\mid (r''(u)w, w) \mid + \mid (Jr''(u)w, w) \mid + \mid (r^{(3)}(u)u_t w, w) \mid \leq \tag{5.14}$$

$$\leq C_0(R) \parallel w \parallel^{2\theta} \parallel w \parallel_V^{2(1-\theta)},$$

where $u_t \equiv -JH'(u) - \gamma u + f$.

We introduce the critical values of the quotient $\parallel \cdot \parallel_V^2 / \parallel \cdot \parallel^2$ i.e.

$$\lambda_k = \underset{F \subset V, \dim F = k}{\text{Max}} \quad \underset{v \in F, v \neq 0}{\text{Min}} \frac{\parallel v \parallel_V^2}{\parallel v \parallel^2}, \tag{5.15}$$

where F denotes a subspace of V. Since $\mathcal{H}$ is infinite dimensional and the imbedding of V into $\mathcal{H}$ is dense and compact, we have

$$\lim_{k \to \infty} \lambda_k = \infty. \tag{5.16}$$

With these notations, we can state our main result.

THEOREM 5.1. We consider a bounded subset X of E which is invariant under the evolution of (1.3) i.e. which satisfy (4.1). We assume that (5.6) and (5.14) hold. Then there exists a constant C_1 such that the uniform Lyapunov exponents $\mu_j(u_0)$, $u_0 \in X$ and $j \geq 1$ satisfy

$$\mu_1(u_0) + ... + \mu_p(u_0) \leq -\gamma p + C_1 \sum_{k=1}^{p} \lambda_k^{-\theta}, \ \forall \ p \geq 1. \tag{5.17}$$

Before giving a rapid sketch of the proof, we deduce from this result the following corollary.

COROLLARY 5.1. Every subset X which is compact in V and satisfy the hypotheses of Theorem 5.1 has finite Hausdorff and fractal dimensions in V.

According to (5.16), we deduce from (5.17) that there exists $p_0 \geq 0$ such that

$$\underset{u_0 \in X}{\text{Sup}} \ \{\mu_1(u_0) + ... + \mu_{p_0}(u_0)\} < 0. \tag{5.18}$$

Then according to Constantin-Foias and Temam [3] and Ghidaglia and Temam [9] (or Temam [14] for a synthesis), the Hausdorff dimension of X is less than $1 + p_0$ and its fractal dimension is finite. This shows the corollary.

We turn now to the proof of Theorem 5.1. We introduce two constants κ and β such that

$$a(w,w) \leq \beta \parallel w \parallel_V^2, \ \forall \, w \in V, \tag{5.19}$$

$$\parallel w \parallel^2 \leq \kappa \parallel w \parallel_V^2, \ \forall \, w \in V. \tag{5.20}$$

Our first goal is to choose μ such that the $q_\mu(t,\cdot)$ are norms on V. More precisely we want that there exist $C_5 > 0$ and C_6 which are independent of t and $u_0 \in X$ such that

$$C_5 \parallel w \parallel_V^2 \leq q_\mu(t,w) \leq C_6 \parallel w \parallel_V^2, \ \forall \, w \in V. \tag{5.21}$$

According to (5.14), there exists C_7 such that

$$\mid (r''(u)w,w) \mid \leq \alpha \parallel w \parallel_V^2 /2 + C_7 \parallel w \parallel^2 .$$

Hence, using (3.5) and (5.3) together with this last estimate, we obtain the first inequality in (5.21) with $C_5 = \alpha/2$ provided we choose $\mu = \lambda + C_7$. The second inequality follows by similar means ; here we have $C_6 = C_0\kappa^\theta + \beta + \lambda + C_7$.

Then we estimate $r_\mu(t,\cdot)$. According to (5.4), (5.14) and (5.20), we find that

$$\mid r_\mu(t,w) \mid \leq C_8 \parallel w \parallel^{2\theta} \parallel w \parallel_V^{2(1-\theta)}, \ \forall \, w \in V \tag{5.22}$$

holds true with a constant C_8 which is independent of t and $u_0 \in X$.

At this point we need a technical result in order to deduce from (5.5), (5.21) and (5.22) an estimate on $\omega_p(L(t,w_0))$, $w_0 \in \mathcal{O}^+(u_0)$. This result is given in Ghidaglia [8,Appendix], and we have

$$\omega_p(L(t,w_0)) \leq \left(\frac{C_6}{C_5}\right)^p \exp\left(-\gamma p + \frac{C_8}{2C_5}\sum_{k=1}^{p} \lambda_k^{-\theta_4}\right) t, \ \forall \, t \geq 0,$$

where the λ_k are given in (5.15). Hence using (4.5), (4.6) and (4.7) we obtain (5.17).

Remark 5.1. The previous proof gives an expression of the constants in (5.17), and this produces an explicit bound on the dimensions.

6. *APPLICATIONS TO NONLINEAR DISPERSIVE EQUATIONS.*

6.1. The K-dV equation and Nonlinear Schrödinger equations.

In the case of Example 1 (with $\mathcal{L} = -\partial^2/\partial x^2$) and Example 2, the only point which remains to be checked is the inequality (5.14). This inequality holds true if we take $E = H_L^2$ and $\theta = 1/2$ in case of Example 1 ([8,eq. (3.21)]) and $\theta = 3/4$ in case of Example 2 ([7,eq. (3.26)]). We notice that the λ_k are explicitly known and $\lambda_k \sim \lambda_1 k^2$, this allows to compute a p_0 for which (5.18) holds. Then we deduce from Corollary 5.1 the following result.

THEOREM 6.1. The global weak-attractor in H_L^2 associated with the KdV equation (2.1) (resp. the nonlinear Schrödinger equation (2.2)) are finite dimensional in the sense of Hausdorff and fractal dimensions.

6.2. Remarks on nonlocal nonlinear dispersive equations.

These equations are of the form (2.1) where the linear operator $\mathcal{L}$ is given in terms of the Fourier multiplier $p(k)$:

$$\hat{L}u(k) = p(k)\hat{u}(k), \tag{6.1}$$

where $\hat{u}(k)$ is the kth Fourier coefficient of u defined by the formula (recall that u is L−periodic).

$$\hat{u}(k) = \frac{1}{L}\int_0^L u(x)\exp\left(-i\frac{2\Pi kx}{L}\right)dx,\ k \in Z,$$

and $p(k) \in \mathcal{R}$ is the symbol of the pseudo-differential operator $\mathcal{L}$.

We assume that $\mathcal{L}$ is of order 1 i.e. there exists positive numbers R, α and β such that

$$\alpha \mid k \mid \leq p(k) \leq \beta \mid k \mid,\ \forall\, k \in Z,\ \mid k \mid \geq R. \tag{6.2}$$

It follows then that the hamiltonian H given in (2.3) can be written as (3.4) with

$$V = H_L^{1/2}, \tag{6.3}$$

$$a(u,v) = \sum_{k\in Z} p(k)\hat{u}(k)\hat{v}(k). \tag{6.4}$$

Here $H_L^s, s > 0$ denotes the usual fractional Sobolev space

$$H_L^s = \{v \in L^2(0,L), \Sigma(1+k^2)^s \mid \hat{v}(k)\mid^2 < 0\}. \tag{6.5}$$

Local in time existence of smooth solutions (for e.g. $u_0 \in E = H_L^2$) to equation (1.1) in that case is shown in Saut [13]. However global in time existence of smooth solutions seems to be connected with the knowledge of "enough" invariants (or nearly invariant quantities). Such a goal is achieved in Abdelhouab *et al* [1] in the three following cases :

$$p(k) = 2\Pi \mid k \mid \ \text{(Benjamin – Ono equation)}, \tag{6.6}$$

$$p_\delta(k) = 2\Pi k\coth(2\Pi\delta k) - (1/\delta),\ \delta > 0 \tag{6.7}$$
$$(Intermediate\ Long\ Wave\ Equation),$$

$$p^s(k) = 2\Pi(\sqrt{k^2+1}-1)\ \text{(Smith equation)}. \tag{6.8}$$

It is likely that for the perturbed Benjamin-Ono equation and the perturbed Intermediate Long-Wave Equation, a global weak-attractor in $E = H_L^2$ exists, by using the methods of [8]. We do not know whether or not the corresponding result for the perturbed Smith equation holds true and this is due to the fact that very few invariants for the unperturbed equation are known.

Concerning the result on the bound on the uniform Lyapunov exponents (i.e. (5.17)), we have to check the inequality (5.14). This inequality holds provided we take $\theta = 1/2$ and we deduce from Corollary 5.1 the following result which applies to the three previous cases (6.6), (6.7) and (6.8).

THEOREM 6.2. Every subset X which is bounded in H_L^2 and invariant under the evolution of (1.3) with $\mathcal{L}$ given in (6.1) where p satisfy (6.2), has finite Hausdorff and fractal dimension in $H_L^{1/2}$.

REFERENCES.

[1] L. Abdelhouab, J.L. Bona, M. Felland and J.C. Saut, Non-local models for nonlinear dispersive waves, *Physica D,* to appear.

[2] M.J. Ablowitz and H. Segur, *Solitons and the Inverse Scattering Transform,* SIAM, Philadelphia, 1981.

[3] P. Constantin, C. Foias and R. Temam, Attractors representing turbulent flows, *Memoirs of A.M.S.,* 53 n° 314, 1985.

[4] I.D. Chueshov, Equilibrium statistical solutions for dynamical systems with an infinite number of degrees of freedom, *Math. USSR Sbornik,* 58 (1987) 397-406.

[5] L. Friedlander, An invariant measure for the equation $u_{tt} - u_{xx} + u^3 = 0$, *Comm. Math. Phys.,* 98 (1985) 1-16.

[6] J.M. Ghidaglia, Comportement de dimension finie pour les équations de Schrödinger faiblement amorties, *C.R. Acad. Sci. Paris,* série I, 305 (1987) 291-294.

[7] J.M. Ghidaglia, Finite dimensional behavior for weakly damped driven Schrödinger equations, *Annales de l'I.H.P., Analyse non Linéaire,* to appear and Preprint, Orsay.

[8] J.M. Ghidaglia, Weakly damped forced Korteweg-de Vries equations behave as a finite dimensional dynamical system in the long time, *J. Diff. Equ.,* (1989), in press and Preprint, Orsay.

[9] J.M. Ghidaglia and R. Temam, Attractors for damped nonlinear hyperbolic equations, *J. Math. Pures Appl.,* 66 (1987) 273-319.

[10] J.M. Ghidaglia and R. Temam, Periodic dynamical system with application to Sine-Gordon equations : estimates on the fractal dimension of the universal attractor, *Contemporary Math.,* to appear.

[11] S. Maache, in preparation.

[12] R. Rudnicki, Invariant measures for the flow of a first order partial differential equation, *Ergod. Th. & Dynam. Sys.,* 5 (1986) 437-443.

[13] J.C. Saut, Sur quelques généralisations de l'équation de Korteweg-de Vries, *J. Math. Pures Appl.*, 58 (1979) 21-61.

[14] R. Temam, *Infinite Dimensional Dynamical Systems in Mechanics and Physics*, Springer, New-York, 1988.

THE CAUCHY PROBLEM FOR THE GENERALIZED KORTEWEG-DE-VRIES EQUATION*

J. Ginibre

Laboratoire de Physique Théorique et Hautes Energies**

Université de Paris Sud, 91405 Orsay Cedex, France

This lecture is devoted to a brief (and partial) survey of the Cauchy problem for the generalized Korteweg-de Vries (GKdV) equation, and to the presentation of some recent results on that problem obtained in collaboration with Y. Tsutsumi and G. Velo. We refer to [12][16] for a general survey and to [8][9][10] for a detailed exposition of the latter results.

The GKdV equation can be written as

$$\partial_t u + D^3 u = D V'(u) \qquad (1)$$

where u is a real function defined in space time $\mathbb{R} \otimes \mathbb{R}$, $D = d/dx$, the prime denotes the derivative and $V \in \mathcal{C}^1(\mathbb{R}, \mathbb{R})$ with $V(0) = V'(0) = 0$. The ordinary KdV equation is the special case $V'(u) = u^2$ and the modified KdV equation is the special case $V'(u) = u^3$. The equation (1) is the Euler-Lagrange equation of a variational problem with Lagrangian density

$$\mathcal{L}(v) = (1/2)\ \partial_t v\, Dv - (1/2)\,(D^2 v)^2 - V(Dv) \qquad (2)$$

where v is a function of space time. In fact, the equation associated with (2) is

$$\partial_t D v + D^4 v = D V'(Dv) \qquad (3)$$

which coincides with (1) for $Dv = u$. The equation (3) satisfies in general three

---

* Lecture delivered at the Workshop on "Integrable Systems and Applications", Ile d'Oléron, June 1988.

** Laboratoire associé au Centre National de la Recherche Scientifique.

invariance laws which give rise to three conserved quantities through the Noether theorem : the fact that (2) or (3) involve only derivatives of v but not v itself implies invariance under the change $v \to v + c$. Equivalently, the equation (1) takes the form of a conservation law

$$\partial_t u + D\, J(u) = 0$$

with $J(u) = D^2 u - V'(u)$, thereby giving rise to the conservation law

$$\int u(t)\, dx = C.$$

More important is the invariance of (1) (2) (3) under space time translations, giving rise to the conservation of the momentum, namely

$$\| u(t) \|_2^2 = C \tag{4}$$

and of the energy

$$E(u(t)) = (1/2) \| Du(t) \|_2^2 + \int dx\, V(u(t)) = C\,. \tag{5}$$

The positivity properties of the invariants (4) and (5) make them important in the study of the Cauchy problem. The invariants (4) and (5) are common to the equation (1) and to the non linear Schrödinger (NLS) equation.

$$i\, \partial_t u + (1/2) D^2 u - f(u) = 0 \tag{6}$$

where now u is a complex function, $V(u) = V(|u|)$ and $f(z) = \partial V/\partial \bar{z}$, although (4) has a different origin in the two cases. As a consequence, some of the methods and estimates that are useful for the NLS equation are also relevant for the equation (1).

The equation (1) is invariant under simultaneous but not independent changes $x \leftrightarrow -x$, $t \leftrightarrow -t$. In what follows, we shall eventually make assumptions on the initial data that are not invariant under the change $x \leftrightarrow -x$, thereby leading to results valid for a preferred time direction (say for increasing time). Corresponding results for the opposite time direction are then obtained by changing $x \leftrightarrow -x$ in the assumptions on the initial data.

A large amount of work has been devoted to the Cauchy problem for the equation (1) with initial data $u(t=0,x) = u_0(x)$ at time zero. The existence problem has been treated either for the ordinary KdV equation with $V'(u) = u^2$ [2] [3] [4] [5] [14] [19] [20] [21] or for more general V' [1] [9] [11] [12] [17] [18] [22] [23] [24] [25]. The available results include in particular the existence of global weak solutions in $L^\infty(\mathbb{R}, L^2)$ for initial data $u_0 \in L^2$ and in $L^\infty(\mathbb{R}, H^1)$ for initial data $u_0 \in H^1$. They require suitable growth restrictions on the function V at infinity. There exists also a wealth of results concerning more regular solutions, for instance solutions that are continuous functions of time with values in Sobolev spaces H^s for $s > 3/2$ corresponding to initial data in H^s. There is comparatively less information available on the uniqueness problem. There exists a well known result according to which the solution is unique if $u_0 \in H^s$ with $s > 3/2$ [2] [11] [19]. More recently new uniqueness results were proved for u_0 in weighted L^2 spaces with either exponential [12] or polynomial [8] [14] weight, either in the special case $V'(u) = u^2$ [12] [14] or in the general case [8], and for u_0 in (possibly weighted) H^1 spaces in the general case [8]. The various results can be classified according to the methods that have been used for their derivation.

(1) Existence results by the inverse scattering method.

It is well known that the equation (1) with $V'(u) = u^2$ can be solved by the inverse scattering method, and that method has been used in particular for a mathematical treatment of the Cauchy problem [4] [5]. Since it is well documented in the literature, we shall not elaborate on it any further. When applied to the Cauchy problem, it has the advantage of requiring little local regularity on u_0, typically $u_0 \in L^1_{loc}$. It has the drawback of requiring a fairly strong decrease of u_0 at infinity, at least $|x|\, u_0 \in L^1$. It is restricted to the integrable cases $V'(u) = u^2$ and to a much smaller extent $V'(u) = u^3$.

(2) Existence results by a compactness method.

That method has been used by many authors. It yields in general the global existence (in time) of solutions, but no uniqueness. It consists of four main steps.

(1) One first regularizes the original equation, and one proves the existence of solutions of the simpler regularized equation.

(2) One derives a priori estimates of the solutions of the regularized equation in some space X of functions of space time which is the dual of a Banach space. Those estimates have to be uniform with respect to the regularization.

(3) One removes the regularization by a limiting procedure, using the fact that closed balls in the dual of a Banach space are weak-* compact.

(4) One proves that the limit points of sequences of solutions of the regularized equation are in fact solutions of the original equation.

Several methods of regularization have been used in the literature. We quote three of them. The Faedo-Galerkin (FG) method consists in projecting the equation (1) on finite dimensional subspaces of L^2, namely in replacing (1) by

$$\partial_t u + P D^3 u = P D V'(u) \tag{7}$$

and replacing the initial condition by $u(0) = P u_0$, where P is taken from an increasing sequence of orthogonal projectors in L^2 converging strongly to $\mathbb{1}$. The parabolic method consists in adding to the equation (1) a dissipative term, namely in replacing (1) by

$$\partial_t u + D^3 u + \eta A u = DV'(u) \tag{8}$$

where A is a positive self adjoint operator in L^2, for instance $A = -D^2$ or $A = D^4$, and η is positive for positive times and negative for negative times. The convolution method consists in changing the non linear term in the equation according to

$$\partial_t u + D^3 u = \varphi * D V'(\varphi * u) \tag{9}$$

and replacing the initial condition by $u(0) = \varphi * u_0$ where φ is a smooth approximation of a δ function and $*$ denotes the convolution in $\mathbb{R}$.

The choice of the space X is dictated by the possibility of obtaining a priori estimates uniformly with respect to the regularization, and therefore ultimately by the conservation laws which provide the basis for such estimates. Natural choices are therefore $X = L^\infty(\mathbb{R}, L^2)$ associated with the conservation of the L^2-norm (4), and $X = L^\infty(\mathbb{R}, H^1)$ associated with the conservation of the L^2-norm (4) and of the energy (5). (Here we denote by $L^q(\mathbb{R}, B)$ the space of L^q functions of the time variable with values in the space B of functions of the space variable).It is important for the second step of the method that the relevant conservation laws be preserved or at least not two badly damaged by the regularization. The FG regularization preserves L^2-norm conservation, but damages the conservation of the energy in an unretrievable way. The parabolic regularization replaces L^2-norm conservation by L^2-norm decay. In fact, from (8) one obtains easily

$$\| u(t) \|_2^2 + 2\eta \int_0^t d\tau \, \| A^{1/2} u(\tau) \|_2^2 = \| P u_0 \|_2^2 \leqslant \| u_0 \|_2^2 \tag{10}$$

(where $\| \cdot \|_r$ denotes the norm in $L^r = L^r(\mathbb{R})$) so that solutions u of (1) with $u_0 \in L^2$

are estimated in $L^{\infty}(\mathbb{R}, L^2)$ uniformly in η, and in addition in $L^2_{loc}(\mathbb{R}^+, \mathscr{D}(A^{1/2}))$ for fixed η (but without uniformity). In the energy conservation law, the parabolic regularization introduces additional terms which are unpleasant to estimate. The convolution regularization (9) preserves both conservation laws (4) and (5), since it preserves the space time translation invariance of the equation and its variational character. In fact it is obtained by simply replacing $V(Dv)$ by $V(\varphi * Dv)$ in the lagrangian density (2).

In the fourth step of the method, when dealing with low regularity solutions, namely solutions in $L^{\infty}(\mathbb{R}, L^2)$, an important role is played by the smoothing properties of the equation (1), whereby under suitable circumstances, solutions with initial data in the Sobolev space $H^s \equiv H^s(\mathbb{R})$ tend to lie not only in $L^{\infty}_{loc}(\mathbb{R}, H^s)$ but also in $L^2_{loc}(\mathbb{R}, H^{s+1}_{loc})$ [8][12][14]. The basis for that property is the fact that the operator D^3 in the linear part of the equation (1) tends to produce commutators of a definite sign. In fact, if h is a smooth function,

$$[D^3, h] = 3\, D\, h' D + h''' \tag{11}$$

and if h is non decreasing, the first and (more singular) term in the right hand side is a negative operator. The simplest instance of that property arises for $s = 0$. Proceeding formally and using (11) and integration by parts, one obtains from the equation (1)

$$\begin{aligned} \partial_t < u, hu > + 3 < Du, h' Du > - < u, h''' u > &= 2 < u, h\, DV'(u) > \\ &= -2 \int h' (u\, V'(u) - V(u)) \end{aligned} \tag{12}$$

(where $< .,. >$ denotes the scalar product in L^2) and by integration

$$\begin{aligned} &< u, h\, u > (t) + \int_0^t d\tau \, \{ 3 < Du, h' Du > - < u, h''' u > \} (\tau) \\ &= < u_0, h\, u_0 > - 2 \int_0^t d\tau \int dx\, h'(u\, V'(u) - V(u)) (\tau, x) \, . \end{aligned} \tag{13}$$

In particular for $h \geqslant 0$, $h' \geqslant 0$, h' with compact support, $h'(x) = 1$ for x in some interval J, (13) provides an a priori estimate of u in $L^2_{loc}(\mathbb{R}, H^1(J))$ in terms of $\| u_0 \|_2$ in so far as the integral in the right hand side can be suitably controlled.

We have used the preceding ideas and methods in [9] to derive the existence of solutions in $L^{\infty}(\mathbb{R}, L^2) \cap L^2_{loc}(\mathbb{R}, H^1_{loc})$ for initial data $u_0 \in L^2$ and of solutions in

$L^\infty(\mathbb{R}, H^1) \cap L^2_{loc}(\mathbb{R}, H^2_{loc})$ for initial data $u_0 \in H^1$ together with the appropriate smoothing identities, under weak and natural assumptions on V.

The main existence result for L^2 solutions is as follows.

<u>Proposition 1</u> – Let $V \in \mathcal{C}^1(\mathbb{R}, \mathbb{R})$, $V(0) = V'(0) = 0$ and let V satisfy

$$\forall \rho \in \mathbb{R}, \ |V'(\rho)| \leqslant C(1 + |\rho|^5) \tag{14}$$

$$\lim_{|\rho| \to \infty} |\rho|^{-6} (\rho V'(\rho) - V(\rho))_- = 0 \tag{15}$$

(where $\lambda_\pm = \mathrm{Max}(\pm\lambda, 0)$). Let $u_0 \in L^2$. Then the equation (1) with initial data $u(0) = u_0$ has a solution u satisfying

$$u \in L^\infty(\mathbb{R}, L^2) \cap L^2_{loc}(\mathbb{R}, H^1_{loc}). \tag{16}$$

Furthermore, for any $h \in \mathcal{C}^3(\mathbb{R}, \mathbb{R}^+)$ with compactly supported h' and $h' \geqslant 0$, u satisfies the inequality

$$< u, h u > (t) + \int_0^t d\tau \{ 3 < Du, h' Du > - < u, h''' u > \} (\tau)$$

$$\leqslant < u_0, h u_0 > - 2 \int_0^t d\tau \int dx \ h' (u V'(u) - V(u)) (\tau, x) \tag{17}$$

for all $t \in \mathbb{R}^+$ and a similar inequality for $t \in \mathbb{R}^-$. In particular $\| u(t) \|_2 \leqslant \| u_0 \|_2$ for all $t \in \mathbb{R}$.

The main existence result for H^1 solutions is as follows.

<u>Proposition 2</u> – Let $V \in \mathcal{C}^1(\mathbb{R}, \mathbb{R})$, $V(0) = V'(0) = 0$ and let V satisfy

$$\forall \rho \in [-1, 1], \ V_-(\rho) \leqslant C \rho^2, \tag{18}$$

$$\lim_{|\rho| \to \infty} |\rho|^{-6} V_-(\rho) = 0. \tag{19}$$

Let $u_0 \in H^1$ satisfy $V_+(u_0) \in L^1$. Then the equation (1) with initial data $u(0) = u_0$ has a solution u satisfying

$$u \in \mathcal{C}(\mathbb{R}, L^2) \cap L^\infty(\mathbb{R}, H^1) \cap L^2_{loc}(\mathbb{R}, H^2_{loc}) \tag{20}$$

and $V(u) \in L^\infty(\mathbb{R}, L^1)$. Furthermore, for any $h \in \mathcal{C}^3(\mathbb{R}, \mathbb{R}^+)$ with compactly supported h' and $h' \geqslant 0$, u satisfies (13), and in addition the inequality

$$\{ < Du, h\, Du > + 2 \int dx\, h\, V(u) \}(t) + \int_0^t d\tau \ \{ 3 < D^2u , h' D^2u >$$

$$- < Du, h''' Du > + < V'(u) , h' V'(u) > - 4 < D^2u, h' V'(u) >$$

$$+ 2 \int dx\, h''' V(u) \}(\tau) \leqslant < Du_0, h\, Du_0 > + 2 \int dx\, h\, V(u_0) \tag{21}$$

for all $t \in \mathbb{R}^+$, and a similar inequality for $t \in \mathbb{R}^-$. In particular $E(u(t)) \leqslant E(u_0)$ for all $t \in \mathbb{R}$.

The proof of Proposition 1 uses a parabolic regularization (8) with $A = D^4$ and an auxiliary FG regularization to prove the existence of solutions of (8). The assumptions (14) (15) on V serve in particular to ensure that the inequality (17) implies an a priori estimate of the solutions in $L^2_{loc}(\mathbb{R}, H^1_{loc})$. Similarly the proof of Proposition 2 uses a convolution regularization (9) and an auxiliary FG regularization. The assumptions (18) (19) on V serve to ensure that the inequality (21) implies an a priori estimate of the solutions in $L^2_{loc}(\mathbb{R}, H^2_{loc})$.

(3) <u>Existence and uniqueness results by a contraction method</u>.

That method yields the local existence (in time) of solutions, and their uniqueness. It does not yield global existence unless it is supplemented by additional information in the form of a priori estimates. It consists in devising an iteration scheme for solving the equation, and showing that the map $u^{(n)} \to u^{(n+1)}$, where $u^{(n)}$ is the n-th iterate, is contracting in a suitable norm $\| \cdot \|_*$, so that

$$\| u^{(n+1)} - u^{(n)} \|_* \leqslant \delta \| u^{(n)} - u^{(n-1)} \|_* \leqslant \delta^n \| u^{(1)} - u^{(0)} \|_* \tag{22}$$

for some $\delta < 1$, thereby ensuring the convergence of the iteration. The simplest iteration scheme, namely

$$\partial_t u^{(n+1)} + D^3 u^{(n+1)} = D V'(u^{(n)}) \tag{23}$$

which is successful for other semi linear equations such as the non linear Schrödinger equation or the non linear Klein-Gordon equation, does not work in the present case because the derivative in the interaction term $DV'(u)$ makes that term too singular. A more elaborate iteration scheme consists in taking [11][12]

$$(\partial_t + D^3 - V''(u^{(n)})\, D\,)\, u^{(n+1)} = 0 \qquad (24)$$

so that $u^{(n+1)}(t) = U_n(t,0)\, u_0$, where $U_n(t,s)$ is the two parameter group of operators which solves the linear equation (with time dependent interaction term)

$$\begin{cases} (\partial_t + D^3 - V''(u^{(n)}(t))\, D\,)\, U_n(t,s) = 0 \\ U_n(s,s) = \mathbb{1} \quad . \end{cases} \qquad (25)$$

That method has been used to derive the existence and uniqueness of local solutions in $\mathcal{C}(., H^s)$ for $s > 3/2$ [11][12]. It requires fairly strong smoothness assumptions on V. Since a priori estimates of solutions that suffice to ensure global existence exist only for integer values of s, that method yields global existence and uniqueness in $\mathcal{C}(\mathbb{R}, H^s)$ for $s \geqslant 2$, but not for smaller values of s.

(4) <u>Uniqueness results by partial contraction</u>

In that method, uniqueness of solutions in a normed function space X (of functions of space time) is obtained by finding an auxiliary norm $\| \cdot \|_*$ on X, weaker than the X norm, and such that for any two solutions u_1 and u_2 in X with the same initial data $u_1(t_0) = u_2(t_0) = u_0$, the difference $w = u_1 - u_2$ satisfies a linear inequality

$$\|w\|_* \leqslant M\,(\mathrm{Max}\,(\|u_1\|_X, \|u_2\|_X))\, \|w\|_* \qquad (26)$$

where $M(.,.)$ can be taken < 1. This implies $w = 0$ and therefore $u_1 = u_2$. Since the uniqueness problem is local in time and since the equation (1) is invariant under time translation, it is sufficient to take $t_0 = 0$ and to consider solutions defined in a time interval $[0,T]$ with T small, and in practice one can take $M(.) < 1$ by taking T sufficiently small. An elementary version of that argument shows the uniqueness of solutions of (1) satisfying $u \in L^\infty_{loc}(L^2 \cap L^\infty)$ and $Du \in L^1_{loc}(L^\infty)$ [18], and a fortiori the uniqueness of solutions in $\mathcal{C}(H^s)$ for $s > 3/2$, namely the uniqueness result of the previous contraction method. In fact, if u_1 and u_2 are two such solutions, then w satisfies the equation

$$\partial_t w + D^3 w = D(\tilde{V}''\, w) \qquad (27)$$

where

$$\tilde{V}'' = \int_0^1 d\lambda\, V''(\lambda\, u_1 + (1-\lambda)\, u_2) \tag{28}$$

and therefore

$$\partial_t \| w \|_2^2 = 2 < w, D(\tilde{V}''\, w) > = < w, (D\, \tilde{V}'') w > \tag{29}$$

so that

$$\| w(t) \|_2^2 \leqslant \| w(0) \|_2^2 \exp\left(\int_0^t d\tau \| D\, \tilde{V}''(\tau) \|_\infty \right) = 0 \tag{30}$$

since $w(0) = 0$, provided $D\, \tilde{V}'' \in L^1_{loc}(L^\infty)$ and in particular if $V \in \mathcal{C}^3$, if $u \in L^\infty_{loc}(L^\infty)$ and $Du \in L^1_{loc}(L^\infty)$. Note in passing that this argument is not restricted to the KdV equation and applies (at least formally) to any equation obtained by replacing D^3 by any antiselfadjoint operator $-L = L^*$, for instance to the Benjamin-Ono equation.

A more elaborate argument of the same type was used more recently in weighted L^2 spaces [12][14] in the special case $V'(u) = u^2$. More precisely it was proved that for $h^{1/2} u_0 \in L^2$ with h a suitable weighting factor, the equation (1) with $V'(u) = u^2$ has a unique solution u such that $h^{1/2} u \in L^\infty_{loc}(\mathbb{R}^+, L^2)$. That result was obtained in [12] with h an exponential weight, namely $h(x) = 1 + e^x$, and in [14] with h satisfying a power law, namely $h(x) = (1 + x_+)^{3/2}$. Note that the weighting factor h is not invariant under the change $x \leftrightarrow -x$. Correspondingly, the results hold only for $t \geqslant 0$. Similar results hold for $t \leqslant 0$ with x changed to $-x$ in the assumptions on u_0.

In [8] we untertook a systematic study of the uniqueness of solutions of the equation (1) for general V and for u_0 in suitably weighted L^2 or H^1 spaces according to

$$(1 + x_+)^{\beta/2}\, u_0 \in L^2 \tag{31}$$

and possibly

$$(1 + x_+)^{\gamma/2}\, Du_0 \in L^2 \tag{32}$$

for suitable $\beta, \gamma \in \mathbb{R}^+$. When combined with Propositions (1) and (2), the uniqueness results yield existence and uniqueness in suitable function spaces. Here for simplicity we give only a partial statement of the final results, concentrating on the assumptions on V and on the associated assumptions on u_0 (namely on β and possibly γ), which are fairly simple. For the case of L^2 solutions, we leave aside the description of the relevant spaces, which is rather complicated, and we only state the most interesting properties of the solutions. Those properties however are in general neither necessary nor sufficient to ensure uniqueness. We refer to the original papers [8][9][10] for a precise description of the spaces where uniqueness holds. For the case of H^1 solutions, we give a full but simplified statement of the uniqueness results, and we refer again to the original papers for a finer description of the uniqueness spaces.

We consider first the case of weighted L^2 solutions. As above the weighting of u_0 is asymmetric in x and correspondingly the results hold for non negative times only.

<u>Proposition 3</u>. Let $V \in \mathcal{C}^2(\mathbb{R}, \mathbb{R})$ with $V(0) = V'(0) = 0$ and let V satisfy

$$\forall\ \rho \in \mathbb{R},\quad |V''(\rho)| \leqslant C|\rho|^p \tag{33}$$

for some $p \in (0,7/2)$. Let u_0 satisfy (31) with $\beta = 1/p - 1/4$ for $0 < \beta \leqslant 2$ and $\beta = 1/4$ for $2 \leqslant p < 7/2$. Then the equation (1) with initial data $u(0) = u_0$ has a unique global solution u for $t \geqslant 0$. The solution u satisfies (16) with $\mathbb{R}$ replaced by $\mathbb{R}^+$ and

$$(1+x_+)^{\beta/2}\, u \in L^\infty_{loc}(\mathbb{R}^+, L^2)\ . \tag{34}$$

In the special case $V'(u) = u^2$ of the ordinary KdV equation where $p = 1$, our result $\beta = 3/4$ improves over that of [14] by a factor 2.

We next consider the case of weighted H^1 solutions.

<u>Proposition 4</u>. Let $V \in \mathcal{C}^2(\mathbb{R}, \mathbb{R})$ with $V(0) = V'(0) = 0$, with V'' absolutely continuous and V''' locally bounded. Let V satisfy (19) and

$$\forall\ \rho \in [-1,1]\ ,\quad |V''(\rho)| \leqslant C|\rho|^p \tag{35}$$

for some $p \geqslant 1$. Let u_0 satisfy (31) and (32) with $\beta \geqslant 0$, $0 \leqslant \gamma \leqslant \beta + 2$, $\beta \geqslant \gamma\ (4-p)_+/(4+p)$ and in addition either

$$(p+1)(\beta+\gamma) \geqslant 1 \tag{36}$$

or

$$(3-p)_+\,\beta + (p-1)(2 + 3\,\mathrm{Min}(\beta,\gamma)) > 1\ . \tag{37}$$

Then the equation (1) with initial data $u(0) = u_0$ has a unique global solution u satisfying (20) with $\mathbb{R}$ replaced by $\mathbb{R}^+$, and satisfying

$$(1+x_+)^{\gamma/2}\, Du \in L^\infty_{loc}(\mathbb{R}^+, L^2)\ . \tag{38}$$

The solution u satisfies in addition (34) and

$$\chi_+ Du \in L^6_{loc}(\mathbb{R}^+, L^\infty) \tag{39}$$

(where χ_+ is the characteristic function of $\mathbb{R}^+$ in the space variable).

Note that for $\beta = \gamma = 0$, the condition (37) reduces to $p > 3/2$. In particular under the assumptions of Proposition 4 on V and with $p > 3/2$, the equation (1) has a unique global solution for $u_0 \in H^1$. In addition in that case, uniqueness (and existence) holds in the space defined by (20). On the other hand, in the special case $V'(u) = u^2$ of the ordinary KdV equation where $p=1$, the assumptions on u_0 reduce to $\beta \geqslant 3\gamma/5$ and $\beta + \gamma \geqslant 1/2$. They are satisfied for instance by $\beta = 1/2, \gamma = 0$ or by $\beta = 3/16, \gamma = 5/16$.

The proofs of the uniqueness results, which form the main part of Propositions 3 and 4, make extensive use of three types of estimates associated with the linear part of the equation (1) and more precisely with the linear one parameter group

$$U(t) = \exp(-tD^3) \tag{40}$$

which solves the Cauchy problem for the inhomogeneous linear equation

$$\partial_t u + D^3 u = f \tag{41}$$

with initial condition $u(0) = u_0$ through the formula

$$u(t) = U(t)\, u_0 + \int_0^t d\tau\; U(t-\tau)\, f(\tau) \ . \tag{42}$$

The group $U(.)$ is unitary in L^2. For each t, $U(t)$ can be represented as the convolution with the function

$$S_t(x) = (3t)^{-1/3}\, Ai(x(3t)^{-1/3}) \tag{43}$$

where Ai is the classical Airy function

$$Ai(x) = (2\pi)^{-1} \int d\xi\, \exp(i\,\xi^3/3 + i\,x\,\xi) \tag{44}$$

which satisfies the estimates

$$\begin{cases} |\,Ai(x)\,| \leqslant C(1+x_-)^{-1/4} \exp(-c\, x_+^{3/2}) \\ |\,DAi(x)\,| \leqslant C(1+x_-)^{1/4} \exp(-c\, x_+^{3/2}) \ . \end{cases} \tag{45}$$

In particular $S_t \in L^\infty(\mathbb{R})$ with norm $C\,t^{-1/3}$. From this fact and from the unitarity of U in L^2, one derives easily the first type of relevant estimates, in complete analogy with similar estimates known to hold for other equation such as the Schrödinger equation [7] [13] [26]. One obtains

Lemma 1. For all r_j, q_j with $2 \leqslant r_j \leqslant \infty$ and $3/q_j = 1/2 - 1/r_j$ for $i = 1,2$, and for any interval $I \subset \mathbb{R}$, the following estimate holds

$$\| \int_I d\tau\, U(\cdot - \tau) f(\tau) ; L^{q_1}(\mathbb{R}, L^{r_1}) \| \leqslant C \| f ; L^{\bar{q}_2}(I, L^{\bar{r}_2}) \| \tag{46}$$

(where $1/\ell + 1/\bar{\ell} = 1$ for any ℓ, $1 \leqslant \ell \leqslant \infty$).

The second type of estimates on the linear group $U(t)$ follows directly from the pointwise estimate (45) and can be stated as follows.

Lemma 2. Let h be a continuous function satisfying

$$\begin{cases} h(x) = \exp(\alpha x) & \text{for some } \alpha > 0 \text{ and all } x \leqslant 0 \\ 0 \leqslant h(y) \leqslant h(x) \leqslant \exp[\alpha(x-y)]\, h(y) & \text{for all } x \geqslant y \, . \end{cases} \tag{47}$$

Then for all $T \geqslant 0$, the following estimate holds

$$\| h^{1/2} \int_0^{\cdot} d\tau\, U(\cdot - \tau)\, Df(\tau) ; L^q([0,T], L^\infty) \|$$

$$\leqslant C(T) \| h^{1/2} (1+x_+)^{1/4} f ; L^{\ell}([0,T], L^1) \| , \tag{48}$$

for all q, ℓ with $0 < 1/q = 1/\ell - 1/4 < 3/4$ or for $q = \infty$, $\ell > 4$, and with $C(.)$ uniformly bounded on the compact subsets of $\mathbb{R}^+$.

Finally the third type of estimates on the linear group expresses the smoothing properties of the equation (41), in close analogy with (17) (21).

Lemma 3. Let $h \in \mathcal{C}^3(\mathbb{R}, \mathbb{R}^+)$ satisfy $h' \geqslant 0$ and $h''' \leqslant c\, h$. Then for all $T \geqslant 0$ the following estimate holds

$$\| h'^{1/2} \int_0^{\cdot} d\tau\, U(\cdot - \tau)\, Df(\tau) ; L^2([0,T], L^2) \| \leqslant 2^{-1/2} g(T)$$

$$\leqslant 2^{-1/2} e^{cT/2} \| h^{1/2} f ; L^1([0,T], L^2) \| \tag{49}$$

where

$$g(T) = \int_0^T d\tau \exp[c(T-\tau)/2] \; \| h^{1/2} f(\tau) \|_2 . \tag{50}$$

With the estimates of the previous lemmas available, the uniqueness proofs in Propositions 3 and 4 essentially follow the method sketched at the beginning of this section. In the case of L^2 solutions, the equation (27) for w together with the initial condition $w(0) = 0$ is rewritten in integral form as

$$w(t) = \int_0^t d\tau \; U(t-\tau) \, D(\tilde{V}'' \, w)(\tau) \quad . \tag{51}$$

One introduces an auxiliary norm $\| \, . \, \|_*$ which in the simpler situation where $p \leqslant 2$ in (33) reduces to

$$\| w \|_* = \| \mathrm{Max}(e^{\alpha x}, 1) w \; ; \; L^q([0,T], L^r) \| \tag{52}$$

with $\alpha > 0$ and suitable q and r, and one derives the basic inequality (26) from (51) by the use of an interpolation between Lemmas 2 and 3. The proof requires that the solutions satisfy (34), and that property in turn follows from the smoothing properties of the equation (1) in the form of the inequality (17). In the more complicated situation where $p > 2$ in (33), a different norm $\| \, . \, \|_*$ is required, as well as additional estimates which generalize those of Lemma 1.

In the case of H^1 solutions, a slight refinement of the argument leading to (29)(30), using the norm (52) with $q = \infty$, $r = 2$ instead of simply the norm of w in $L^\infty(L^2)$, shows that the condition $Du \in L^1_{loc}(L^\infty)$ can be weakened to $\chi_+ \, Du \in L^1_{loc}(L^\infty)$ in the subsequent uniqueness proof. One then estimates $\chi_+ \, Du$ from the integral form of the equation (1), namely

$$u(t) = U(t) \, u_0 + \int_0^t d\tau \;\; U(t-\tau) \, DV'(u(\tau)) \tag{53}$$

by using the pointwise estimates on the propagator $S_t(x)$ that follow from (45), either in a direct way, or through the use of Lemma 2. The proof requires that the solutions satisfy (38) and that property again follows from the smoothing properties of the equation (1), now in the form of the inequality (21). We refer to [8] for the details of the proofs.

REFERENCES

[1] J.L. Bona and J.C. Saut, (a) Singularités dispersives de solutions d'équations de type Korteweg-de Vries, C.R. Acad. Sc. Paris, 303 (1986), 101-103.
(b) Dispersive blow up solutions of non linear dispersive wave equations, article in preparation.

[2] J.L. Bona and L.R. Scott, Solutions of the Korteweg-de Vries equation in fractional order Sobolev spaces, Duke Math. J., 43 (1976), 87-99.

[3] J.L. Bona and R. Smith, The initial value problem for the Korteweg-de Vries equation, Phil. Trans. Royal Soc. London A, 278 (1975), 555-601.

[4] A. Cohen Murray, Solutions of the Korteweg-de Vries equation from irregular data, Duke Math. J., 45 (1978), 149-181.

[5] A. Cohen and T. Kappeler, Solutions to the Korteweg-de Vries equation with initial profile in $L^1_1(\mathbb{R}) \cap L^1_N(\mathbb{R}^+)$, SIAM J. Math. Anal., 18 (1987), 991-1025.

[6] P. Constantin and J.C. Saut, Local smoothing properties of dispersive equations, J. Amer. Math. Soc. 1 (1988), 413-439.

[7] J. Ginibre and G. Velo, Scattering theory in the energy space for a class of non linear Schrödinger equations, J. Math. Pures Appl., 64 (1985), 363-401.

[8] J. Ginibre and Y. Tsutsumi, Uniqueness of solutions for the generalized Korteweg-de Vries equation, SIAM J. Math. Anal., in press.

[9] J. Ginibre, Y. Tsutsumi and G. Velo, Existence and uniqueness of solutions for the generalized Korteweg-de Vries equation, preprint, Orsay (1988).

[10] J. Ginibre, Y. Tsutsumi et G. Velo, Existence et unicité des solutions de l'équation de Korteweg-de Vries généralisée, C.R. Acad. Sc. Paris, in press.

[11] T. Kato, On the Korteweg-de Vries equation, Manuscripta Math. 28 (1979), 89-99.

[12] T. Kato, On the Cauchy problem for the (generalized) Korteweg-de Vries equation, Studies in Applied Mathematics, Adv. Math. Supplementary Studies, 18 (1983), 93-128.

[13] T. Kato, On non linear Schrödinger equations, Ann. IHP (Phys. Théor.), 46 (1987), 113-129.

[14] S.N. Kruzhkov and A.V. Faminskii, Generalized solutions of the Cauchy problem for the Korteweg-de Vries equation, Math. USSR Sbornik, 48 (1984), 391-421.

[15] J.L. Lions, Quelques méthodes de résolution des problèmes aux limites non linéaires, Dunod and Gauthier-Villars, Paris, 1969.

[16] R.M. Miura, The Korteweg-de Vries equation : a survey of results, SIAM Review, 18 (1976), 412-459.

[17] T. Mukasa and R. Iino, On the global solutions for the simplest generalized Korteweg-de Vries equation, Math. Japonica, 14 (1969), 75-83.

[18] J.C. Saut, Sur quelques généralisations de l'équation de Korteweg-de Vries, J. Math. Pures Appl. 58 (1979), 21-61.

[19] J.C. Saut and R. Temam, Remarks on the Korteweg-de Vries equation, Israel J. Math., 24 (1976), 78-87.

[20] S. Tanaka, Korteweg-de Vries equation : construction of solutions in terms of scattering data, Osaka J. Math. ,11 (1974), 49-59.

[21] R. Temam, Sur un problème non linéaire, J. Math. Pures Appl. 48 (1969), 159-172.

[22] M. Tsutsumi, On global solutions of the generalized Korteweg-de Vries equation, Publ. RIMS, Kyoto Univ., 7 (1972), 329-344.

[23] M. Tsutsumi, T. Mukasa and R. Iino, On the generalized Korteweg-de Vries equations, Proc. Jap. Acad., 46 (1970), 921-925.

[24] M. Tsutsumi and T. Mukasa, Parabolic regularization of the generalized Korteweg-de Vries equation, Funkcialaj Ekvacioj, 14 (1971), 89-110.

[25] Y. Tsutsumi, On uniqueness of solutions in the energy space for the modified KdV equation (unpublished).

[26] K. Yajima, Existence of solutions of Schrödinger evolution equations, Commun. Math. Phys., 110 (1987), 415-426.

Effective Stability in Hamiltonian Systems in the Light of Nekhoroshev's Theorem

Antonio Giorgilli
Dipartimento di Matematica dell'Università, Via Saldini 50, 20133 MILANO, Italy

Abstract.

The methods of classical perturbation theory are revisited in the light of a rigorous algebraic approach and of Nekhoroshev's theorem on stability over exponentially large times. The applications to the restricted three body problem and to a statistical model of a diatomic gas of identical molecules are illustrated, with the aim of giving good estimates for the size of the stability region and for the dependence on the number of degrees of freedom.

1. Introduction

The present lecture is concerned with the study of a near to integrable Hamiltonian system of differential equations. Such a problem is of interest in many fields of mathematical physics and astronomy. My plan is to illustrate recent rigorous results concerning such kind of systems, in the spirit of Nekhoroshev's theorem on stability over exponentially large times.

Let me recall that, according to Poincaré[1], the fundamental problem of dynamics is the study of a canonical system with Hamiltonian

$$H(p,q,\varepsilon) = h(p) + \varepsilon f(p,q,\varepsilon) \ , \tag{1}$$

where $p = (p_1, \ldots, p_n) \in \mathcal{G} \subset \mathbf{R}^n$, with $\mathcal{G}$ an open set, and $q = (q_1, \ldots, q_n) \in \mathbf{T}^n$ are action–angle variables, ε is a real (small) parameter and $H(p,q,\varepsilon)$ is assumed to be an analytic function of p, q and ε.

The dynamical evolution of the unperturbed system, i.e. the one with Hamiltonian $H(p,q,0) = h(p)$, is a well known topic: the system admits n independent prime integrals, namely the actions $p_1, \ldots, p_n$, and the flow in phase space is given by

$$p(t) = p^{(0)} \ , \quad q(t) = \omega(p^{(0)})t + q^{(0)} \ ,$$

where $\omega(p^{(0)}) = \frac{\partial h}{\partial p}(p^{(0)})$ are the unperturbed frequencies and $(p^{(0)}, q^{(0)})$ the initial data. So, the phase space is foliated into invariant tori, and the motion on a torus is either quasi periodic or periodic, according to the possible existence of resonance relations among the frequencies, namely relations like $k \cdot \omega = 0$ with $0 \neq k \in \mathbf{Z}^n$.

The naive approach of perturbation theory is essentially an attempt to extend the same plain picture to the perturbed system, namely the case $\varepsilon \neq 0$. From a technical viewpoint, one usually looks for a near to identity canonical transformation which removes the dependence of the transformed Hamiltonian on the angle variables. Fortunately, as was proven by Poincaré himself[1], such a naive program fails. Indeed, he proved that under nondegeneracy conditions on the unperturbed Hamiltonian $h(p)$, namely $\det\left(\frac{\partial^2 h}{\partial p_j \partial p_l}\right) \neq 0$, the system (1) does not generally admit analytic prime integrals independent of the Hamiltonian.

On the other hand, it is well known that the series produced by perturbative methods, although generally non convergent, are very useful in practical applications. The relevance of such a fact is stressed by Poincaré, who also discusses the possibility that these series are in fact asymptotic expansions[1]. A rigorous approach on this direction can already be found in some works of Moser[2] and Littlewood[3], and a general result is given by Nekhoroshev's theorem[4]. Roughly speaking, one renounces to look for information on the dynamical evolution of the system for infinite times, being satisfied with results concerning a time interval like

$$T = T_* \exp\left(\frac{\varepsilon_*}{\varepsilon}\right)^a , \tag{2}$$

with suitable constants T_*, ε_* and a. Due to the exponential dependence on the inverse of the perturbative parameter ε, such a time interval can become very large, and possibly exceed any realistic time. In the very words of Littlewood[3], "while not eternity, this is a considerable slice of it".

Now, it is natural to raise the question whether such results can be effectively applied to physical systems, mainly in the case of statistical models, with a large number of degrees of freedom. A first problem here concerns the size of the constants T_*, ε_* and a in the exponential estimate (2): when explicitly computed, these constants turn out to be ridiculously small. For example, if one considers the solar system the natural perturbation parameter is the ratio between Jupiter's mass and the mass of the Sun; if one tries to apply the Nekhoroshev's result, as in the original formulation, one finds that the mass of Jupiter should be several orders of magnitude less than that of a proton. A second remark concerns the dependence of the same constants on the number of degrees of freedom: for example, still using the original Nekhoroshev's formulation, one has $a \sim 1/n^2$. This fact led some authors to conclude that Nekhoroshev's like results are not applicable to statistical systems, where the number n of degrees of freedom goes to infinity[5].

Of course, one is not authorized to immediately conclude that Nekhoroshev's theorem is a wonderful mathematical result, but completely useless for physics. Indeed, the explicit values of the constants are estimated via a set of inequalities which are often far from being optimal, and could hardly be optimized without more specific assumptions on the physical system. The aim of the present lecture is precisely to discuss some recent theoretical improvements which give a positive, although partial answer to the question raised above.

2. Model problems

Let me illustrate the theory by making explicit reference to two physical examples: the stability of an elliptic equilibrium point, which is a classical topic, and the problem of relaxation times in statistical systems, which has recently been investigated[6,7].

2.1. The stability of an elliptic equilibrium.

As a first model, let's consider a system of harmonic oscillators described by the Hamiltonian

$$H(x,y) = \frac{1}{2}\sum_{l=1}^{n} \omega_l \left(x_l^2 + y_l^2\right) + \sum_{s>0} H_s(x,y) , \tag{3}$$

where H_s is a homogeneous polynomial of degree $s+2$ in the canonical variables $(x,y) \in \mathbf{R}^{2n}$, and $\omega \in \mathbf{R}^n$ is the vector of the harmonic frequencies, which are assumed to be all nonvanishing. This is a classical problem, and can be considered as the simple version of the "fundamental problem of dynamics". The reason is that the frequencies of the unperturbed system are independent of the initial point, so that the formal approach is definitely simpler. Specific models of such a kind are the nonlinear chain, like the celebrated FPU model[8], and the Lagrangian triangular equilibrium points L_4 and L_5 of the circular restricted problem of three bodies.

For the FPU model, namely a chain of $n+2$ identical point masses on a straight line, connected with nonlinear springs and with fixed ends, the harmonic frequencies are given by $\omega_l = 2\alpha \sin \frac{l\pi}{2(n+1)}$, $1 \leq l \leq n$, (α being a constant).

For the Lagrangian points, denoting by m and M the masses of the primaries, and introducing the dimensionless parameter $\mu = m/(M+m)$, the triangular points turn out to be elliptic (and so linearly stable) for $0 < 27\mu(1-\mu) < 1$, but, unlike the FPU model, the frequencies have different signs. For example, considering the point L_4 for the Sun–Jupiter case, and allowing the asteroid to move in space one gets $\mu \sim 0.95387 \times 10^{-3}$, $\omega_1 \sim 0.99676$, $\omega_2 \sim -0.80464 \times 10^{-1}$ and $\omega_3 = 1$.

In these models two problems naturally arise.

i. *Stability of the equilibrium:* this is easily solved for the FPU model, since the Hamiltonian has a minimum in the equilibrium, but remains unsolved, up to now, for the L_4 point, because the harmonic frequencies have different sign, so that the equilibrium is a saddle point for the Hamiltonian; in the latter case strong results have been obtained in the framework of KAM theory, but definite conclusions can only be drawn in the planar case.

ii. *Freezing of the harmonic actions:* since the pioneering work of Fermi, Pasta and Ulam a lot of numerical experiments produced evidence that, at least for low total energy, there is no equipartition of energy among the normal modes. Such a fact is relevant for the foundations of classical statistical mechanics.

A possible approach, which is valid for both these problems, consists in relaxing the definition of stability: instead of looking for a confinement of orbits (or for freezing of the actions) over an infinite time, one tries to prove that this happens up to a finite, but large time.

2.2. Relaxation times in statistical systems

The second model has been investigated in refs. [6] and [7]. One considers a canonical system with analytic Hamiltonian

$$H(p,x,\pi,\xi) = \hat{h}(p,x) + h_\omega(\pi,\xi) + f(p,x,\pi,\xi) \ , \tag{4}$$

where

$$h_\omega(\pi,\xi) = \frac{1}{2}\sum_{l=1}^{\nu}\left(\pi_l^2 + \omega_l^2\xi_l^2\right) \ , \quad (\pi,\xi) \in \mathbf{R}^{2\nu}$$

is the Hamiltonian of a system of harmonic oscillators, $\hat{h}(p,x)$ is the Hamiltonian of a generic n–dimensional system, and $f(p,x,\pi,\xi)$ a coupling term which is assumed to be of order ξ, and so to vanish for $\xi = 0$. More specific models are the realization of physical constraints and a statistical system like a diatomic gas of identical molecules.

In the case of constraints the Hamiltonian $\hat{h}(p,x)$ describes the motion of the constrained system, while $h_\omega(\pi,\xi)$ describes the vibrations of the constraints. One is then interested in investigating the relations between the time evolution of the constrained system and that of the whole system in the limit $\omega \to \infty$.

The statistical model of a diatomic gas can be studied in the same spirit: the Hamiltonian $\hat{h}(p,x)$ describes the translational and rotational degrees of freedom, while $h_\omega(\pi,\xi)$ describes the internal vibrations. Assuming that the molecules interact via a regular short range potential, one is still interested in understanding the dynamical evolution in the limit of large ω's. The peculiarity of such a system with respect to that of constraints lies in the fact that the number ν of molecules is large, but all the frequencies are equal, so that the system is completely resonant.

The identification of a perturbative parameter in these systems proceeds as follows. Write $\omega = \lambda\Omega$ with Ω of the same order of the inverse of a typical time scale of the constrained system (for example the characteristic time for the collision of two molecules, which is non zero if the interaction potential is regular) and large λ; then transform the variables according to $\pi = \pi'\sqrt{\lambda\Omega}$ and $\xi = \xi'/\sqrt{\lambda\Omega}$, and assume the total energy of the constraints (or of the internal vibrations) to be finite, so that the variables (π', ξ') turn out to be confined in a disk of size $1/\sqrt{\lambda}$. Then the Hamiltonian can be given the form, omitting primes,

$$H(p,x,\pi,\xi,\lambda) = \hat{h}(p,x) + \lambda h_\Omega(\pi,\xi) + \frac{1}{\lambda} f_\lambda(p,x,\pi,\xi) \tag{5}$$

with

$$h_\Omega(\pi,\xi) = \frac{1}{2}\sum_{l=1}^{\nu} \Omega_l \left(\pi_l^2 + \xi_l^2\right) \tag{6}$$

(here, a straightforward computation would give $\lambda^{-1/2}$ in front of f, but f itself turns out to be of order $\lambda^{-1/2}$, since it vanishes for $\xi = 0$). Thus, the problem looks similar to the fundamental problem of dynamics, corresponding to the Hamiltonian (1), since λ^{-1} plays the role of a small parameter; at the same time it resembles the problem of the elliptic equilibrium, due to the form of h_Ω. In fact, it is definitely different from both the models above, because $\hat{h}$ is not restricted to represent an integrable system: one needs a perturbative scheme for a nonintegrable system.

The approach taken here consists in considering the whole system as composed of two separate subsystems $\hat{h}(p,x)$ and $h_\Omega(\pi,\xi)$, each with its own internal, possibly chaotic, dynamics, and in looking only for a bound on the energy exchange between these two subsystems over a large time scale, in the spirit of Nekhoroshev's theorem.

3. The perturbation scheme

My aim is now to describe a general perturbation scheme that can be applied to both the models above. The main tool will be an algebraic approach to perturbation theory based on explicit recursive formulae, like in the Lie transform method described in refs. [9–10]. Such a scheme is made rigorous by adding explicit estimates on the convergence properties of the series so produced; in doing this, particular attention is paid to obtaining good estimates, mainly for what concerns the dependence on the number of degrees of freedom. The general formulation of the present scheme can be found in ref. [11].

3.1. Algebraic framework

The first step consists in translating the concept of "perturbation order" into an algebraic structure. This is given by introducing a function space $\mathcal{P}$ which is an algebra with respect to sum, scalar multiplication and product, and a sequence $\{\mathcal{P}_s\}_{s\geq 0}$ of subspaces of $\mathcal{P}$ such that:

i. $\mathcal{P}_s$ is a linear space for $s \geq 0$, and $\mathcal{P} = \oplus_{s\geq 0}\mathcal{P}_s$
ii. for any $f \in \mathcal{P}_s$ and $g \in \mathcal{P}_r$ one has $fg \in \mathcal{P}_{s+r}$
iii. there exists $2n$ nonnegative integers $(a_1, \ldots, a_n, b_1, \ldots, b_n)$ such that for any $f \in \mathcal{P}_s$ one has $\frac{\partial f}{\partial p_j} \in \mathcal{P}_{s-a_j}$ and $\frac{\partial f}{\partial q_j} \in \mathcal{P}_{s-b_j}$ for $1 \leq j \leq n$. Here, p, q are the canonical coordinates.

The properties above essentially ask that the algebraic structure of the classes $\mathcal{P}_s$ be compatible with the operations needed in order to build a perturbative scheme, namely sums, derivatives and Poisson brackets. Moreover, one can expand any function $f \in \mathcal{P}$ as $f = \sum_{s\geq 0} f_s$ with $f_s \in \mathcal{P}_s$, like in a power series development in a small parameter. The main, and important, difference with respect to the usual one–parameter scheme is that nonhomogeneous expansions are also allowed. Indeed, introducing for $s \geq 0$ the space $\mathcal{P}^{[s]} = \bigoplus_{l\geq s} \mathcal{P}_l$, one can consider an expansion like $f = \sum_{s\geq 0} f^{[s]}$ with $f^{[s]} \in \mathcal{P}^{[s]}$. Such an expansion is clearly not unique, but has the advantage that one is allowed to control the minimum order of each term. A more detailed control is possible by introducing for $0 \leq s \leq r$ the space $\mathcal{P}^{[s,r]} = \bigoplus_{s\leq l\leq r} \mathcal{P}_l$, so that also the maximum order of each term is bounded. Such a generalization is the technical tool which allows to build up a perturbation scheme for the model problem proposed in sect. 2.2. Indeed, in that case one has to deal with at least two parameters—the dimensionless quantity λ^{-1} and the size of the (π, ξ) variables. In such a case the straightforward application of the usual one–parameter expansion does not work. More generally, one can imagine cases when the identification of an expansion parameter is not trivial at all; a possible way out is to decide *a priori*, on the basis of heuristic considerations, *what is small and why*, and to check the consistency of that choice on the basis of rigorous estimates on the expansions. This is possible in the present general scheme.

In order to produce rigorous estimates one must take into account the domains where the functions are defined, and to introduce a suitable norm which allows to estimate the size of the functions.

The choice of the domains is the usual one: starting with a suitable subset M_0 of the phase space, one builds the domain

$$\mathcal{D}_R(M_0) = \bigcup_{(p,q)\in M_0} \Delta_R(p,q) \tag{7}$$

where $\Delta_R(p,q)$ is the polydisk

$$\Delta_R(p,q) = \{(p',q') \in \mathbf{C}^{2n} \ :\ |p_j - p'_j| \leq \varrho_j\,,\ |q_j - q'_j| \leq \sigma_j\,,\ 1 \leq j \leq n\}\,. \tag{8}$$

Here, $R \equiv (\varrho_1, \ldots, \varrho_n, \sigma_1, \ldots, \sigma_n)$ is a vector with positive components. For brevity, $\mathcal{D}_R(M_0)$ will be simply denoted by $\mathcal{D}_R$ in what follows.

Having so defined the domain, one considers the function space $\mathcal{F}_R \subseteq \mathcal{P}$ of these functions which are analytic in the interior of $\mathcal{D}_R$ and bounded on $\mathcal{D}_R$, and introduces a norm $\|\cdot\|_R$ on $\mathcal{F}_R$ with the following properties:

i. if $\mathcal{D}_{R'} \subseteq \mathcal{D}_R$ then $\|f\|_{R'} \leq \|f\|_R$;

ii. for any $(p,q) \in \mathcal{D}_R$ one has $|f(p,q)| \leq \|f\|_R$;

iii. there exists a positive constant C such that for any $\chi \in \mathcal{F}_{(1-d')R}$ with $0 \leq d' < 1$, for any $f \in \mathcal{F}_R$ and for any positive $d \leq 1 - d'$ one has

$$\|L_\chi f\|_{(1-d)R} \leq \frac{C}{(d+d')d} \|\chi\|_{(1-d')R} \|f\|_R \ ;$$

here, $L_\chi f$ denotes the Lie derivative of f, i.e., in fact, the Poisson bracket $\{\chi, f\}$. Let me add some comment. In order to bound the size of a function, the natural choice would be to use the supremum norm over the domain; this is indeed the usual method. However, when looking for good estimates, mainly with respect to the number n of degrees of freedom, this turns out to be a bad choice. The reason is that at some step of the perturbation procedure one necessarily has to consider an expansion of a function either in powers of the coordinates or in Fourier components, and to use general theorems in order to get bounds on the coefficients of that expansion; then, after some operation on the coefficients one comes back to the supremum norm by summing up all the contributions. This, of course, introduces a very bad dependence on n. The only way out is to change the norm, still maintaining some compatibility properties with the supremum norm, like ii. above.

Let's come now to the examples. In the case of the elliptic equilibrium, the natural (and classical) choice is to identify $\mathcal{P}_s$ with the linear space of homogeneous polynomials of degree $s+2$ in the canonical variables, so that the Hamiltonian (3) is characterized by $H_s \in \mathcal{P}_s$. Choosing M_0 as the origin, the domain is naturally defined as a polydisk. Writing a polynomial $f \in \mathcal{P}_s$ as $f = \sum_{j,k} f_{jk} x^j y^k$, with $f_{jk} \in \mathbf{C}$, a suitable norm is then $\|f\|_R = \sum_{j,k} |f_{jk}| \varrho^j \sigma^k$. Then all the properties above are satisfied with $a_j = b_j = 1$ for $1 \leq j \leq n$ and with $C = 1$.

The model of constraints is more complex. Here, one must take into account the fact that the variables (π, ξ) play an active role in determining the perturbation order, since they are confined in a neighbourhood of the origin of size $\varepsilon = \lambda^{-1/2}$, while the variables (p, x) should be considered essentially as parameters. The algebraic framework is then built up by defining $\mathcal{P}_s$ as the space of homogeneous polynomials of degree $s+2$ in π, ξ, ε, whose coefficients are analytic functions of (p, x). The domain has now the form $\mathcal{D}_R = \mathcal{G}_{R'} \times \Delta_{R''}$, where $\mathcal{G}_{R'}$ is the domain of the variables (p, x) and $\Delta_{R''}$ a polydisk centered on the origin for the variables (π, ξ) (here, I set $R = (R', R'')$, with analogous settings for ϱ and σ). Writing then $f \in \mathcal{P}_s$ as

$$f = \sum_{l+|j|+|k|=s+2} \varepsilon^l f^{(l)}_{jk}(p, x) \pi^j \xi^k \ ,$$

the norm is defined as

$$\|f\|_R = \sum_{l,j,k} \varepsilon^l \left| f^{(l)}_{jk} \right|_{R'} \varrho''^j \sigma''^k \ , \qquad \left| f^{(l)}_{jk} \right|_{R'} = \sup_{(p,x) \in \mathcal{G}'_R} \left| f^{(l)}_{jk}(p, x) \right| \ .$$

Then, all the properties above are satisfied with $a_j = b_j = 0$ for the variables p, x, with $a_j = b_j = 1$ for the variables π, ξ, and with $C = 2$.

3.2. Near to identity canonical transformations

The next step consists in defining a near to identity canonical transformation via a recursive algebraic algorithm.

Considering a *generating sequence* $\chi = \{\chi_s\}_{s \geq 1}$ of functions $\chi_s \in \mathcal{P}^{[s]}$, one defines the operator

$$T_\chi = \sum_{s \geq 0} E_s \ ,$$

where

$$E_0 = \mathrm{Id} \ , \quad E_s = \sum_{j=1}^{s} \frac{j}{s} L_{\chi_j} E_{s-j} \ .$$

Such an operator turns out to be linear, invertible, and to preserve products and Poisson brackets, i.e. $T_\chi(f \cdot g) = (T_\chi f)(T_\chi g)$ and $T_\chi\{f, g\} = \{T_\chi f, T_\chi g\}$. Moreover, any near to identity canonical transformation $(p, q) = \mathcal{C}(p', q')$ can be given the form

$$p_l = T_\chi p'_l \ , \quad q_l = T_\chi q'_l \ , \quad 1 \leq l \leq n \ , \tag{9}$$

with a suitable generating sequence χ, and explicit recursive formulae can be found for the inverse operator T_χ^{-1} and for the generating sequence, φ say, of the composition of two transformations, T_χ and T_ψ say, so that $T_\varphi = T_\chi \circ T_\psi$.

Coming now to a rigorous viewpoint, one proves that, *under the hypothesis that there exist real constants $\beta \geq 0$ and $\Phi > 0$ such that $N_{R,R}(\chi_s) \leq \frac{\beta^{s-1}}{s}\Phi$, for any positive $d < 1/2$, and with the condition*

$$\left(\frac{2Ce^2\Phi}{d^2} + \beta\right) \leq 1$$

the canonical transformation defined by T_χ is analytic, and one has

$$\mathcal{D}_{(1-2d)R} \subset T_\chi(\mathcal{D}_{(1-d)R}) \subset \mathcal{D}_R \ .$$

The condition above on Φ, β and d turns out, in the examples discussed above, to be a condition on the size of both the parameter ε and the size R of the domain.

The present approach is more effective when compared with the classical one involving generating functions in mixed variables. Let me add two remarks to illustrate this fact.

A first remark is that the transformation is given by an explicit recursive expression: no inversion is needed, like in the classical method, and the definition can be easily translated into an explicit recursive algorithm. The same holds for the inverse transformation.

A second remark is concerned with the fact that, given a function $f(p,q)$, the canonical transformation (9) transforms it to $f'(p',q') = (f \circ T_\chi)(p',q')$: a standard result in the Lie series theory, namely the exchange theorem (see ref. [9]), states that $f'(p',q')$ coincides with $T_\chi f$, i.e. that $f \circ T_\chi = T_\chi f$. The actual consequence is that no substitution is needed in order to determine the transformed function f', since it is given by an explicit algorithm. This is, according to Gröbner, the most fascinating aspect in the Lie transform theory.

Let me also stress that the present algebraic approach has some advantages with respect to the usual approach to Lie transforms, which makes essential use of the flow generated by a nonautonomous canonical system. Indeed, in the latter method one must choose a single perturbation parameter, to be identified with the time variable of the canonical flow, thus making not so natural the extension of the method to the case of many perturbation parameters. The usual approach could, of course, be recovered, but let me stress that only the algebraic algorithm is used, in fact, in practical applications, and that all the relevant properties can be proven by purely algebraic methods. The reference to a canonical flow looks like a psychological support more than like an essential mathematical tool.

3.3. Normal form of the Hamiltonian

The development of the theory now follows more closely the usual scheme. Let me make reference, for definiteness, to the Hamiltonian of the model of constraints. With the algebraic framework described in sect. 3.1, the Hamiltonian can be expanded in the form

$$H = H_0 + \sum_{s\geq 1} H_s \ ,$$

$$H_0 = h_\Omega(\pi,\xi) + \varepsilon^2 \hat{h}(p,x) \ , \quad H_s \in \mathcal{P}_{2s+2} \ ,$$

with $h_\Omega(\pi,\xi)$ and $\hat{h}(p,x)$ as in (6); one looks then for a truncated generating sequence $\chi^{(r)} = \{\chi_s\}_{s=1}^r$ which gives the Hamiltonian the form

$$H^{(r)} = Z_0 + \sum_{s=1}^{r} Z_s + \mathcal{R}^{(r)} \ ,$$

where $Z_0 = H_0$, Z_s is in normal form and $\mathcal{R}^{(r)} \in \mathcal{P}^{[2r+2]}$ is a nonnormalized remainder. By normal form it is simply meant here that $L_{h_\Omega} Z_s = 0$. This is a classical and well known problem, so let me stress only the relevant differences with the usual procedure. These differences are mainly concerned with two problems: first, the Poisson bracket between homogeneous functions necessarily produces a nonhomogeneous result, and, second, the unperturbed Hamiltonian H_0 is not integrable.

Trying to separate the homogeneous components is impractical, so, let's proceed as follows. By substituting the explicit definition of T_χ and the expansion above of the Hamiltonian in the expression of the normal form, and acting exactly as in the usual case of homogeneous expansions (for example the case of the elliptic equilibrium, as discussed in ref. [12]), one gets the infinite system of equations $L_{H_0}\chi_s + Z_s = \Psi_s$, with known Ψ_s, to be recursively solved for $s \geq 1$ with respect to χ_s and Z_s with the condition $L_{H_0} Z_s = 0$. Such a system can be consistently used in the nonhomogeneous case too, since, as is easily checked, the equation at order s involves only terms belonging to the space $\mathcal{P}^{[2s+2]}$, so that the minimum order of each term increases with s. At this point the second problem arises. Indeed, due to the lack of knowledge about $\hat{h}$, it is in general impossible to solve the equation above. The key point is instead to notice that $L_{H_0}\chi_s = L_{h_\Omega}\chi_s + L_{\varepsilon^2\hat{h}}\chi_s$, and that, if $\chi_s \in \mathcal{P}^{[2s+2]}$, then one has $L_{h_\Omega}\chi_s \in \mathcal{P}^{[2s+2]}$, while $L_{\varepsilon^2\hat{h}}\chi_s \in \mathcal{P}^{[2s+4]}$. Such an elementary remark allows to shift $L_{\varepsilon^2\hat{h}}\chi_s$ to the next order, where it becomes a known term, so that the equation above takes the simpler form

$$L_{h_\Omega}\chi_s + Z_s = \Psi_s \ , \tag{10}$$

with known Ψ_s, and the (p, x) variables play essentially the role of parameters. The solution proceeds then, at least formally, as usual: the vector Ω determines a resonance module $\mathcal{M}_\Omega \in \mathbf{Z}^\nu$ defined as $\mathcal{M}_\Omega = \{k \in \mathbf{Z}^\nu : k \cdot \Omega = 0\}$; this module in turn determines a splitting $\Psi_s = \overline{\Psi}_s + \tilde{\Psi}_s$, where $\overline{\Psi}_s$ collects the resonant terms of Ψ_s and $\tilde{\Psi}_s$ the remaining ones; finally one puts $Z_s = \overline{\Psi}_s$ and uses $\tilde{\Psi}_s$ in order to determine χ_s.

From a rigorous viewpoint, this is not enough if one looks for good estimates. Indeed, as is well known, in solving the equation (10) above there appear small denominators of the form $k \cdot \Omega$ with $k \in \mathbf{Z}^\nu \setminus \mathcal{M}_\Omega$. A key point in removing, as far as possible, the dependence of the results on the number n of degrees of freedom, is that at any order one has to deal only with a finite number of small denominators. This is obtained by carefully controlling how the nonhomogeneity of the expansions propagate through the successive application of Poisson brackets. Indeed, one can check that at any order s of the perturbative procedure one has $\Psi_s \in \mathcal{P}^{[2s+2,4s]}$, and so also Z_s and χ_s belong to the same space. This furnishes a bound on the maximum order of each term, and so ensures that only a finite number of small denominators do appear, in this case those with $0 < |k| \leq 4s$. So, in order to control the action of the small denominators, it is enough to determine a nonincreasing sequence $\{\alpha_s\}_{s\geq 1}$ of positive constants such that $|k \cdot \Omega| \geq \alpha_s$ for $0 < |k| \leq 4s$ and $k \notin \mathcal{M}_\Omega$.

Giving rigorous estimates for the generating sequence and the normal form is now a technical matter: the formal normalization algorithm is translated into a set of recursive estimates. The result is that, *assuming that there exist real positive constants E_0, σ and E such that $\|\hat{h}\|_R \leq E_0$ and $\|H_s\|_R \leq \sigma^{s-1}E$, one proves that for any positive $d < 1$ one has $\|\chi_s\|_{(1-d)R} \leq \frac{\beta^{s-1}}{s}\Phi$ with $\beta = C_0 r/\alpha_r$ and $\Phi = 4E/\alpha_r$; the constant C_0 depends on E_0, E, σ and R.* Thus, the generating sequence satisfies the conditions for the convergence of the canonical transformation stated in sect. 3.2, provided the constants β and Φ are small enough. This can be satisfied if R' is chosen to be of order ε, namely if the initial condition are such the the energy of the subsystem h_Ω is small. Setting now $d = 1/4$, and recalling that $\varepsilon^2 = \lambda^{-1}$ for the original Hamiltonian (5) of the model of constraints, one gets the final result that *under the hypotheses above there exist real positive constants* λ_*, C_1, C_2 *and* C_3 *depending on* E_0, E, σ *and* R *such that for any* $\lambda > \lambda_*$ *the normalized Hamiltonian is analytic in the domain* $\mathcal{D}_{R/2}$, *and one has the estimates*

$$\begin{aligned} |T_\chi h_\Omega - h_\Omega| &< C_1 \lambda^{-2} \\ \left|T_\chi \hat{h} - \hat{h}\right| &< C_2 \frac{\lambda^{-2}}{\alpha_r} \\ \left|\mathcal{R}^{(r)}\right| &< C_3 \left(\frac{r}{\alpha_r}\right)^r \left(\frac{\lambda_*}{\lambda}\right)^r . \end{aligned} \tag{11}$$

Here, the dependence on the normalization order r has been put into evidence.

The estimates above have been adapted to the case of constraints, but essentially the same results are obtained in the case of the elliptic equilibrium, with the same dependence on the normalization order r. The size of the constants β and Φ in this case depends only on the size of the domain, which is a polydisk centered on the origin. A detailed proof for the case of constraints can be found in ref. [7]; the case of elliptic equilibrium has been investigated in ref. [12].

3.4. Exponential estimates of Nekhoroshev type

Forget now, for a moment, the remainder $\mathcal{R}^{(r)}$, and consider the dynamical evolution of the truncated Hamiltonian $Z = Z_0 + \sum_{s=1}^{r} Z_s$. Let $I = I_0 + I_1 + \ldots$ be a prime integral for Z (in general, given $I_0 = \sum_{l=1}^{\nu} \mu_l(\pi_l^2 + \xi_l^2)$ with $\mathcal{M}_\Omega \perp \mu \in \mathbf{R}^\nu$, then $I = T_\chi I_0$ is a prime integral; for example, $T_\chi h_\Omega$ is such an integral). Then, if one observes the dynamical evolution of I_0 one sees that its value changes over a time scale of order λ, i.e. a quite short time; however, since I is constant, this change in time is bounded by $|I - I_0|$, evaluated over the domain where the motion takes place. If one can guarantee that this domain is contained in $\mathcal{D}_{R/2}$, then the change of I_0 is bounded for all times, being of order λ^{-1}. Such a change is due to the *deformation* of coordinates induced by the near to identity canonical transformation.

Taking now into account the whole Hamiltonian $H^{(r)}$, one sees that I is an approximate integral whose time derivative is of the same order λ^{-r} of the remainder $\mathcal{R}^{(r)}$. Thus, superimposed to the deformation, there may be a diffusion (the Arnold diffusion) due to the *noise* induced by the remainder. Such a diffusion is slow, since it can become of the same order λ^{-1} of the deformation only after a time interval of order λ^{r-1}, but its possible existence does not allow to draw any conclusion about the motion for infinite times.

The bound above on the noise has an evident shortcoming: it depends on the normalization order r, which is clearly an extraneous element introduced by the perturbation scheme. Thus, one is naturally led to remove such an element by looking for a choice of r, say $r_{\rm opt}$, which minimizes the size of the remainder.

A general method is the following. Recalling the usual diophantine theory, choose the sequence $\{\alpha_s\}_{s\geq 1}$ as $\alpha_s = \gamma s^{-\tau}$, with suitable constants $\gamma > 0$ and $\tau \geq 0$ (it is known that for $\tau > \nu - 1$ such a condition is satisfied by a set of Ω's of large measure). Replace now this value in the bound (11) of the remainder, so that

$$|\mathcal{R}^{(r)}| < C' r^{r(\tau+1)} \left(\frac{\lambda_*}{\lambda}\right)^r ,$$

with a suitable constant C', and minimize it by setting

$$r = r_{\mathrm{opt}} = \left[\frac{1}{e}\left(\frac{\lambda}{\lambda_*}\right)^{\frac{1}{\tau+1}}\right] \tag{12}$$

(here, $[\cdot]$ denotes the integer part). Thus, the optimal normalization order r_{opt} is determined as a function of the perturbation parameter λ. By substituting this value in the estimate for the remainder one gets

$$|\mathcal{R}_\lambda| < \mathcal{A}\lambda \exp\left[-\frac{\tau+1}{e}\left(\frac{\lambda}{\lambda_*}\right)^{\frac{1}{\tau+1}}\right] , \tag{13}$$

where $\mathcal{R}_\lambda$ denotes the remainder when the normalization order is chosen according to (12), and $\mathcal{A}$ is a constant.

The conclusion is that *the change in time of a prime integral is of the same order of the deformation up to a time* $|t| \leq \min(T, T_0)$, *where*

$$T = T_* \lambda^{-1} \exp\left[\frac{\tau+1}{e}\left(\frac{\lambda}{\lambda_*}\right)^{\frac{1}{\tau+1}}\right] , \tag{14}$$

and T_0 *is the escape time of the orbit from the domain* $\mathcal{D}_{R/2}$, *where the estimates hold.* The time T_0 can usually be determined by independent considerations; for example, it is actually infinite if the motion takes place on a compact surface of constant energy which is contained in $\mathcal{D}_{R/2}$.

4. Application to the model problems

Let me now conclude by coming back to the specific problems illustrated in sect. 2, and quoting some recent results. As said in the introduction, the aim is to show that the application of the previous scheme to these models can substantially improve the original Nekhoroshev's results, at least in some cases.

4.1. Nonlinear chains

Assuming that the frequencies ω are nonresonant, the normalized Hamiltonian admits n prime integrals of the form $\Phi^{(l)} = I_l + \Phi_1^{(l)} + \ldots$, which are perturbations of the harmonic actions $I_l = \frac{1}{2}\sum_l \omega_l(x_l^2 + y_l^2)$. If the initial data $I_l(0)$ are small enough, then one has

$$|I_l(t) - I_l(0)| < C I_{\mathrm{max}}^{3/2} ,$$

with a suitable constant C and $I_{\mathrm{max}} = \max_l(I_l(0))$, up to a time

$$T = T_* \exp\left[\frac{\tau+1}{e}\left(\frac{I_*}{I_{\mathrm{max}}}\right)^{\frac{1}{2(\tau+1)}}\right] .$$

Unfortunately, here one has $\tau > n-1$, which is a bad dependence on n; however, this result has not been optimized.

Such results can be obtained by the methods described in the present lecture. A substantially simpler, but unfortunately less general, method can be found in ref. [13]. An extension to the case of an infinite system of coupled oscillators, but with a particular class of initial conditions, can be found in ref. [14].

4.2. The Lagrangian point L_4

Considering the spatial, circular restricted three body problem in the Sun–Jupiter case, one can look for a radius R_0 such that an asteroid starting in a neighbourhood of radius R_0 of the Lagrangian point L_4 is confined to a neighbourhood of radius $2R_0$ for a time T of the order of the age of the universe.

Due to the particular simplicity of the problem, one can improve the choice of the optimal normalization order by explicitly computing the constant α_r in (11), and performing optimization by computer. This gives for R_0 a value of a few kilometers, which is not far from being a realistic result. Let me stress, in such a connection, that the use of the computer is only limited to the evaluation of the constants which appear in the theory.

4.3. The diatomic gas

In this model, the only prime integral for the normalized Hamiltonian is h_Ω. Such a result is in a sense optimal, since, due to the complete resonance, essentially no bound can be put on the exchange of energy among the internal vibration of the molecules. So, the best we can do is to prove that the subsystem of the translational and rotational degrees of freedom of the molecules and the subsystem of the internal vibrations evolve in fact independently, each with its own internal, possibly chaotic dynamics, and that a significant transfer of energy between these two subsystems takes an exponentially large time.

The results of sect. 3.4 allow to conclude, in this case, that the energy exchange is of order λ^{-1} up to a time

$$T = T_* \exp\left(\frac{\lambda}{\lambda_*}\right) .$$

Note that the exponent a which appears in (2), and is of order $1/n$ in the case of the nonlinear chain, here is exactly 1. This is precisely due to the resonance. Indeed, due to the fact that $\Omega_1 = \ldots = \Omega_\nu$, one has that the expression $k \cdot \Omega$ either vanishes, but the corresponding term goes into the normal form, or is a multiple of Ω_1, so that there are no small denominators at all. The price paid is the fact that no information is available on the internal evolution of the subsystem of the vibrations.

A dependence on n still remains in the constants T_* and λ_*, which turn out to be typically of order $1/n^2$. However one should note that this is too pessimistic, since the perturbation scheme, being global in (a subset of) the phase space, is developed as if all the molecules were simultaneously colliding at any time: a foolish picture. It seems quite reasonable to hope that such a dependence can be removed in the framework of a statistical approach, which for example excludes the initial data leading to multiple collisions. Some more indications can be found in ref. [7].

[1] Poincaré, H.: *Les méthodes nouvelles de la mécanique céleste*, Gauthier–Villars, Paris (1892).

[2] J. Moser: *Stabilitätsverhalten kanonisher differentialgleichungssysteme,* Nachr. Akad. Wiss. Göttingen, Math. Phys. K1 IIa, nr.6 (1955), 87–120.

[3] J. E. Littlewood: *On the equilateral configuration in the restricted problem of three bodies*, Proc. London Math. Soc.(3) **9** (1959), 343–372;
J. E. Littlewood: *The Lagrange configuration in celestial mechanics*, Proc. London Math. Soc.(3) **9** (1959), 525–543.

[4] Nekhoroshev N. N.: *Exponential estimate of the stability time of near–integrable Hamiltonian systems*, Russ. Math. Surveys **32** N.6, 1–65 (1977).
Nekhoroshev N. N.: *Exponential estimate of the stability time of near–integrable Hamiltonian systems, II*, Trudy Sem. Petrovs. N.5, 5–50 (1979).

Benettin, G., Galgani, L. and Giorgilli, A.: *A proof of Nekhoroshev's theorem for the stability times in nearly integrable Hamiltonian systems*, Cel. Mech **37**, 1–25 (1985).
Benettin, G. and Gallavotti, G.: *Stability of motion near resonances in quasi integrable Hamiltonian systems*, J. Stat. Phys. **44**, 293–338 (1986).
Gallavotti, G.: *Quasi-integrable mechanical systems*, in K. Osterwalder and R. Stora (eds.), *Les Houches, Session XLIII* (1984).

[5] Zaslavski, H. M.: *Stochasticity of dynamical systems*, Nauka, Moscow (1984).
Chirikov, B. V.: Phys.Rep. **52**, 263 (1979).

[6] Benettin, G., Galgani, L. and Giorgilli, A.: *Realization of holonomic constraints and freezing of high frequency degrees of freedom in the light of classical perturbation theory, part I*, Comm. Math. Phys., **113**, 87–103 (1987).

[7] Benettin, G., Galgani, L. and Giorgilli, A.: *Realization of holonomic constraints and freezing of high frequency degrees of freedom in the light of classical perturbation theory, part II*, Comm. Math. Phys, to appear.

[8] Fermi, E., Pasta, J., and Ulam, S.: *Los Alamos Report* No. LA-1940 (1955), later published in *Fermi, E.: Collected Papers* (Chicago, 1965), and *Lect. Appl. Math.* **15**, 143 (1974).

[9] Gröbner, W.: *Die Lie-Reihen und Ihre Anwendungen*, VEB Deutscher Verlag der Wissenschaften (1967).

[10] Hori, G.: *Theory of general perturbation with unspecified canonical variables*, Publ. Astron. Soc. of Japan **23** 567–587 (1966).
Deprit, A.: *Canonical transformations depending on a small parameter*, Cel. Mech. **1**, 12–30 (1969).

[11] Colombo, R. and Giorgilli, A.: *A rigorous algebraic approach to Lie transforms in Hamiltonian mechanics*, preprint (1988).

[12] Giorgilli, A., Delshams, A., Fontich, E., Galgani, L. and Simó, C.: *Effective stability for a Hamiltonian system near an elliptic equilibrium point, with an application to the restricted three body problem*, J. Diff. Eqs., **77** N.1, 167–198 (1989).

[13] Giorgilli, A.: *Rigorous results on the power expansions for the integrals of a Hamiltonian system near an elliptic equilibrium point*, Ann. I.H.P, **48** N.4, 423–439 (1989).

[14] Benettin, G., Fröhlich, J. and Giorgilli, A.: *A Nekhoroshev-type theorem for Hamiltonian systems with infinitely many degrees of freedom*, Comm. Math. Phys., to appear.

This article was processed by the author using the TEX Macropackage from Springer-Verlag.

Analysis of the Linearization Around a Critical Point of an Infinite Dimensional Hamiltonian System

Manoussos Grillakis*

Chapter 1. Introduction

Partial differential equations that conserve energy can often be written as infinite dimensional Hamiltonian systems of the following general form

$$\frac{du}{dt} = JE'(u) \tag{1.1}$$

where $J: X^* \to X$ is a symplectic matrix (i.e., $JJ^* = -1$) and $E: X \to R$ is a C^2 functional defined on some Hilbert space X. Examples of partial differential equations that can be put in this form are the nonlinear Klein-Gordon equation

$$u_{tt} + i\omega u_t - \Delta u + f(x, |u|^2)u = 0 \tag{1.2}$$

and the nonlinear Schrödinger equation

$$iu_t - \Delta u + f(x, |u|^2)u = 0 \tag{1.3}$$

A critical point of equation (1.1) is a point $\phi \in X$ such that $E'(\phi) = 0$. One is interested in the stability of such a critical point and as a first step towards that goal I would like to consider the linearization of equation (1.1) around ϕ, i.e.,

$$\frac{dv}{dt} = JE''(\phi)v + JO(\|v\|^2). \tag{1.4}$$

In this paper I will study the spectrum of the operator $A := JE''(\phi)$, for the following cases.

(1) Nonliner Schrödinger equation:

$$\vec{u} = (Re\, u, Im\, u) = (u, v);\ X = H^1(\mathbf{R}^n) \times H^1(\mathbf{R}^n)$$

$$J = \begin{bmatrix} 0 & 1 \\ -1 & 0 \end{bmatrix};\ E(\vec{u}) = \frac{1}{2}\int \{|\nabla u|^2 + |\nabla v|^2 + F(x, u^2 + v^2)\}\, dx$$

$$\frac{\partial F(x,s)}{\partial s} = f(x,s).$$

---

* Partially supported by NSF Grant DMS 8610730.

Now (1.2) can be written as

$$\frac{d\vec{u}}{dt} = JE'(\vec{u})$$

Linearizing around a critical point $\vec{\phi} = (\phi, 0)$, I have

$$A = -\Delta + f(x, \phi^2) + 2f'(x, \phi^2)\phi^2$$

$$B = -\Delta + f(x, \phi^2) \qquad (1.7)$$

$$f'(x, s) = \frac{\partial f}{\partial s}(x, s)$$

$$JE'' = \mathcal{A} = \begin{bmatrix} 0 & B \\ -A & 0 \end{bmatrix} \qquad (1.8)$$

(2) Klein-Gordon equation:

$$\vec{u} = (Re\, u, Im\, u_t, Im\, u, Re\, u_t) := (u, z, v, w)$$

$$X = H^1(\mathbf{R}^n) \times L^2(\mathbf{R}^n) \times H^1(\mathbf{R}^n) \times L^2(\mathbf{R}^n)$$

$$J = \begin{bmatrix} 0 & J_2 \\ J_2 & 0 \end{bmatrix}; \; J_2 = \begin{bmatrix} 0 & 1 \\ -1 & 0 \end{bmatrix}$$

$$E(\vec{u}) = \frac{1}{2} \int \{w^2 + z^2 + |\nabla u|^2 + |\nabla u|^2 + F(u^2 + v^2)\} dx$$

Now (1.3) can be written as $d\vec{u}/dt = JE'(\vec{u})$, notice that equation (1.3)is invariant under the group $\{e^{is}: s \in \mathbf{R}\}$, denote by $\{T(s): s \in \mathbf{R}\}$ this group in the present setting. Suppose that we are looking vor periodic solutions of the form $T(\omega t)\vec{\varphi}(x)$, then the transformation $\vec{u} \to T(\omega t)\vec{u}$ gives the equation

$$\frac{d\vec{u}}{dt} = J(E'(\vec{u}) - \omega Q'(\vec{u})) \qquad (1.3a)$$

where $Q(\vec{u}) = \frac{1}{2}\langle J^{-1}T'(0)\vec{u}, \vec{u}\rangle = \frac{1}{2}\int(vw - uz)dx$. Hence a periodic solution of (1.3) reduces to a steady-state solution of (1.3a) with the reduced energy $E(\vec{u}) - \omega Q(\vec{u})$.

Linearizing around a critical point $\vec{\phi} = (\phi, \omega\phi, 0, 0)$ I have

$$JE'' = \begin{bmatrix} 0 & J_2 B \\ J_2 A & 0 \end{bmatrix} \qquad (1.9)$$

$$A = \begin{bmatrix} A_0 & -\omega \\ -\omega & 1 \end{bmatrix}, \quad B = \begin{bmatrix} B_0 & \omega \\ \omega & 1 \end{bmatrix}$$

$$A_0 = -\Delta + f(x, \phi^2) + 2f'(x, \phi^2)\phi^2 \qquad (1.10)$$

$$B_0 = -\Delta + f(x, \phi^2)$$

Finally I want to study operators that are of the form either (1.6) or (1.9). In general both A, B are going to have nonempty kernel, a fact that is actually useful in studying the stability, see (2). Let $\zeta = \rho + i\omega \in \mathbf{C}$ be an eigenvalue with (p, q) the corresponding eigenfunction for operator (1.7) or (1.9), then (1.7) gives

$$Bq = \zeta p$$
$$-Ap = \zeta q$$

which can be written

$$(R - zS)p = 0$$
$$p \in X := \{\ker A \cup \ker B\}^{\perp}; \quad z = -\zeta^2$$
$$R = \text{ restriction of } A \text{ on } X$$
$$S = \text{ restriction of } B^{-1} \text{ on } X$$

Similarly (1.9) gives

$$J_2 Bq = \zeta p$$
$$J_2 Ap = \zeta q$$

which reduces to

$$(R - zS)p = 0$$
$$p \in X\colon \{\ker A \cup \ker (J_2^* B^{-1} J_2)\}^{\perp}; \quad z = -\zeta^2$$
$$R = \text{ restriction of } A \text{ on } X$$
$$S = \text{ restriction of } J_2^* B^{-1} J_2 \text{ on } X$$

In both cases the problem is reduced to studying the following operator

$$K(z) = R - zS \tag{1.11}$$

where R, S are selfadjoint operators without kernel. In what follows I am going to study the operator of the form (1.11) keeping in mind that eigenvalues of (1.11) translate into eigenvalues of $\mathcal{A}$ as follows.

(a) $z \in \mathbf{R}^- \Rightarrow \mathcal{A}$ has a pair of real eigenvalues.

(b) $z \in \mathbf{R}^+ \Rightarrow \mathcal{A}$ has a pair of imaginary eigenvalues.

(c) $z \in \mathbf{C}; Im\, z \neq 0 \Rightarrow \mathcal{A}$ has a quadruplet of eigenvalues.

Assumptions concerning (R, S):

(AI) $S, S: X \to X$ are selfadjoint operators with empty kernel.

(AII) There exists strictly positive selfadjoint operator H such that $R = H + W, S^{-1} = H + V$ and W, V are compact relative to H.

Definition: For a pair (R, S) of selfadjoint operators that satisfy (AI), (AII) define the resolvent $\rho(R, S) \subset \mathbb{C}$ to be $\{z \in \mathbb{C}\colon K^{-1}(z)$ exists and is bounded$\}$ and the spectrum $\sigma(R, S) = \mathbb{C} \backslash \rho(R, S)$.

Remarks: 1) It is easy to show, using (AII) and Weyl's essential spectrum theorem, that

$$\sigma_{\mathrm{ess}}(A) = \sigma_{\mathrm{ess}}(R, S) = \sigma_{\mathrm{ess}}(H^2) \tag{1.12}$$

where $\sigma_{\mathrm{ess}} = \sigma - \sigma_{\mathrm{disc}}$; see (5) for the definitions.

2) Assumptions (AI), (AII) are quite general and are satisfied by all the examples that I know, one might think of the assumptions as follows: $H = -\Delta + m$; $m > 0$, $X = L^2(\mathbb{R}^n)$ and $W = W(x)$, $V = V(x)$ potentials that are smooth bounded and decay exponentially at infinity.

Definition: The pair (R, S) satisfying (AI, II) is said to have *unstable* eigenvalue $z \in \mathbb{C}$ iff $z \in \sigma_{\mathrm{disc}}(R, S)$ such that

(a) $Re\, z < 0$ or $Im\, z \neq 0$

(b) $z = \lambda \in \mathbb{R}^+$ and there exists finite dimensional subspace E invariant under $S^{-1}R$ such that

$$R \mid_E = \begin{bmatrix} \lambda & 1 & \dots & 1 \\ & \cdot & & \cdot \\ & & \cdot & \cdot \\ 0 & & & \cdot \\ & & & \lambda \end{bmatrix} S \mid_E .$$

Otherwise (R, S) is called stable.

Notice that existance of an unstable eigenvalue for (R, S) translates into existence of an eigenvalue of A that will make the critical point of (1.1) unstable. A concept related to stability is that of *structural* stability.

Definition: The pair (R, S) satisfying (AI, II) is called structurally stable if it is stable and moreover there exists $\varepsilon > 0$ such that for every $\widetilde{W}, \widetilde{V}$ operators compact relative to H such that

$$\|H^{-1}\widetilde{W}\| < \varepsilon, \ \|H^{-1}\widetilde{V}\| < \varepsilon, \text{ if } \widetilde{R} = R + \widetilde{W}, \widetilde{S}^{-1} = S^{-1} + \widetilde{V}$$

then the pair $(\widetilde{R}, \widetilde{S})$ is also stable. Otherwise (R, S) is called structurally unstable.

Definition: Let $\lambda > 0$ be a stable eigenvalue for (R, S) with corresponding eigenfunction, $p \in X$, i.e., $(R - \lambda S)p = 0$. Define the sign of λ to be

$$s(\lambda) = \text{sign}\langle Sp, p\rangle = \text{sign}\langle Rp, p\rangle \tag{1.13}$$

where $\langle \cdot \ , \ \cdot \rangle$ is the inner product of X.

Some explanations are in order here. The sign associated with the above eigenvalue is a special case of the sign of eigenvalues for a symplectic operator T introduced by M. Krein, see (1, 6). The situation is as follows: Let T: $X \to X$ be a symplectic operator acting on a finite dimensional space X and J the symplectic matrix on X. Denote by $\langle \cdot \ , \ \cdot \rangle$ the inner product on X and $[\cdot \ , \ \cdot] = \langle J\cdot \ , \ \cdot \rangle$ a skew product coming from the symplectic structure on X. T symplectic means that $[T\xi, T\eta] = [\xi, \eta]$, $\forall \xi, \eta \in X$. It is easy to see that the eigenvalues of T exist in quadruplets $\{z, \overline{z}, z^{-1}, \overline{z}^{-1}\}$. If $Tp = e^{i\omega}p$, $T\overline{p} = \overline{e}^{i\omega}\overline{p}$, is a pair of eigenvalues on the unit circle let π_ω be the real invariant space corresponding to $e^{\pm i\omega}$ then

$$s(\omega) - \text{sign}[T\xi, \xi] \ \forall \xi \in \pi_\omega \tag{1.14}$$

It is easy to see that this is a well defined quantity, see (6) for more details.

The correspondence with the present situation is $T = e^{\mathsf{A}}$ and if I denote by $\mathsf{H} = E''(\phi)$ then if $\mathsf{A}p = i\omega p$, $\mathsf{A}\overline{p} = -i\omega\overline{p}$,

$$s(\omega) = \text{sign}\langle \mathsf{H}p, \overline{p}\rangle. \tag{1.15}$$

On the other hand if $\mathsf{A} = \left[\begin{smallmatrix} 0 & B \\ -A & 0 \end{smallmatrix}\right]$ it is straightforward to compute first that $(R - \omega^2 S)p = 0$ and

$$s(\omega) = \text{sign}\{\langle Rp, p\rangle + \omega^2\langle Sp, p\rangle\} = 2\omega^2\langle Sp, p\rangle.$$

The computation is similar if $\mathsf{A} = \begin{bmatrix} 0 & J_2 B \\ J_2 A & 0 \end{bmatrix}$ and gives the same answer about $s(\omega)$.

Definition: For R, S selfadjoint operators satisfying (AI, II) let

(a) $N(R,S) = \#$ negative eigenvalues of R, S.

(b) $C(R,S) = \{p \in X \colon \langle Rp, p\rangle \leq 0 \text{ or } \langle Sp, p\rangle \leq 0\}$;
$\partial C(R,S) = \{p \in X \colon \langle RP, p\rangle = 0 \text{ or } \langle Sp, p\rangle = 0\}$.

The following was proved in (3, 4).

(*) If $N(R) \neq N(S)$ then (R,S) has at least one real negative eigenvalue. Hence it is unstable.

In view of (*) the interesting case to be examined is when $N(R) = N(S) = n$. In the analysis in the next chapter, although it applies in general when $N(R) \neq N(S)$, we are going to be concerned mainly with $N(R) = N(S)$. Now going back to the sign $s(\lambda)$ of the eigenvalues M. Krein observed, see (1, 6), that when two eigenvalues of opposite sign collide they may leave the unit circle, thus creating a quadruplet of unstable eigenvalues, while if two eigenvalues of the same sign collide they just pass through each other staying on the unit circle. Since A is the infinitisimal generator of T and given the correspondence between eigenvalues of A and (R,S) the schematic picture is as follows.

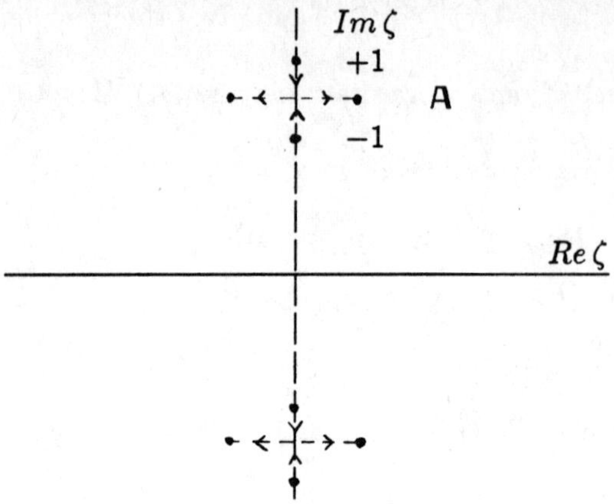

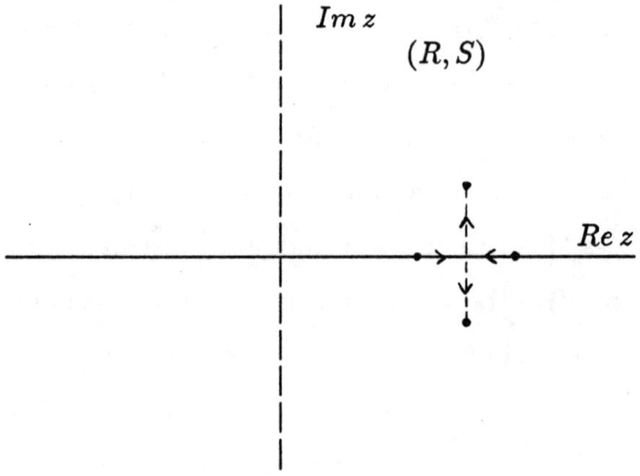

In this paper I want to develop a method for locating eigenvalues of the pair (R, S) tht are either positive with negative sign or unstable. It is not surprising that the analysis here is similar to the one of Hill's equation (7). Noticc also that the continuous part of the spectrum of the pair (R, S) can be thought of as having positive sign and there is the possibility of an eigenvalue with opposite sign embedded in the continuous spectrum; naturally the question arises if such a situation is structurally stable or not.

The existence of eigenvalues of A of negative sign is also connected with the following interesting phenomenon. Assume that we add a small dissipation in equation (1.1) as follows,

$$\frac{du}{dt} = (J - \varepsilon 1)E'(u) \tag{1.16}$$

where 1 is the identity operator. This corresponds, for example for (1.3), to the equation

$$iu_t + \varepsilon u_t - \Delta u + f(|u|^2)u = 0. \tag{1.17}$$

One can compute formally:

$$dE(u)|dt = -\varepsilon\|E'(u)\|^2 \leq 0.$$

If $E'(\phi)$ is a critical point then the linearization of (1.16) gives

$$\mathsf{A}_\varepsilon = (J - \varepsilon 1)\mathsf{H} = \mathsf{A} - \varepsilon \mathsf{H}$$

where $E''(\phi) = \mathsf{H}$.

Now let $\zeta(\varepsilon) \in \mathbf{C}$ be an eigenvalue of A_ε with corresponding eigenfunction $p(\varepsilon)$ such that if $\zeta(\varepsilon) = \rho(\varepsilon) + i\omega(\varepsilon)$ then $\rho(0) = 0$ and $\omega(0) = \omega_0 \neq 0$ then differentiating with respect to ε at $\varepsilon = 0$, I get

$$(\mathsf{A}_0 - \xi(0))\dot{p}(0) - \mathsf{H}p(0) = (\dot{\rho}(0) + i\dot{\omega}(0))p(0).$$

Taking inner product with $\overline{p}(0)$, given that $(\mathsf{A}_0 - \xi(0))p(0) = 0$, I have

$$\left.\frac{d\rho}{d\varepsilon}\right|_{\varepsilon=0} = -\langle \mathsf{H}p(0), \overline{p}(0)\rangle. \tag{1.18}$$

Hence if $s(\omega_0) = -1$ then the eigenvalue is moving to the right half plane where $Re\,\zeta > 0$ causing an exponential instability (i.e., initial data arbitrarily close to ϕ grow exponentially in time with exponent of order ε). Now from the analysis in Chapter 2 it will follow that if $N(R) = N(S) = \eta$ then A has either unstable eigenvalues or exactly n pairs of eigenvalues of negative sign.

The structure of the rest of the paper is as follows: In Chapter 2 a method is developed for locating eigenvalues that are either unstable or of negative sign. Part of the difficulty here comes from the existence of continuous spectrum. Next it is shown that if an eigenvalue of negative sign is embedded in the continuous spectrum the operator A is structurally unstable in the sense that there exists an arbitrarily small perturbation of A that has complex eigenvalues. Finally, certain instability criteria are developed. In Chapter 3 the results from Chapter 2 are applied to some specific examples. For the rest of the paper summation over repeated indices is assumed.

Chapter 2.

In this chapter I want to analyze the following eigenvalue problem:

(E) Find $p \in X + iX$, $z \in \mathbf{C}$ such that

(i) $(R - zS)p = 0$

(ii) $\langle Sp, \overline{p} \rangle \leq 0$

where R, S are selfadjoint operators that satisfy assumptions AI, AII, and $\overline{p}$ is the complex conjugate of p.

Pick any set of vectors $\{\psi_i,\ i = 1, \ldots, n\} \subset X$ where $n = N(S)$ with the following properties.

$$\begin{aligned} &(i) \qquad \sigma_{ij} = \langle S^{-1}\psi_i, \psi_j \rangle \quad \text{is strictly negative definite} \\ &(ii) \qquad \langle \psi_i, \psi_j \rangle = \delta_{ij} \end{aligned} \tag{2.1a}$$

Definition: 1)

$$\begin{aligned} \mathbf{P}\colon &= \{p \in X\colon \langle p, \psi_i \rangle = 0 \ \forall i = 1, \ldots, n\} \\ \pi\colon &= \text{orthogonal projection on } \mathbf{P} \\ R_1\colon &= \pi R\colon \mathbf{P} \to \mathbf{P} \text{ and } S_1\colon = \pi S\colon \mathbf{P} \to \mathbf{P} \\ \rho_{ij}\colon &= \langle R^{-1}\psi_i, \psi_j \rangle, r_{ij} = \langle R\psi_i, \psi_j \rangle, s_{ij} = \langle S\psi_i, \psi_j \rangle \\ u_i\colon &= \pi R\psi_i = R\psi_i - r_{ij}\psi_j \\ v_i\colon &= \pi S\psi_i = S\psi_i - s_{ij}\psi_j. \end{aligned}$$

Notice that with the above decomposition I have the following isomorphism $X + iX \simeq \mathbf{C}^n \oplus \mathbf{P} + i\mathbf{P}$. If $\alpha \in \mathbf{C}^n$ I will write $\alpha = (\alpha_1, \alpha_2, \ldots, \alpha_n)$ where $\alpha_i \in \mathbf{C}\ \forall i = 1, \ldots, n$. I would like to define an inner product on $\mathbf{C}^n$.

Definition: 2) If $\alpha, \beta \in \mathbf{C}^n$ let $\alpha \cdot \overline{\beta} = \Sigma_{i=1}^n \alpha_i \overline{\beta}_i$ and define the inner product to be in the realizatin of $\mathbf{C}^n$

$$(\alpha, \beta) = Re(\alpha \cdot \overline{\beta}) \tag{2.2}$$

Also in the same spirit if $p, q \in X + iX$ define

$$(p, q) = Re\langle p, \overline{q} \rangle. \tag{2.3}$$

Let $S^{2n-1} = \{\alpha \in \mathbf{C}^n\colon (\alpha,\alpha) = |\alpha|^2 = 1\}$. Fix $\alpha \in S^{2n-1}$ and assume that $\alpha_i\psi_i + p \in \mathbf{C}^n \oplus \mathbf{P} + i\mathbf{P}$, $z \in \mathbf{C}$ is a solution of problem (E), then:

$$\left\{\begin{aligned} R(p+\alpha_k\psi_i) - zS(p+\alpha_i\psi_i) = 0 \\ \langle S(p+\alpha_i\psi_i), \overline{p}+\overline{\alpha}_i\psi_i\rangle \leq 0 \end{aligned}\right\}.$$

These equations can be rewritten as follows:

$$(R_1 - zS_1)p + \alpha_i(u_i - zv_i) = 0 \tag{2.4}$$

$$r_{ij}\alpha_j - zs_{ij}\alpha_j + \langle p, u_i - zv_i\rangle = 0 \tag{2.5}$$

$$(Sp,p) + 2(p,\alpha_i v_i) + s_{ij}(\alpha_i,\alpha_j) \leq 0 \tag{2.6}$$

Equation (2.4) can be written as

$$R_1(p+\alpha_i S_a^{-1}v_i) - zS_1(p+\alpha_i S_1^{-1}v_1) + \alpha_i(u_i - R_1S_1^{-1}v_i) = 0 \tag{2.4a}$$

In order to bring equations (2.4), (2.5), (2.6) in a form that is easier to analyze I will need some formulas.

Lemma 2.1: *The following formulas hold.*

$$\left\{\begin{aligned} R_1 &= R - \langle\cdot, u_i\rangle\psi_i \\ R_1^{-1} &= R^{-1} - \langle\cdot, \pi R^{-1}\psi_i\rangle\rho^{ij}R^{-1}\psi_j \end{aligned}\right\} \tag{2.5}$$

$$\left\{\begin{aligned} S_1 &= S - \langle\cdot, v_i\rangle\psi_i \\ S_1^{-1} &= S^{-1}\langle\cdot, \pi S^{-1}\psi_i\rangle\sigma^{ij}S^{-1}\psi_j \end{aligned}\right\} \tag{2.6}$$

$$\left\{\begin{aligned} R_1^{-1}u_i &= \psi_i - \rho^{i\ell}R^{-1}\psi_\ell \\ \langle R_1^{-1}u_i, u_j\rangle &= -\rho^{ij} + r_{ij} \end{aligned}\right\} \tag{2.7}$$

$$\left\{\begin{aligned} S_1^{-1}v_i &= \psi_i - \sigma^{i,\ell}S^{-1}\psi_\ell \\ \langle S_1^{-1}v_i, v_j\rangle &= -\sigma^{ij} + s_{ij} \end{aligned}\right\} \tag{2.8}$$

Proof: I will give the proof only for R_1, the rest being similar.

$$\begin{aligned} R_1^{-1}R_1p &= R_1^{-1}(Rp - \langle p, u_i\rangle\psi_i) \\ &= p - \langle p, u_i\rangle R^{-1}\psi_i - \langle(Rp - \langle p, u_k\rangle\psi_k), \phi R^{-1}\psi_i\rangle\rho^{ij}R^{-1}\psi_j \\ &= p - \langle p, u_i\rangle R^{-1}\psi_i + \langle p, u_k\rangle\langle\psi_k, R^{-1}\psi_i\rangle\cdot\rho^{ij}R^{-1}\psi_j = p \end{aligned}$$

and this proves (2.5).

$$\begin{aligned}
R_1^{-1}u_i &= R_1^{-1}(R\psi_i - r_{ij}\psi_j)\\
&= \psi_i - r_{ij}R^{-1}\psi_j - \langle (R\psi_i - r_{ij}\psi_j), R^{-1}\psi_k\rangle \rho^{k\ell}R^{-1}\psi_\ell\\
&= \psi_i - r_{ij}R^{-1}\psi_j - \delta_{ik}\rho^{k\ell}R^{-1}\psi_\ell + r_{ij}\rho_{jk}\rho^{k\ell}R^{-1}\psi_\ell\\
&= \psi_i - r_{ij}R^{-1}\psi_j - \rho^{i\ell}R^{-1}\psi_\ell + r_{ij}R^{-1}\psi_j\\
&= \psi_i - \rho^{i\ell}R^{-1}\psi_\ell\\
\langle R_1^{-1}u_i, u_j\rangle &= \langle \psi_i - \rho^{i\ell}R^{-1}\psi_\ell,\ R\psi_j - r_{jk}\psi_k\rangle\\
&= r_{ij} - \rho^{i\ell}\delta_{\ell j} - r_{jk}\delta_{ik} + r_{jk}\rho_{k\ell}\rho^{\ell i}\\
&= r_{ij} - \rho^{ij} - r_{ij} + r_{ij}\\
&= -\rho^{ij} + r_{ij}
\end{aligned}$$

and this proves (2.7). □

Shifting the origin in equations (2.4, 2.5, 2.6) to $-\alpha_i^{-1}v_i$, i.e., writing $p = p + \alpha_i S_1^{-1}v_i$, they become

$$\begin{aligned}
(R_1 - zS_1)p + \alpha_i(u_i - R_1S_1^{-1}v_i) &= 0\\
(r_{ij} - zs_{ij})\alpha_j = \langle p, u_i - zv_i\rangle &- \langle \alpha_j S_1^{-1}v_j, v_i - zv_i\rangle\\
\langle Sp, \overline{p}\rangle + \sigma^{ij}(\alpha_i, \alpha_j) &\le 0
\end{aligned}$$

where in the third equation I used (2.8). The second equation becomes, using Lemma 2.1,

$$\begin{aligned}
&\rho^{ij}\alpha_j - z\sigma^{ij}\alpha_j + \langle (R_1^{-1}u_j - S_1^{-1}v_j)\alpha_j, u_i\rangle + \langle p, u_i - zv_i\rangle = 0\\
\Rightarrow\ &(\rho^{ij} - z\sigma^{ij})\alpha_j + \langle R_1^{-1}(u_j - R_1S_1^{-1}v_j), (u_i - R_1S_1^{-1}v_i)\rangle\alpha_j\\
&\qquad + \langle u_j - R_1S_1^{-1}v_j, S_1^{-1}v_i\rangle\alpha_j\\
&\qquad + \langle (R_1 - zS_1)p, S_1^{-1}v_i\rangle + \langle p, u_i - R_1S_1^{-1}v_i\rangle = 0\\
\Rightarrow\ &(\rho^{ij} - z\sigma^{ij})\alpha_j + \langle R_1^{-1}(u_i - R_1S_1^{-1}v_i), u_j - R_1S_1^{-1}v_j\rangle\alpha_j + \langle p, u_i - R_1S_1^{-1}v_i\rangle = 0
\end{aligned}$$

Definition: 3) Define the following vectors in $\mathbf{P}$ and $\mathbf{P} + i\mathbf{P}$

$$b_i := u_i - R_1S_1^{-1}v_i;\ \ b(\alpha) := \alpha_i b_i \tag{2.9}$$

$$g_i := S_1^{-1/2}(u_i - R_1S_1^{-1}v_i);\ \ g(\alpha) = \alpha_i g_i \tag{2.10}$$

Using Lemma 2.1, it is easy to express b_i and g_i in terms of R, S and ψ_i for $i = 1, \ldots, n$.

Corollary 2.1: *The following formulas hold:*

$$b_i = \sigma^{i\ell}(RS^{-1}\psi_\ell - \langle S^{-1}\psi_\ell, R\psi_k\rangle\psi_k) \tag{2.11}$$

$$S_1^{-1/2}g_i = \sigma^{i\ell}(S^{-1}RS^{-1}\psi_\ell - \langle S^{-1}\psi_\ell, \psi_k\rangle\sigma^{km}S^{-1}\psi_m) \tag{2.12}$$

Now I can write equations (2.4, 2.5, 2.6) as follows.

$$(R_1 - zS_1)p + \alpha_j b_j = 0 \tag{2.13}$$

$$(\rho^{ij} - z\sigma^{ij} + \langle R_1^{-1}b_i, b_j\rangle)\alpha_j + \langle p, b_i\rangle = 0 \tag{2.14}$$

$$(S_1 p, p) + \sigma^{ij}(\alpha_i, \alpha_j) \leq 0 \tag{2.15}$$

I want to solve the system of equations (2.13, 2.14, 2.15) for $p \in \mathbf{P} + i\mathbf{P}$, $z \in \mathbf{C}$ and $\alpha \in \mathbf{C}^n$ but a few remarks are in order. First notice from (2.15) and the fact that $\sigma^{ij} = (\sigma_{ij})^{-1}$ is strictly negative definite it follows that $S_1 > 0$. Second notice that if $(R - zS)p = 0$ and $Im\, z \neq 0$ then $\langle Rp, \overline{p}\rangle = 0 = \langle Sp, \overline{p}\rangle$; hence all the complex eigenvalues live on the following set. $M = \{p \in X + iX \colon (Rp,p) = 0 = (Sp,p);\ \|p\| = 1\}$, on M one can define a vector field as follows:

$$\begin{aligned} f(p) :&= Rp - z(p)SP \\ z(p) &= \frac{\langle Rp, S\overline{p}\rangle}{\|Sp\|^2} \end{aligned} \tag{2.16}$$

It is easy to check that $(f(p), Rp) = (f(p), Sp) = (f(p), p) = (f(p), ip) = 0$ hence $f(p)$ is a vector field on the "manifold" M modulo the $e^{i\theta}, \theta \in [0, 2\pi]$ action which leaves M invariant. One possible approach would be to examine the topological properties of M so that the Euler characteristic $\chi(M \mid e^{i\theta})$ in an appropriate sense will measure exactly how many nonreal eigenvalues problem (E) has. In this paper I will prefer an analytic approach to the prolem which although it is less intuitive is easier to compute; however I would like to keep some of the intuition that comes from the geometry of the set M.

Definition: 4) Define the following sets.

a) $C(R) \colon = \{p \in X + iX \colon (Rp,p) = 0\}$;

$C(S) \colon = \{p \in X + iX \colon (Sp,p) = 0\}$

b) $M \colon = C(R) \cap C(S) \cap \{p \in X + iX \colon \|p\| = 1\}$

c) $S(\alpha) = \{e^{i\theta}(\alpha_i\psi_i): \theta \in [0, 2\pi]\}$, where $\alpha \in S^{2n-1}$

d) $E_R(\alpha) = \{p \in \mathsf{P} + i\mathsf{P}: (R(p+\psi), p+\psi) = 0$, where $\psi \in S(\alpha)\}$;
$E_S(\alpha) = \{p \in \mathsf{P} + i\mathsf{P}: (S(p+\psi), p+\psi) = 0$, where $\psi \in S(\alpha)\}$
$T(\alpha) = E_R(\alpha) \cap E_S(\alpha)$.

For the rest of the chapter let $\widehat{\ }$ denote a set modulo the $e^{i\theta}$, $\theta \in [0, 2\pi]$ action. $\widehat{S}^{2n-1} \simeq \mathsf{C}P(n-1)$ and I can write $\widehat{M} = \{(\alpha, \widehat{T}(\alpha)): \alpha \in \mathsf{C}P(n-1)\}$ thus the topology of $\widehat{M}$ can be understood if for fixed $\alpha \in S^{2n-1}$ one understands how $\widehat{T}(\alpha)$ looks like. Going back to equations (2.13, 2.14, 2.15), I want to make the folliwng construction.

Theorem 2.1: *There exist continuous functions*

(a) $z: S^{2n-1} \to \mathsf{C}$;

(b) $y: S^{2n-1} \to S(\alpha) \oplus \mathsf{P} + i\mathsf{P}$;

(c) $h: S^{2n-1} \to \mathsf{C}^n$

such that

(i) $(R + zS)y = h_i\psi_i$; $(Sy, y) \leq 0$, $\forall\, \alpha \in S^{2n-1}$

(ii) $z(e^{i\theta}\alpha) = z(\alpha)$; $\forall\, \theta \in [0, 2\pi]$; $\forall\, \alpha \in S^{2n-1}$

(iii) $y(e^{i\theta}\alpha) = e^{i\theta}y(\alpha)$, $h(e^{i\theta}\alpha) = e^{i\theta}h(\alpha)$; $\forall\, \theta \in [0, 2\pi]$, $\forall\, \alpha \in S^{2n-1}$

(iv) $(h(\alpha), e^{i\theta}\alpha) = 0$, $\forall\, \theta \in [0, 2\pi]$, $\forall\, \alpha \in S^{2n-1}$.

Remark: Notice that because of (iv), $h(\alpha)$ is a vector field on $\mathsf{C}P(n-1)$.

Remark: The choice of the vectors ψ_i $i = 1, \ldots, n$ is only subject to the restrictions (2.1a) hence I can always choose them so that $\ker(R_1) = 0$.

Since $S_1 > 0$, R_1 has a spectral decomposition with respect to S_1, to be precise define

$$H_1 = S_1^{-1/2} R_1 S_1^{-1/2} \tag{2.17}$$

then

$$R_1 - zS_1 = S^{1/2}(H_1 - z)S^{1/2} \tag{2.18}$$

Let $\{dP_\xi;\ \xi \in \mathbf{R}\}$ be the spectral decomposition of the selfadjoint operator H_1, then (2.13) can be written

$$p = \int\limits_{\mathbf{R}} \frac{1}{z-\xi} S_1^{-1/2} dP_\xi S_1^{-1/2}(\alpha_j b_j) + k \tag{2.19}$$

$$k \in \ker(R_1 - zS_1)$$

also

$$\langle S_1 p, \overline{p}\rangle = \int_{\mathbf{R}} \frac{1}{|z-\xi|^2} \langle dP_\xi \alpha_j g_j, \alpha_j g_j\rangle \tag{2.20}$$

$$\langle R_1^{-1} b_i, b_j\rangle = \langle H_1^{-1} g_i, g_j\rangle = \int\limits_{\mathbf{R}} \frac{1}{\xi} \langle dP_\xi g_i, g_j\rangle \tag{2.21}$$

$$\langle p, b_i\rangle = \int\limits_{\mathbf{R}} \frac{1}{z-\xi} \langle dP_\xi(\alpha_j g_j), g_i\rangle \tag{2.22}$$

where recall that $g_i = S_1^{-1/2} b_i$. From the above formulas it is obvious that I need some definitions.

Definition: 5) *Define the measures on the real line.*

(a) $d\nu_{ij}(\xi) = \langle dP_\xi g_i, g_j\rangle$; $d\nu_g(\xi) = d\nu_{ij}(\xi)(\alpha_i, \alpha_j)$

Using these measures define the matrices and the corresponding functions

(b) $\chi_{ij}(z){:} = \int_{\mathbf{R}} \frac{1}{|z-\xi|^2} d\nu_{ij}(\xi)$; $\chi(\alpha, z) = \chi_{ij}(z)(\alpha_i, \alpha_j)$.

(c) $\phi_{ij}(z) = \int_{\mathbf{R}} \frac{1}{z-\xi} d\nu_{ij}(\xi)$; $\phi(\alpha, z) = \phi_{ij}(z)(\alpha_i, \alpha_j)$.

(d) $\tau_{ij} = \int_{\mathbf{R}} \frac{1}{\xi} d\nu_{ij}(\xi)$; $\tau(\alpha) = \tau_{ij}(\alpha_i, \alpha_j)$.

(e) $\sigma(\alpha) = -\sigma^{ij}(\alpha_i, \alpha_j)$; $\rho(\alpha) = -\rho^{ij}(\alpha_i, \alpha_j)$.

and finally the function

$$f(\alpha, z) = \sigma(\alpha) z + \tau(\alpha) + \phi(\alpha, z) \tag{2.27}$$

Now equations (2.13, 2.14, 2.15) can be rewritten

$$\left\{ \begin{array}{c} p = \int\limits_{\mathbf{R}} \frac{1}{z-\xi} dP_\xi(g(\alpha)) + k;\ k \in \ker(R_1 - zS_1) \\ \chi(\alpha, z) + (S_1 k, k) \le \sigma(\alpha) \\ \rho^{ij}\alpha_j - [z\sigma^{ij} + \tau_{ij} + \phi_{ij}(z)]\alpha_j = 0 \end{array} \right\} \tag{2.28}$$

The problem of understanding how $\widehat{T}(\alpha)$ looks like should follow from the following. Let

$$E_R(f){:}\ = \{p \in \mathsf{P} + i\mathsf{P}{:}\ (R_1(p+\alpha_j R_1^{-1} b_j), p+\alpha_j R_1^{-1} b_j) = f\}$$

Problem (T): Find $f \in \mathbf{R}$ such that $\widehat{E}_R(f)$ is tangent to $\widehat{E}_S(\alpha)$.

Problem (T) is equivalent to the eigenvalue problem: Find $\lambda \in \mathbf{R}$ such that

$$(R_1 - \lambda S_1)p = -\alpha_j b_j; \ (S_1 p, p) = \sigma(\alpha) \tag{2.29}$$

In order to solve (2.29) p is given by (2.23) while

$$\chi(\alpha, \lambda) + (S_1 k, k) = \sigma(\alpha) \tag{2.30}$$

$$f = \lambda\sigma(\alpha) + \tau(\alpha) + \phi(\alpha, \lambda) \tag{2.31}$$

For $\alpha \in S^{2n-1}$ let me call

$$\begin{aligned} \Lambda_d(\alpha) &: = \{\lambda \in \mathbf{R} : \chi(\alpha,\lambda) < \sigma(\alpha)\} \cap \sigma_{\text{disc}}(R_1, S_2) \\ \Lambda_c(\alpha) &: = \{\lambda \in \mathbf{R} : \chi(\alpha,\lambda) < \sigma(\alpha)\} \cap \sigma_{ac}(R_1, S_1) \\ \Lambda(\alpha) &: = \{\lambda \in \mathbf{R} : \chi(\alpha,\lambda) > \sigma(\alpha)\} \end{aligned} \tag{2.31a}$$

In general the spectrum of (R_1, S_1) consists of discrete, absolutely continuous and singular parts hence the measure $d\nu_g(\xi)$ can be decomposed accordingly. It turns out that in all the examples the singular part of the spectrum is empty, see the appendix. Henceforth I am going to assume that $d\nu_g$ consists only of discrete and absolutely continuous parts thus simplifying the statements that follow.

Lemma 2.2: *$\forall \alpha \in S^{2n-1}$ $\Lambda(\alpha) = \cup_{i=1}^m \Lambda_i$, where $\Lambda_i = (\lambda_i^\ell, \lambda_i^u)$ and $\lambda_i^{\ell,u}$ satisfy the inequalities*

$$\lambda_1^\ell < \lambda_1^u \le \lambda_2^\ell < \lambda_2^u \le \cdots \le \lambda_m^\ell < \lambda_m^u \tag{2.32}$$

Proof: Decompose the measure $d\nu_g(\xi) = \sum b_i \delta(\lambda_i - \xi) + a(\xi)d\xi$, where $b_i > 0$, $\lambda_i \in \mathbf{R}$ and $a(\xi)$ is a positive measurable function. Notice that WLOG I can take $a(\xi)$ to be continuous, otherwise approximate $a(\xi)$ with a sequence of positive continuous functions and take the limit. Now

$$\chi(\alpha, \lambda) = \sum \frac{b_i}{(\lambda - \lambda_i)^2} + \int a(\xi)^2 d\xi$$

hence if $\lambda \in \operatorname{ess\,supp}\{d\nu_g\}$ then $\chi(\alpha,\lambda) = +\infty$ and inequalities (2.32) follow. □

Remark: Notice that if $\lambda_i^u = \lambda_{i+1}^\ell$ then $\Lambda_i \cup \Lambda_{i+1}$ merge and form a common interval. Also in (2.32) λ_m^u or m can be $+\infty$. $\chi(\alpha, \lambda)$ is a function that looks like the figure below.

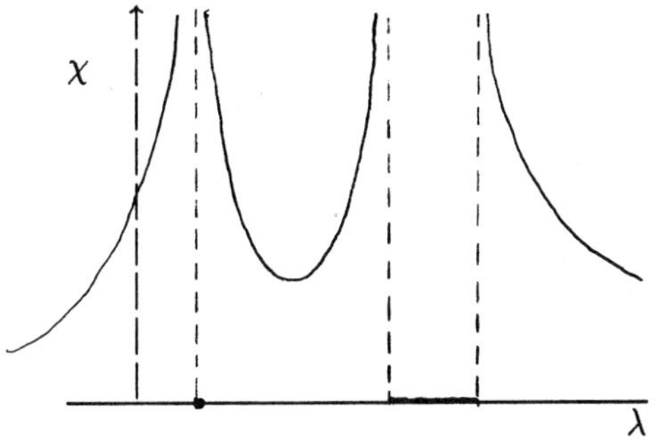

Let me call

(a) $f_i^{\ell,u}(\alpha) = f(\alpha, \lambda_i^{\ell,u})$, $F_i = (f_i^\ell, f_i^u)$, $F(\alpha) := \cup_i F_i$

(2.32a)

(b) $F_d(\alpha) := f(\alpha, \Lambda_d(\alpha))$, $F_c(\alpha) := f(\alpha, \Lambda_c(\alpha))$

Lemma 2.3: *The following inequalities hold*

$$f_1^\ell < f_1^u \le f_2^\ell < f_2^u \le \cdots \le f_m^\ell < f_m^u \tag{2.33}$$

Proof: If $\lambda \in (\lambda_i^u, \lambda_{i+1}^\ell)$ for some $i = 1, \ldots, m$ then $\frac{df}{d\lambda} = \sigma(\alpha) - \chi(\alpha, \lambda) > 0$ therefore $f(\alpha, \lambda^u) < f(\alpha, \lambda_{i+1}^\ell)$. If $\lambda \in (\lambda_i^\ell, \lambda_i^u)$ then

$$f(\alpha, \lambda_i^u) - f(\alpha, \lambda_i^\ell) = (\lambda_i^u - \lambda_i^\ell)\left[\sigma(\alpha) - \int \frac{1}{(\lambda_i^u - \xi)(\lambda_i^\ell - \xi)} d\nu_g(\xi)\right] > 0$$

since

$$\int \frac{1}{(\lambda_i^u - \xi)(\lambda_i^\ell - \xi)} d\nu_g < \left(\int \frac{1}{(\lambda_i^u - \xi)^2} d\nu_g \int \frac{1}{(\lambda_i^\ell - \xi)^2} d\nu_g\right)^{1/2} \le \sigma(\alpha).$$

The function $f(\alpha, \lambda)$looks like the figure below.

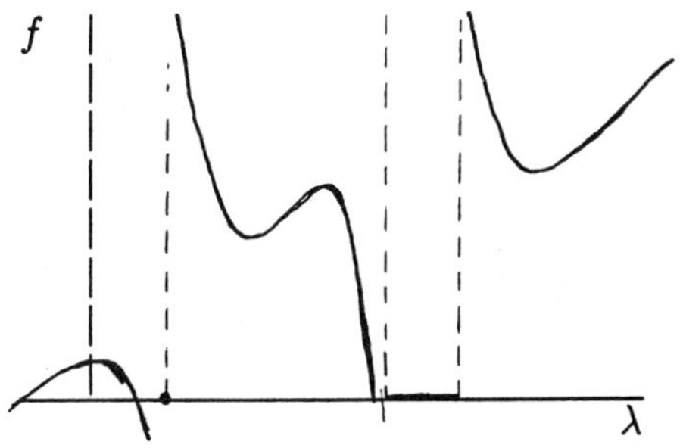

I can state now a theorem characterizing the topology of the set $\widehat{T}(\alpha)$; before I do that however I would like to state precisely what I mean by topology in infinite dimensions.

Definition: 6) Given $\{\pi_N, \mathbf{P}_N\,;\ N \in \mathbf{N}\}$ a sequence of finite dimensional subspaces of $\mathbf{P}$ and the corresponding projections such that (i) $\mathbf{P}_N \subset \mathbf{P}_{N+1}$, (ii) dim $\mathbf{P}_N = N$, (iii) $\cup_N \mathbf{P}_N = \mathbf{P}$, I will say that:

(a) $\widehat{T}(\alpha) \simeq$ sphere, if $\exists\, N_0$ such that $\forall N > N_0$, $\widehat{T}_N(\alpha) \simeq S^{2N_2}$ where $\widehat{T}_N(\alpha)$ is the natural restriction of $\widehat{T}(\alpha)$ on $\mathbf{P}_N + i\mathbf{P}_N$.

(b) $\widehat{T}(\alpha) \simeq$ torus, if $\exists\, N_0$ such that $\forall N > N_0$, $\widehat{T}_N(\alpha) \simeq S^{2k-1} \times S^{2(N-k)-1}$ for some integer k.

(c) $\widehat{E}_R(\alpha)$ and $\widehat{E}_S(\alpha)$ are tangent if problem (T) has a solution.

(d) $\widehat{E}_R(\alpha)$ and $\widehat{E}_S(\alpha)$ are weakly tangent if there exists a sequence $\{p_i\}$, $i \in \mathbf{N}$ such that $\{p_i\}$ converges weakly but not strongly and

(1) $(R_1 - \lambda_0 S_1)p_i \to -\alpha_j b_j$ for$\lambda_0 \in \lambda_c$

(2) $(R_1(p + i + R_1^{-1}(\alpha_j b_j)), p_i + R_1^{-1}(\alpha_j b_j)) = \rho(\alpha)$

(3) $(S_1 p_i, p_i) = \sigma(\alpha)$.

Theorem 2.2:

(a) *If* $\rho(\alpha) \in F = \cup_i F_i$ *then* $\widehat{T}(\alpha) \simeq$ *sphere;*

(b) *If* $\rho(\alpha) \notin F \cup F_d \cup F_c$ *then* $\widehat{T}(\alpha) \simeq$ *torus;*

(c) *If* $\rho(\alpha) \in \partial F \cup F_d$ *then* $\widehat{E}_R(\alpha)$ *and* $\widehat{E}_S(\alpha)$ *are tangent;*

(d) *If* $\rho(\alpha) \in F_c$ *then* $\widehat{E}_R(\alpha)$ *and* $\widehat{E}_S(\alpha)$ *are weakly tangent.*

Proof: First notice that (c) follows from equations (2.28) and (2.30), (2.31). Also for (d), pick $\{k_i\}$ in (2.28) such that $(R_1 - \lambda S_1)k_i \to 0$ and $\langle S_1 k_i, k\rangle = \sigma(\alpha) - \chi(\alpha, \lambda)$, so it remains to show (a) and (b).

(a) Let $\rho(\alpha) \in F_i$ for some $i \in \{1, \ldots, m\}$. Consider $g_t = tg(\alpha)$ and the corresponding $\lambda_i^{\ell,u}(t)$ defined in Lemma 2.2. WLOG I can assume that supp$\{d\nu_g\} = \sigma(R_1, S_1)$. As t increases it is easy to see from the definition of $\chi(\alpha, \lambda)$ that $\lambda_i^{k,u}(t) \uparrow$ and $\lambda_i^{\ell}(t) \downarrow$ until they meet and disappear so that $\Lambda_i(t)$ and $\Lambda_{i+1}(t)$ merge into one interval. Let

$\widehat{T}_t = \widehat{E}_R(\rho_t) \cap \widehat{E}_S(\alpha)$ where $\rho_0 = \rho(\alpha)$ and ρ_t changes continuously in such a way that $\rho_t \neq f_i^{\ell,u}(t)$, $\forall i = 1, \ldots, m$ (this is possible in view of the way that $\lambda_i^u(t)$ and$\lambda_{i+1}^{\ell}(t)$ change). Finally for t large enough I will have

$$\Lambda(t) = (\lambda_1^{\ell}(t), \lambda_m^u(t)), \ \ \rho_t \in \operatorname{int} F = (f_1^{\ell}(t), f_m^u(t)) \tag{2.34}$$

But (2.34) implies, using Morse theory, that $\widehat{T}_t \simeq$ sphere and $\widehat{T}_t$ is a continuous deformation of $\widehat{T}(\alpha)$.

(b) In order to prove (b) it is enough to observe that if $\rho(\alpha) = f_i^{\ell,u}$ then $\widehat{T}(\alpha) = \widehat{E}_R(\alpha) \cap \widehat{E}_S(\alpha) \simeq \Sigma(2N, 2m)$ for some $m < N$ where $\Sigma(2N, 2m)$ is a sphere S^{2N} with $S^{2m} \subset S^{2N}$ and S^{2m} identified to a point. If I change ρ continuously, as ρ crosses from int(F) to the complement of $F \cup F_d \cup F_c$, I have the following changes in the topologyof $\widehat{T}(\alpha)$.

$$S^{2N} \rightarrow \sum(2N, 2m) \rightarrow S^{2m+1} \times S^{2(N-m)-1}$$

The figure below shows possible changes for S^2.

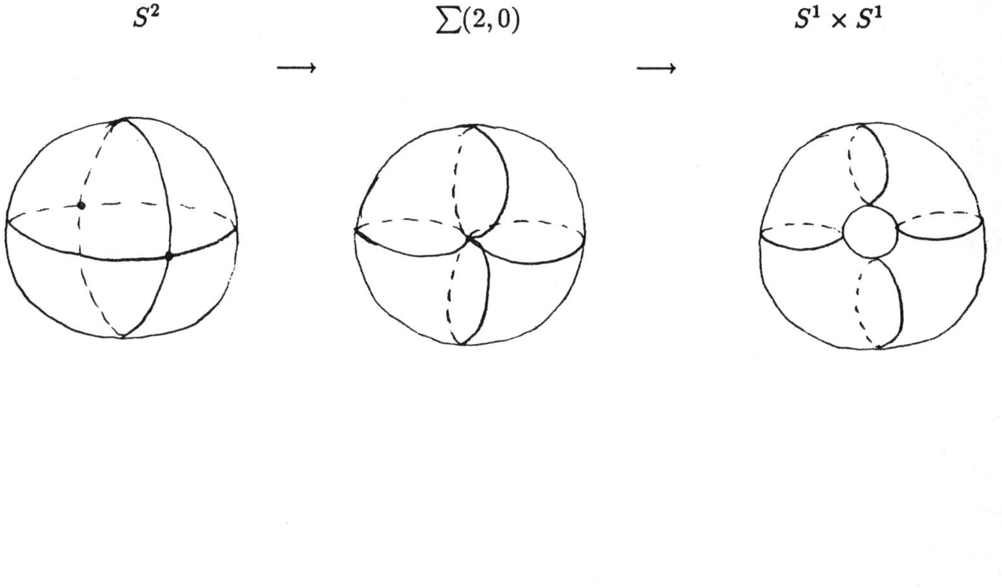

□

Proof of Theorem 2.1: I want to find$\{z(\alpha), y(\alpha) = \alpha_i \psi_i + p(\alpha)\}$ where$p(\alpha) \in \mathbf{P} + i\mathbf{P}$ such that

$$\left\{ \begin{array}{c} (R_1 - z(\alpha)S_1)p(\alpha) = -\alpha_j b_j \\ \rho(\alpha) = \sigma(\alpha)z + \tau(\alpha) + \phi(\alpha, z) \\ (S_1 p(\alpha), p(\alpha)) \leq \sigma(\alpha) \end{array} \right\} \tag{2.35}$$

Set

$$h(\alpha) = (\rho^{ij} - z\sigma^{ij} + \tau_{ij} + \phi_{ij})\alpha_j$$

It follows from (2.35), (2.36) that

$$(h(\alpha), e^{i\theta}\alpha) = 0\,, \quad \forall \theta \in [0, 2\pi] \quad \forall \alpha \in S^{2n-1}.$$

Choose:

$$p = \int_{\mathbf{R}} \frac{1}{z - \xi} dP_\xi(g(\alpha))$$

and the problem reduces to solving the equations

$$\left\{ \begin{array}{c} \rho(\alpha) = \sigma(\alpha)z + \tau(\alpha) + \phi(\alpha, z) \\ \int_{\mathbf{R}} \frac{1}{|z-\xi|^2} d\nu_g \le \sigma(\alpha) \end{array} \right\} \tag{2.37}$$

Case (A): $\rho(\alpha) \in (\cup F_i)^c$

Then $\rho(\alpha) \in [\rho_i^u, \rho_{i+1}^\ell]$ and since $f(\alpha, \lambda)$ is strictly increasing in the interval $[\lambda_i^u, \lambda_{i+1}^\ell]$ there exists unique $\lambda = \lambda(\alpha)$ such that

$$f(\alpha, \lambda(\alpha)) = \rho(\alpha)\,; \quad \chi(\alpha, \lambda(\alpha)) < \sigma(\alpha)$$

Choose:

$$z(\alpha) = \lambda(\alpha)\,; \quad p(\alpha) = \int_{\mathbf{R}} \frac{1}{\lambda(\alpha) - \xi} dP_\xi(g(\alpha)).$$

Case (B): $\rho(\alpha) \in F_i = (\rho_i^\ell, \rho_i^u)$ for some $i \in \{1, \dots, m\}$.

Call $z = \lambda + i\mu$, if $\rho(\alpha) = z\sigma(\alpha) + \tau(\alpha) + \phi(\alpha, z)$ then

$$\rho(\alpha) = \lambda\sigma(\alpha) + \tau(\alpha) + \int_{\mathbf{R}} \frac{\lambda - \xi}{(\lambda - \xi)^2 + \mu^2} d\nu_g \tag{2.38a}$$

$$\sigma(\alpha) = \int_{\mathbf{R}} \frac{1}{(\lambda - \xi)^2 + \mu^2} d\nu_g \tag{2.38b}$$

In order to solve the above system let first $\mu^2(\lambda)$ be the unique solution of (2.38b) for $\lambda \in (\lambda_i^\ell, \lambda_i^u)$. Observe that $\mu^2(\lambda) \to 0$ as $\lambda \to \lambda_i^{\ell,u}$. Differentiating with respect to λ

$$2\int \frac{(\lambda - \xi)}{[(\lambda - \xi)^2 + \mu^2]^2} d\nu_g + \frac{d\mu^2}{d\lambda} \int \frac{1}{[(\lambda - \xi)^2 + \mu^2]^2} d\nu_g = 0.$$

Let $\alpha(\lambda) = \lambda\sigma(\alpha) + \int \frac{\lambda-\xi}{(\lambda-\xi)^2+\mu^2} d\nu_g$ then

$$\begin{aligned}\frac{d\alpha}{d\lambda} &= \sigma(\alpha) + \int \frac{1}{|z-\xi|^2} d\nu_g - 2\int \frac{(\lambda-\xi)^2}{[(\lambda-\xi)^2+\mu^2]^2} d\nu_g - \frac{d\mu^2}{d\lambda}\int \frac{\lambda-\xi}{[(\lambda-\xi)^2+\mu^2]^2} d\nu_g \\ &= 2\mu^2 \int \frac{1}{[(\lambda-\xi)^2+\mu^2]^2} d\nu_g \\ &\quad + 2\Big(\int \frac{\lambda-\xi}{[(\lambda-\xi)^2+\mu^2]^2} d\nu_g\Big)^2 \Big/ \int \frac{1}{[(\lambda-\xi)^2+\mu^2]^2} d\nu_g > 0.\end{aligned}$$

Since $\frac{d\alpha}{d\lambda} > 0$ and $f(\alpha, \lambda_i^\ell) < \rho(\alpha) < f(\alpha, \lambda_i^u)$ there exist unique up to complex conjugation $z(\alpha) = \lambda(\alpha) + i\mu(\alpha)$ such that$f(\alpha, z(\alpha)) = \rho(\alpha)$. Choose $z(\alpha) = \lambda(\alpha) + i\mu(\alpha)$; $p(\alpha) = \int_{\mathbf{R}} \frac{1}{z(\alpha)-\xi} dP_\xi(g(\alpha))$ then

$$(S_1 p(\alpha), p(\alpha)) = \int_{\mathbf{R}} \frac{1}{|z(\alpha)-\xi|^2} d\nu_g = \sigma(\alpha) \tag{2.39}$$

Remark: In view of Theorem 2.2 if $\Im z(\alpha) = 0$ then either $\widehat{E}_R(\alpha)$, $\widehat{E}_S(\alpha)$ are tangent or $\widehat{T}(\alpha) \cong$ torus, while $\Im z(\alpha) \neq 0$ implies that $\widehat{T}(\alpha) \cong$ sphere. In Case (B) of the previous theorem when $\Im z(\alpha) = \mu(\alpha) \neq 0$ it is obvious that I can choose $\lambda(\alpha) \pm i\mu(\alpha)$ as solutions therefore I can construct two vector fields, let $h(\alpha)$ correspond to $\mu(\alpha) > 0$, while $h^*(\alpha)$ to $\mu(\alpha) < 0$.

Definition: 7) *Define* $\mathcal{E} = \{\alpha \in S^{2n-1}\colon z(\alpha) \in \Lambda_c(\alpha)\}$, $\mathcal{R} = S^{2n-1} \backslash \mathcal{E}$, $\mathcal{E}_d = \{\alpha \in S^{2n-1}\colon z(\alpha) \in \Lambda_d(\alpha)\}$, $\mathcal{E}_c = \{\alpha \in S^{2n-1}\colon z(\alpha) \in \Lambda_c(\alpha)\}$.

If$\alpha \in \mathcal{R}$ then $\{z(\alpha), p(\alpha), h(\alpha)\}$ are smooth functions with respect to α. The set $\mathcal{E}$ can be characterized as follows, let $\lambda_i^{\ell,u}(\alpha)$ $i = 1, \ldots, m$ be the zeros of the equation $\chi(\alpha, \lambda) = \sigma(\alpha)$ and define the functions

$$d_i^{\ell,u}(\alpha) = f(\alpha, \lambda_i^{\ell,u}(\alpha)) - \rho(\alpha) \quad i = 1, \ldots, m \tag{2.40}$$

then

$$\mathcal{E} = \{\alpha \in S^{2n-1}\colon d_i^{\ell,u}(\alpha) = 0 \quad \text{for some} \quad d_i^{\ell,u}\}. \tag{2.40a}$$

The function $z(\alpha) = \lambda(\alpha) + i\mu(\alpha)$ satisfies the equation

$$-\rho(\alpha) + z\sigma(\alpha) + \tau(\alpha) + \int \frac{1}{z-\xi} d\nu_g = 0. \tag{2.41}$$

Let $\alpha_j = x_j + iy_j$ and denote

$$\partial_j = \frac{1}{2}\left(\frac{\partial}{\partial x_j} - i\frac{\partial}{\partial y_j}\right) ; \; \overline{\partial}_j = \frac{1}{2}\left(\frac{\partial}{\partial x_j} + \frac{\partial}{\partial y_j}\right).$$

If $\alpha \in \mathcal{R}$, differentiating (2.41) I have

$$h_j(\alpha) + [\sigma(\alpha) - \chi(\alpha, z)]\overline{\partial}_j\lambda - M(\alpha)\overline{\partial}_j z = 0$$
$$\overline{h_j^*}(\alpha) + [\sigma(\alpha) - \chi(\alpha, z)]\partial_j\lambda - M(\alpha)\partial_j z = 0$$

where

$$M(\alpha) = -i\mu \int \frac{\overline{z} - \xi}{|z - \xi|^4} d\nu_g \tag{2.42}$$

If$\mu = 0$ then $M = 0$, $h^* = h$ and

$$h_j - \langle Sy(\alpha), \overline{y}(\alpha)\rangle \overline{\partial}_j \lambda = 0 \tag{2.43}$$

where $y(\alpha) = p(\alpha) + \alpha_i\psi_i$. If on the other hand $\mu \neq 0$ then $\langle Sy, \overline{y}\rangle = \chi(\alpha, z) - \sigma(\alpha) = 0$ and

$$\begin{aligned} h_j - M(\alpha)\overline{\partial}_j z &= 0 \\ \overline{h_j^*} - M(\alpha)\partial_j z &= 0 \end{aligned} \tag{2.44}$$

Because of Theorem 2.1 for every $\alpha \in S^{2n-1}$ I have the relation

$$(R - z(\alpha)S)y(\alpha) = h_j(\alpha)\psi_j \tag{2.45}$$

Taking the $\langle \cdot, \cdot \rangle$ inner product with $y(\alpha)$ I have $\langle Ry, y\rangle - z\langle Sy, y\rangle = h_j(\alpha)\alpha_j$, now differentiating I get

$$\begin{aligned} 2h_j &= I(\alpha)\partial_j z + \partial_j(h_\ell \alpha_\ell) \\ 0 &= I(\alpha)\overline{\partial}_j z + \overline{\partial}_j(h_\ell \alpha_\ell) \end{aligned} \tag{2.46}$$

where $I(\alpha) = \langle Sy(\alpha), y(\alpha)\rangle$. Finally let $\alpha \in \mathcal{E}$, then $d_i^{\ell,u}(\alpha) = 0$ for some $d_i^{\ell,u}$ and $z(\alpha) = \lambda_i^{\ell,u}(\alpha)$. Differentiating $d_i^{\ell,u}$ I have

$$\overline{\partial}_j d_i^{\ell,u}(\alpha) = h_j(\alpha) \tag{2.47}$$

where $d_i^{\ell,u}(\alpha) = 0$. The last relation implies that $\mathcal{E}$ is a smooth hypersurface of codimension one unless $h(\alpha) = 0$.

Theorem 2.3: *The problem (E) has exactly n pairs of eigenvalues and corresponding eigenvectors $\{z_i, y_i, \overline{z}_i, \overline{y}_i\,;\, i = 1,\ldots,n\}$ where $n = N(S)$. Moreover (E) will have an eigenvalue of the form*

$$\begin{pmatrix} \lambda & 1 \\ 0 & \lambda \end{pmatrix}; \quad \lambda \in \mathbf{R}$$

if and only if there exist $\alpha_0 \in S^{2n-1}$ such that (i) $d(\alpha_0) = 0$ and (ii) $\overline{\partial}_j d(\alpha_0) = 0$ $j = 1,\ldots,n$ for some of the functions $d_i^{\ell,u}(\alpha)$.

Proof: The vector field h: $\mathbf{C}P(n-1) \to \mathbf{C}^n$ has n critical points (i.e., points where $h = 0$) since the Euler chracteristic of $\mathbf{C}P(n-1)$ is

$$\chi(\mathbf{C}P(n-1)) = \sum_{i=0}^{2(n-1)} (-1)^i \operatorname{rank} H_i = \sum_{k=0}^{n-1} (-1)^{2k} = n.$$

Without loss of generality I can assume that there are no critical points of n in $\mathcal{E}$. Indeed if for some $\alpha_0 \in \mathcal{E}$ I have $d(\alpha_0) = 0$ and $\overline{\partial}_j d(\alpha_0) = 0$ then from Sard's theorem I can find $\varepsilon \in \mathbf{R}$ orbitrarily small such that $d(\alpha) = \varepsilon$ is a regular value of $d(\alpha)$. Now consider the matrix $\rho_\varepsilon^{ij} = \rho^{ij} - \varepsilon\delta_{ij}$ and set

$$h_i^\varepsilon(\alpha) = [\rho_\varepsilon^{ij} - z^\varepsilon(\alpha)\sigma^{ij} + \tau_{ij} + \phi_{ij}(z^\varepsilon(\alpha))]\alpha_j$$

where $z^\varepsilon(\alpha)$ solves the equation $\rho_\varepsilon(\alpha) = f(\alpha, z)$ with $f(\alpha, z) = z\sigma(\alpha) + \tau(\alpha) + \phi(\alpha, z)$. $h^\varepsilon(\alpha)$ is a new vector field on $\mathbf{C}P(n-1)$ and the set on which h^ε is not smooth is given by $d(\alpha) = \varepsilon$ which is a smooth hypersurface. Assume that $\alpha_0 \in \mathcal{R}$ such that $h(\alpha_0) = 0$ and $\beta_j \partial_j h_i(\alpha_0) = 0$ (i.e., α_0 is a degenerate critical point) for some $\beta \in \mathbf{C}^n$. Differentiating (2.45) I have

$$(R - zS)\beta_j \partial_j y = (\beta_j \partial_j z) S y + \beta_j \partial_j h_i \psi_i$$

From (2.46) I have $0 = I(\alpha)(\beta_j \partial_j z) + (\beta_j \partial_j h_i)\alpha_i \Rightarrow \beta_j \partial_j z = 0$ hence

$$(R - zS)(\beta_j \partial_j y) = 0$$

and $h(\alpha)$ vanishes on the set spanned by $\{\alpha_0, \beta\}$. This argument shows that there are at least n eigenvalues. If $(R - z_i X) y_i = 0$ $i = 1, 2$ and $z_1 \neq \{z_2, \overline{z}_2\}$ then$\langle S y_1, \overline{y}_2 \rangle = 0$, hence if there exist more than n distinct pairs, counting multiplicity, that solve problem (E) then

S will be negative definite on the space spanned by$\{y_i,\ i=1,\ldots,n_1\}$, $n_1>n$ arriving at a contradiction.

Finally, if $(R-\lambda S)y=0$ and $\langle Sy,y\rangle=0$ then by setting $t=(R-\lambda S)^{-1}Sy$ and $E_\lambda=\mathrm{span}\{t,y\}$ I have that $S^{-1}R$ leaves E_λ invariant and

$$R\mid_{E_\lambda}=\begin{pmatrix}\lambda & 1\\ 0 & \lambda\end{pmatrix}S\mid_{E_\lambda}$$

□

Let $\Omega=\{\alpha\in S^{2n-1}\colon Im\,z=0\}$, then because of (2.43) $h=0\Leftrightarrow\overline{\partial}_j\lambda=0$, also $\partial\Omega=E=E_i^{\ell,u}$, where $E_i^{\ell,u}=\{\alpha\in S^{2n-1}\colon d_i^{\ell,u}(\alpha)=0\}$. If $\eta(\alpha)$ is the unit outward normal on $\partial\Omega$ then it is easy to see that

$$\begin{aligned}\alpha_0\in E_i^{\ell}&\Rightarrow(h(\alpha_0),\eta(\alpha_0))<0\\ \alpha_0\in E_i^{u}&\Rightarrow(h(\alpha_0),\eta(\alpha_0))>0\end{aligned}\tag{2.48}$$

Now I would like to examine what happens if there is an eigenvalue of negative sign embedded in the continuous spectrum, i.e.,

$$(R-\lambda_0 S)y_0=0\,,\ \lambda_0\in\sigma_c(R,S)\,,\ \langle Sy_0,y_0\rangle<0\tag{2.49}$$

By writing $y_0=p_0+\alpha_i^0\phi_i$ (2.49) becomes

$$\begin{gathered}(R_1-\lambda_0 S_1)p_0=-\alpha_j b_j\\ [\rho^{ij}-\lambda_0\sigma^{ij}+\tau_{ij}+\phi_{ij}(\lambda_0)]\alpha_j^0=0\\ \chi(\alpha_0,\lambda_0)<\sigma(\alpha_0)\end{gathered}\tag{2.50}$$

I want to show that this eigenvalue is structurally unstable and for this reason I will have to study how the continuous spectrum changes under perturbations. Since in practice it is always the operators R,S^{-1} that are given, see (ch. 3), I will consider perturbations of the form

$$R(\delta)=R+\delta\widetilde{W}\,;\ S(\delta)=(S^{-1}+\delta\widetilde{V})^{-1}$$

Assumption $(*)$: Assume that $\widetilde{W},\widetilde{V}$ are bounded operators that satisfy

$$\sup_{0<|\varepsilon|<1;\,\lambda\in\sigma_{ac}(R,S)}|\varepsilon|\,\left\||B|^{1/2}K_1(\lambda+i\varepsilon)\right\|^2<+\infty\tag{2.51}$$

where $B = \widetilde{W} - \langle \cdot, W\psi_i\rangle\psi_i$ or $B = S\tilde{V}S - \langle \cdot, S\tilde{V}S\psi_i\rangle\psi_i$ and $K_1(z) = (R_1 - zS_1)^{-1}$.

This assumption is equivalent to assuming that $|B|^{1/2}$ is smooth with respect to the continuous part of the spectrum of $K_1(z)$. For the definition of H-smoothness and its implications see (5) vol.IV. Since I have the freedom to choose $\{\psi_i\, i = 1, \ldots, n\}$ in any dense subset of X, it is easy to see that (2.51) is equivalent to

$$\sup_{0<|\varepsilon|<1;\ \lambda\in\sigma_{ac}(R,S)} |\varepsilon|\ \left\||B|^{1/2}K(\lambda + i\varepsilon)\right\|^2 < +\infty \tag{2.51a}$$

where $B = |\widetilde{W}|^{1/2}$ or $B = |SVS|^{1/2}$ and $K(z) = (R - zS)^{-1}$. The reason for assumption $(*)$ is Stone's formula for the spectral projection which in this case has the form

$$\frac{1}{2}\langle p, (\Pi_{(a,b)} + \Pi_{[a,b]})\, p\rangle = \lim_{\varepsilon\downarrow 0} \pi^{-1} \int_a^b Im\langle p, K_1(\delta, \lambda + i\varepsilon)p\rangle d\lambda$$

where

$$d\Pi_\xi(\delta) = S_1^{-1/2}(\delta) dP_\xi(\delta) S_1^{-1/2}(\delta)$$

$\{dP_\xi(\delta)\}$is the spectral projection of $H_1(\delta) = S_1^{-1/2}(\delta)R_1(\delta)S_1^{-1/2}(\delta)$, and

$$K_1(\delta, z) = (R_1(\delta) - zS_1(\delta))^{-1}$$

If $[a, b] \subset \sigma_{ac}(R(\delta), S(\delta))$ so that $\Pi_{\{a\}}(\delta) = \Pi_{\{b\}}(\delta) = \phi$ then it is easy to show (by expanding in powers of δ) that $\Pi_{(a,b)}(\delta)$ is smooth, in fact analytic, with respect to δ, see (5)vol. IV, pg. 164. Finally in order to motivate the assumption $(*)$ consider the following typical case, $R = -\Delta + m + W(x)$, $S^{-1} = -\Delta + m + V(x)$,$X = L^2(\mathbf{R}^N)$ where $W(x)$, $V(x)$ are bounded potentials decaying exponentially at infinity, then $\widetilde{W}(x)$, $\tilde{V}(x)$ can be any potentials having the same properties as $W(x)$, $V(x)$.

Theorem 2.4: *If there exists $\lambda_0 \in \mathbf{R}^+$ satisfying (2.49) then there exists perturbation $(R(\delta), S(\delta))$ of (R, S) satisfying assumption $(*)$ such that for δ small $(R(\delta), S(\delta))$ has a pair of complex conjugate (nonreal) eigenvalues near λ_0.*

Proof: I will assume for simplicity that $N(S) = 1$, the general case. $N(S) > 1$ follows from the case $N(S) = 1$ combined with the observation made in (2.48) and the fact that if $z(\alpha) = \lambda(\alpha) \in \mathbf{R}$, where $z(\alpha)$ is the function constructed in Theorem (2.1), then the critical points of $h(\alpha)$ coincide with the critical points of the function $\lambda(\alpha)$, see (2.43).

Set

$$\rho(\delta) = -\langle (R + \delta\widetilde{W})^{-1}\psi, \psi\rangle$$
$$d\nu(\delta,\xi) = \langle d\Pi_\xi(\delta)b(\delta), b(\delta)\rangle$$

where $b(\delta)$ is given by (2.11), i.e.,

$$b(\delta) = \frac{1}{\sigma(\delta)}\{R(\delta)S^{-1}(\delta)\psi - \langle S^{-1}(\delta)\psi, R(\delta)\psi\rangle\psi\}$$
$$\sigma(\delta) = -\langle (S^{-1} + \delta\widetilde{V})\psi, \psi\rangle$$

The criterion for deciding if the eigenvalue of $(R(\delta), S(\delta))$ is complex is simply

$$\rho(\delta)^{-1} \in F(\delta) \Rightarrow Im\, z(\delta) \neq 0$$

where $F(\delta)$ is defined by (2.32a).

Let $I = [\lambda_0 - \delta_1, \lambda_0 + \delta_1]$ for some $\delta_1 > 0$ and$\Pi_I(\delta) = \int_I d\Pi_\xi(\delta)$ so that

$$\nu(\delta, I) = \int_I d\nu(\delta,\xi) = \langle \Pi_I(\delta)b(\delta), b(\delta)\rangle.$$

Since $d\Pi_\xi(\delta) = S_1^{-1/2}(\delta)dP_\xi(\delta)S_1^{-1/2}(\delta)$ and $\{dP_\xi(\delta)\}$ is the spectral projection of the selfadjoint operator $H_1(\delta) = S_1^{-1/2}(\delta)R_1(\delta)S_1^{-1/2}(\delta)$, I have that $\Pi_I(\delta)S_1(\delta)\Pi_I(\delta) = \Pi_I(\delta)$ hence

$$\nu(\delta, I) = \langle S_1(\delta)\Pi_I(\delta)b(\delta), \Pi_I(\delta)b(\delta)\rangle. \tag{2.52}$$

Let «′» denote differentiation with respect to δ, then differentiating twice (2.52) and using the fact that $\Pi_I(0)b(0) = 0$ I have

$$\nu(\delta, I) = \langle S_1(0)(\Pi_I'(0)b(0) + \Pi_I(0)b'(0)),$$
$$\Pi_I'(0)b(0) + \Pi_I(0)b'(0)\rangle + 0(\delta^3)$$

hence $\nu(\delta, I)$ will be strictly positive for δ small iff

$$\Pi_I'(0)b(0) + \Pi_I(0)b'(0) \neq 0 \tag{2.53}$$

Let $S_1(0) = S_1$, $R_1(0) = R_1$, $K_1(z) = (R_1 - zS_1)^{-1}$, $\widetilde{W}_1 = \widetilde{W} - \langle\cdot, \widetilde{W}\psi\rangle\psi$ and $(S\widetilde{V}S)_1 = S\widetilde{V}S - \langle\cdot, S\widetilde{V}S\psi\rangle\psi$, and calculate

$$
\begin{aligned}
\Pi_I'(0)b(0) &= \lim_{\varepsilon\downarrow 0}\frac{1}{\pi}\int_I Im[K_1(\lambda+i\varepsilon)(\widetilde{W}_1 + \lambda(S\widetilde{V}S)_1)K_1(\lambda+i\varepsilon)]b(0)d\lambda \\
&= \lim_{\varepsilon\downarrow 0}\frac{1}{\pi}\Big\{\int_I (Im\,K_1(\lambda+i\varepsilon))(\widetilde{W}_1 + \lambda(S\widetilde{V}S)_1)(ReK_1(\lambda+i\varepsilon))b(0)d\lambda + \\
&\quad + \int_I (ReK_1(\lambda+i\varepsilon))(\widetilde{W}_1 + \lambda(S\widetilde{V}S)_1)(Im\,K_1(\lambda+i\varepsilon))b(0)d\lambda\Big\}
\end{aligned}
\tag{2.54}
$$

Since $Im\,K_1(\lambda+i\varepsilon) = S_1^{-1/2}\frac{\varepsilon}{(H_1-\lambda)^2+\varepsilon^2}S_1^{-1/2}$; $H_1 = S_1^{-1/2}R_1S_1^{-1/2}$, the second integrand in the above relation can be bounded by

$$
|\varepsilon|\left\|K_1(\lambda+i\varepsilon)\left(|\widetilde{W}_1|^{1/2} + |S\widetilde{V}S|_1^{1/2}\right)\right\|\left\|S_1^{-1/2}\frac{1}{(H_1-\lambda)^2+\varepsilon^2}S_1^{-1/2}b(0)\right\|
$$

Since $\Pi_I(0)b(0) = 0$, $\|S_1^{-1/2}[(H_1-\lambda)^2+\varepsilon^2]^{-1}S_1^{-1/2}b(0)\|$ is bounded independent of ε $\forall\lambda \in I$, while assumption $(*)$ implies

$$
\sup_{\substack{0<\varepsilon<1\\ \lambda\in I}}\sqrt{\varepsilon}\|K_1(\lambda+i\varepsilon)(|\widetilde{W}_1|^{1/2} + |S\widetilde{V}S|_1^{1/2})\| < +\infty.
$$

Hence the second term in (2.54) goes to zero as $\varepsilon \downarrow 0$ and the relation simplifies to

$$
\Pi_I'(0)b(0) = \lim_{\varepsilon\downarrow 0}\frac{1}{\pi}\int_I d\Pi_\lambda(0)(\widetilde{W}_1 + \lambda(S\widetilde{V}S)_1)a(\lambda)d\lambda
$$

where $a(\lambda) = S_1^{-1/2}(H_1-\lambda)^{-1}S_1^{-1/2}b(0) = (R_1 - \lambda S_1)^{-1}b(0)$, which is well defined and bounded in norm $\forall\lambda \in I$. Also

$$
b'(0) = \frac{1}{\sigma(0)}\left\{-\frac{\sigma'(0)}{\sigma(0)}b(0) + \widetilde{W}_1(S^{-1}\psi) + R_1(\widetilde{V}\psi)\right\}
$$

so that finally I can write (2.53) as follows

$$
\Pi_I(0)\left\{(\widetilde{W}_1 + \lambda_0(S\widetilde{V}S)_1)a(\lambda_0) + \frac{1}{\sigma(0)}[\widetilde{W}_1(S^{-1}\psi) + R_1(\widetilde{V}\psi)]\right\} + 0(\delta_1) \neq 0.
$$

Since I can choose δ_1 small, all I need is to choose $\widetilde{W}$ and $\widetilde{V}$ such that

$$
\widetilde{W}_1(a(\lambda_0) + \frac{1}{\sigma(0)}S^{-1}\psi) + \lambda_0(S\widetilde{V}S)_1a(\lambda_0) + \frac{1}{\sigma(0)}R_1(\widetilde{V}\psi) = d \tag{2.55}
$$

where $d \in \mathbf{P}$ is a vector such that $\Pi_I(d) \neq 0$.

Recall that using $R(\delta)$, $S(\delta)$ I can define the functions

$$\chi(\delta, z) = \int \frac{1}{|z-\xi|^2} d\nu(\delta, \xi)$$

$$f(\delta, z) = z\sigma^{-1}(\delta) + \tau(\delta) + \int \frac{1}{z - xi} d\nu(\delta, \xi)$$

If (2.55) is satisfied then for δ small there exist $\lambda^{\ell,u}(\delta) \in \mathbf{R}$ such that

$$\lambda^{\ell}(\delta) < \lambda_0 - \delta_1 < \lambda_0 + \delta_1 < \lambda^{u}(\delta)$$

$$\chi(\delta, \lambda^{\ell,u}(\delta)) = \sigma^{-1}(\delta),$$

this follows from Lemma 2.2. Moreover, $\rho(0)^{-1} = f(0, \lambda_0)$ and $f(0, \lambda)$ is strictly increasing in a neighborhood of λ_0 (see Lemma 2.3). It is easy to see the following

$$f(\delta, \lambda^{\ell}(\delta)) < f(0, \lambda_0 - \delta_1) < \rho(0)^{-1} < f(0, \lambda_0 + \delta_1) < f(\delta, \lambda^{u}(\delta)).$$

Hence for δ small I will have

$$f(\delta, \lambda^{\ell}(\delta)) < \rho(\delta)^{-1} < f(\delta, \lambda^{u}(\delta))$$

and this last inequality implies that $(\rho(\delta))^{-1} \in F(\delta)$ hence the equation $\rho^{-1}(\delta) = f(\delta, z)$ has a pair of complex conjugate (non real) solutions $\{z(\delta), \overline{z}(\delta)\}$.

Concluding, the problem reduces to finding $\widetilde{W}, \widetilde{V}$ that satisfy (2.55), this can obviously be achieved, for example using finite rank operators, however in specific examples we would like to consider only a special kind of operators imposed by the nature of the problem; to be more specific, consider the following example.

$$R = -\Delta + m + W(x), \ S^{-1} = -\Delta + m + V(x)$$

where $m > 0$ and $W(x)$, $V(x)$ are bounded potentials decaying exponentially at infinity (this is the case if one considers Schrödinger equation in $\mathbf{R}^N$, see the examples). Then I would like to consider as perturbations $\widetilde{W}(x)$, $\widetilde{V}(x)$ potentials with the same properties as $W(x)$, $V(x)$ so that assumption $(*)$ is satisfied. Assume for simplicity that $\widetilde{V}(x) = 0$ then (2.55) reduces to

$$\widetilde{W}(x)p(x) - \langle p(x), \widetilde{W}(x)\psi(x)\rangle_{L^2}\psi(x) = d(x)$$

where $p(x) = a(\lambda_0)(x) + 1/\sigma(0)S^{-1}\psi(x)$. Here $p(x)$, $\psi(x)$, $d(x)$ are given function of x in $L^2(\mathbf{R}^N)$ and a reasonable choice is

$$\widetilde{W}(x) = d(x)/p(x)$$

The only problem is on the set $\{x\colon p(x) = 0\}$ but I can always choose $d(x) \in L^2(\mathbf{R}^N)$ such that $\Pi_I(0)d \neq 0$ and $p(x) = 0 \Rightarrow d(x) = 0$, notice that $p(x)$ can be chosen to be a smooth function. □

I am now in a position to state a theorem about structural stability, from Theorem (2.3) the vector field $h\colon \mathbf{C}P(n-1) \to \mathbf{C}^n$ has n critical points counting multiplicity, call them α_i $i = 1, \dots, n$.

Theorem 2.5: a) *The pair (R, S) is stable iff all the critical points $\{\alpha_i \; i = 1, \dots, n\}$ of the vector field $h(\alpha)$ are in $\mathcal{R}$ and moreover $z(\alpha_i) = \lambda i \in \mathbf{R}^+$.*

b) The pair (R, S) is structurally stable iff it is stable and moreover $\lambda_i \notin \Lambda_d \cup \Lambda_{ac}$, where Λ_d andΛ_{ac} are given in (2.31a).

Proof: The proof of (a) is obvious from Theorem 2.3, while the proof of (b) follows from Theorem 2.4. □

Let $C \subset X$ be a cone in X and denote by $d(C)$ the dimension of the maximal linear subspace of X that is contained in C. The following instability criterion is a generalization of those found in (4,3).

Theorem 2.6: *Problem (E) has $\max\{N(R), N(S)\} - d(C(R) \cap C(S))$ negative real eigenvalues, hence A has that many pairs $\{\pm\lambda_i \; i = 1, \dots, \max\{N(R), N(S)\} - d(C(R) \cap C(S))\}$ of real eigenvalues.*

Recall that $C(R) = \{p \in X\colon \langle Rp, p\rangle < 0\}$ etc.

The last result is a criterion of structural instability for a pair of operators (R, S) that turns out to be useful in applications (see Chap. 3).

Theorem 2.7: *Assume that the essential spectrum of the pair (R, S) consists of absolutely continuous part and is $\sigma_{\mathrm{ess}}(R, S) = \sigma_{ac}(R, S) = [m, +\infty)$, where $m > 0$. There exists positive constant $c = c(m)$ such that if $\sup_{|\alpha|=1} \rho(\alpha)/\sigma(\alpha) > c(m)$ then (R, S) has either an* *unstable eigenvalue, or a positive eigenvalue of negative sign embedded in the continuous spectrum of (R,S). Hence (R,S) is either unstable or structurally unstable.*

Proof: Let me call α^k $k=1,\ldots,n$ the critical points of the vector field $h(\alpha)$. Assume that $z(\alpha^k)=\lambda_k\in(0,m)\ \forall k=1,\ldots,n$, i.e.,

$$\rho^{ij}\alpha_j^k=\lambda_k\left[\sigma^{ij}\alpha_j^k-\int\frac{1}{\xi(\lambda_k-\xi)}d\nu_{ij}(\xi)\alpha_j^k\right]$$

Let me normalize α^k so that $\sigma(\alpha^k)=1$ and denote $\sigma(\alpha^k,\alpha^\ell)=-\sigma^{ij}\alpha_j^k\alpha_i^\ell$; $\rho(\alpha^k,\alpha^\ell)=-\rho^{ij}\alpha_j^k\alpha_i^\ell$. Let $\alpha=b_k\alpha^k$ so that $\sigma(\alpha)=1$, then

$$\rho(\alpha)=b_kb_\ell\rho(\alpha^k,\alpha^\ell)=\left[\lambda_k\sigma(\alpha^k,\alpha^\ell)+\int\frac{1}{\xi(\lambda_i-\xi)}d\nu_{ij}(\xi)\alpha_j^k\alpha_i^\ell\right]b_kb_\ell \tag{2.55}$$

$$\sigma(\alpha^k,\alpha^\ell)\le(\sigma(\alpha^k)\sigma(\alpha^\ell))^{1/2}\le 1 \tag{2.56}$$

Notice that the choice of $\{\psi_i\ i=1,\ldots,n\}$ is only subject to the restrictions (2.1a) hence they can be chosen so that $\ker R_1=\phi$ and this in turn means that there exists $\varepsilon>0$ such that $\nu_g([-\varepsilon,\varepsilon])=0$ for every $g=g(\alpha)$. Here recall that $\rho_{ij}=\langle R^{-1}\psi_i,\psi_j\rangle$ and $\rho(\alpha)=-\rho^{ij}(\alpha_i,\alpha_j)$ so if $\ker R_1\neq\phi$ then $\ker\rho_{ij}\neq\phi$ and $\sup\limits_{|\alpha|=1}\rho(\alpha)=+\infty$. Since $0<\lambda_k<m\ \forall k=1,\ldots,n$ and $|\xi|>\varepsilon$ I have

$$\frac{1}{|\xi|^2}<\left(1+\frac{m}{\varepsilon}\right)^2\frac{1}{|\lambda_k-\xi|^2}\Longrightarrow\int\frac{1}{\xi^2}d\nu_g\le\left(1+\frac{m}{\varepsilon}\right)^2\chi(\alpha,\lambda_k)$$

$$\left|\int\frac{1}{\xi(\lambda_k-\xi)}d\nu_{ij}\alpha_j^k\alpha_i^\ell\right|\le\left(1+\frac{m}{\varepsilon}\right)(\chi(\alpha^k,\lambda_k)\chi(\alpha^\ell,\lambda_\ell))^{1/2}<1+\frac{m}{\varepsilon} \tag{2.57}$$

since $\chi(\alpha^k,\lambda_k)<\sigma(\alpha^k)=1$. Now using (2.56), (2.57)in (2.55) I have $\rho(\alpha)\le c(m,\varepsilon)$ for some constant $c(m,\varepsilon)$ and this gives a contradiction if $\sup\limits_{\sigma(\alpha)=1}\rho(\alpha)>c(m,\varepsilon)$. □

Chapter 3

Applictions

(1) Nonlinear Schrödinger Equation

Consider the equation

$$iu_t - \Delta u + u - g(|u|^2)u = 0;\ u\colon \mathbf{R}\times\mathbf{R}^N \to \mathbf{C} \tag{3.1}$$

Looking for steady-state solutions reduces the problem to the equation

$$-\Delta\varphi + \varphi - g(\varphi^2)\varphi = 0;\ \varphi\colon \mathbf{R}^N \to \mathbf{R}^N \tag{3.2}$$

Let $G(s^2) = \int\limits_0^s g(t^2)tdt$ and $F(s) = 1/2s^2 - G(s^2)$, the following existence theorem holds for equation (3.2) see (9).

Theorem 3.1: *Assume that*

a) *There exists* $s_0 \in \mathbf{R}$ *such that* $F(s_0) < 0$;

b) $|s|^q < G(s^2) < |s|^p$ *for some* $p, q \in (2, 2N/N-2)$.

Then equation (3.2) has infinitely many radial solutions $\varphi_n(|x|)$ $n = 0, 1, \ldots$ *where* n *is the numer of nodes of the function* $\varphi_n(|x|)$; *moreover if* $\lambda_n := |\varphi_n(0)|$ *then* $\lim\limits_{n\to\infty} \lambda_n = +\infty$.

The linearization around $\varphi_n(|x|)$ gives the following operator

$$\mathbb{A}_n = \begin{pmatrix} 0 & B_n \\ -A_n & 0 \end{pmatrix}$$

where $A_n = -\Delta + 1 - g(\varphi_n^2) - 2g'(\varphi_n^2)\varphi_n^2$, $B_n = -\Delta + 1 - g(\varphi_n^2)$. For simplicity I want to consider the special case where $g(s) = |s|^p$, $0 < p < 4/N-2$ so that $A_n = -\Delta + 1 - (p+1)|\varphi_n(x)|^p$; $B_n = -\Delta + 1 - |\varphi_n(x)|^p$. The function $\widehat{\varphi}_n(|x|) = \lambda_n^{-1}\varphi_n(\lambda_n^{-p/2}|x|)$ satisfies the equation

$$-\Delta\varphi + \lambda_n^{-p}\varphi - |\varphi|^p\varphi = 0;\ \varphi(0) = 1 \tag{3.4}$$

Let $W_n(x) = 1 - (p+1)|\varphi_n(x)|^p$, $V_n(x) = 1 - |\varphi_n(x)|^p$ and $\widehat{W}_n(x) = \lambda_n^{-p} - (p+1)|\widehat{\varphi}_n(x)|^p$, $\widehat{V}_n(x) = \lambda_n^{-p} - |\widehat{\varphi}(x)|^p$. It is easy to show that

$$\widehat{\varphi}_n(|x|) \to \widehat{\varphi}_0(|x|) \text{ in } C^1_{\text{loc}}(\mathbf{R}^N) \tag{3.5}$$

where $\widehat{\varphi}_0$ is the radial solution of the equation

$$-\Delta\varphi - |\varphi|^p\varphi = 0 \quad \varphi(0) = 1 \tag{3.6}$$

It can be proved that the solution of (3.6) has infinitely many nodes. In view of (3.5) I have that $\widehat{W}_n \to \widehat{W}_0 := -(p+1)|\widehat{\varphi}_0(x)|^p$ and $\widehat{V}_n \to \widehat{V}_0 := -|\widehat{\varphi}_0(x)|^p$ in $C^i_{\text{loc}}(\mathbf{R}^N)$.

Proposition 3.2: *The operators $-\Delta + \widehat{W}_0$ and $-\Delta + \widehat{V}_0$ have infinitely many negative eigenvalues $\widehat{E}_0 < \widehat{E}_1 \leq \widehat{E} \leq \cdots$. If $\widehat{E}_{n,i}$ is a negative eigenvalue of $-\Delta + \widehat{W}_n$ or $-\Delta + \widehat{V}_n$ then $\lim_{n\to\infty} \widehat{E}_{n,i} = \widehat{E}_i$*

The proof of the above proposition follows easily from Sturm-Liouville theory since $(-\Delta + \widehat{V}_0)\widehat{\varphi}_0 = 0$ and $\widehat{\varphi}_0(|x|)$ has infinitely many nodes. If $\widehat{E}_n$ is an eigenvalue of $-\Delta + \widehat{W}_n$ or $-\Delta + \widehat{V}_n$ then $E_n := \lambda_n^p \widehat{E}_n$ is an eigenvalue of $-\Delta + W_n$ or $-\Delta + V_n$ respectively, hence if $E_{n,i}$ is the ith eigenvalue of $-\Delta + W_n$ or $-\Delta + V_n$ then $E_{n,i} \to -\infty$ as $n \to \infty$. Since $(-\Delta + V_n)\varphi_n = 0$ let $X_n := \varphi_n^{\perp} = \{p \in L^2_r(\mathbf{R}^N)\colon p \perp \varphi_n\}$ and let $R_n = A_n \mid X_n$, $S_n = (-\Delta + V_n)^{-1} \mid X_n$. Choose ψ_j $j = 1, \ldots, n$ to be the eigenfunctions corresponding to negative eigenvalues of S_n since $R_n < S_n^{-1}$ I have that

$$\sup_{|\alpha|=1} \rho_n(\alpha)/\sigma_n(\alpha) \to +\infty \text{ as } n \to \infty \tag{3.7}$$

where $\sigma_n(\alpha) = -(\langle S_n^{-1}\psi_i, \psi_j\rangle)^{-1}\alpha_i\alpha_j$, $\rho_n(\alpha) = -(\langle R_n^{-1}\psi_i, \psi_j\rangle)^{-1}\alpha_i\alpha_j$. Now in view of Theorem 2.7 I have the following result.

Theorem 3.2: *Let $\varphi_n(|x|)$ be the steady-state solution of the equation*

$$iu_t - \Delta u + u - |u|^p u = 0. \tag{3.8}$$

Then for n sufficiently large the operator A_n in (3.3) coming from the linearization around $\varphi_n(|x|)$ has either unstable eigenvalues or is structurally unstable.

The above result holds true for any nonlinearity $g(s)$ as long as the assumptions of Theorem 3.1 and 3.7 are satisfied. it is known that for equation (3.8) the steady states are unstable if $p \in \left(\frac{4}{N}, \frac{4}{N-2}\right)$ see (3) while the question of stability is open for $p \in (0, \frac{4}{N})$ and $n \geq 1$ (for $n = 0$ i.e., if $\varphi_0(x)$ is a positive solution of (3.2) and $p \in (0, 4/N)$ then the solution is stable, see (2)). Theorem 3.2 states that for $p \in (0, 4/N)$ a steady-state solution of (3.8) with large enough number of nodes has a linearlization that has either unstable eigenvalues or is structurally unstable. In view of Theorem 3.1 one would like to show that the steady-states $\varphi_n(x)$ for n large are structurally unstable with respect to the nonlinear term in equation (3.1). Assume that $\varphi_{0,n}(x)$ is a solution of 3.2 for some nonlinearity $g_0(s)$, where n is the number of notes and such that the linearization $\mathcal{A}_{0,n}$ has a pair of eigenvalues $\pm i\lambda_0$ embedded in the continuous spectrum of $\mathcal{A}_{0,n}$ (this is possible in view of Theorem 3.2). One should be able to prove that for every $\varepsilon > 0$ there exists nonlinearity $g_\varepsilon(s)$ such that $\sup_s |g_\varepsilon(s) - g_0(s)| < \varepsilon$ and the solution $\varphi_{\varepsilon,n}(x)$ corresponding to $g_\varepsilon(s)$ gives a linearization $\mathcal{A}_{\varepsilon,n}$ that has a quadruplet of complex eigenvalues near $\pm i\lambda_0$. In order to prove such a statement one can use a smooth path of nonlinearities $g(t,s)$ such that $g(0,s) = g_0(s)$ and then use (2.55) where $d(x)$ is a given function and $\widetilde{W}(x)$, $\widetilde{V}(x)$ in (2.55) can be computed by differentiating with respect to t.

(2) Klein-Gordon Equation

Consider the equation

$$u_{tt} - \Delta u + u - g(|u|^2)u = 0 \tag{3.9}$$

where u: $\mathbf{R} \times \mathbf{R}^N \to \mathbf{C}$. A solution of the form $e^{iwt}\varphi(x)$ satisfies the equation

$$-\Delta\varphi + (1 - w^2)\varphi - g(\varphi^2)\varphi = 0 \tag{3.10}$$

Equation (3.10) is the same as equation (3.2 and Theorem 3.1 about existence of solutions applies in this situation. The linearization around a state $e^{iwt}\varphi_n(x)$ gives the operator

$$\mathcal{A}_n = \begin{pmatrix} 0 & J_2 B_n \\ J_2 A_n & 0 \end{pmatrix}$$

where

$$A_n = \begin{pmatrix} -\Delta + 1 - g(\varphi_n^2) - 2g'(\varphi_n^2)\varphi_n^2 & -w \\ -w & 1 \end{pmatrix}$$

$$B_n = \begin{pmatrix} -\Delta + 1 - g(\varphi_n^2) & w \\ w & 1 \end{pmatrix}$$

$$J_2 = \begin{pmatrix} 0 & 1 \\ -1 & 0 \end{pmatrix}.$$

Let $A_{0n} = -\Delta+1-g(\varphi_n^2)-2g'(\varphi_n^2)\varphi_n^2$; $B_{0n} = -\Delta+1-g(\varphi_n^2)$ and E a negative eigenvalue of either A_n or B_n then

$$(C - w^2)p = \left(1 + \frac{w^2}{1-E}\right) Ep \tag{3.11}$$

where$C = A_{0n}$ or B_{0n}. Hence negative eigenvalues of A_n or B_n correspond to negative eigenvalues of the operators

$$\begin{aligned} &-\Delta + 1 - w^2 - g(\varphi_n^2) \\ &-\Delta + 1 - w^2 - g(\varphi_n^2) - 2g'(\varphi_n^2)\varphi_n^2 \end{aligned} \tag{3.12}$$

Consider the special case where $g(s^2) = |s|^p$ for $p \in \left(0, \frac{4}{N-2}\right)$ then as in the previous example the negative eigenvalues of (3.12) tend to infinity as $n \to \infty$ hence because of (3.11) the negative eigenvalues of A_n and B_n tend to infinity as $n \to \infty$, also it is easy to see that $B_n < A_n$. A theorem similar to Theorem 3.2 holds.

Theorem 3.3: *Let $g(s^2) = |s|^p$ and $\varphi_n(|x|)$ a solution of (3.10) with n nodes. For n large enough the linearized operator A_n has either unstable eiganvalues or an eigenvalue of negative sign embedded in the continuous spectrum of A_n.*

The remarks made in the previous example hold for this case also

(3) Nonlinear Schrödinger Equation with Potential

The last example serves to illustrate that in some cases the linearlized operator has all its spectrum on the imaginary axis (i.e. the linearized equation is stable) while the energy at the critical point is truly a saddle point, i.e. even after considering the obvious conserved quantities the energy is still a saddle point on the hypersurface defined by the conserved quantitites.

Consider the equation

$$iu_t - \Delta u + V(x)u - \lambda|u|^p u = 0 \quad p \in (0, 4/N-2) \tag{3.13}$$

If $\lambda = 0$ the above equation reduces to the linear equation

$$iu_t - \Delta u + V(x)u = 0$$

Assume that $V(x)$ is potential that is decreasing fast enough at infinity, for example exponentially fast, and let $w \in \mathbf{R}^+$ such that

$$-\Delta\varphi_0 + (V(x) + w)\varphi_0 = 0 \tag{3.14}$$

For some function $\varphi_0(x)$ normalized so that $\|\varphi_0\|_{L^2} = 1$, hence the linear equation has a periodic solution of the form $e^{-iwt}\varphi_0(x)$. For $\lambda \in \mathbf{R}^+$ small the following existence theorem holds, see (8).

Theorem 3.4: *Let w be an eigenvalue of equation (3.14) of multiplicity m. There exists $\lambda_0 > 0$ such that if $0 < \lambda < \lambda_0$ the nonlinear equation*

$$-\Delta\varphi_\lambda + (V(x) - w_\lambda)\varphi_\lambda - \lambda|\varphi_\lambda|^p\varphi_\lambda = 0$$

has m solutions, where $w_\lambda \to w$ as $\lambda \to 0$ and $\|\varphi_\lambda - \varphi_0\|_{L^2} \to 0$ as $\lambda \to 0$.

The proof of the above theorem uses bifurcation techniques, see (8). If $\lambda = 0$ the linearization around $e^{iwt}\varphi_0$ gives the operator

$$\mathsf{A}_0 = \begin{pmatrix} 0 & A_0 \\ -A_0 & 0 \end{pmatrix}$$

where $A_0 = -\Delta + V(x) + w$. The operator A_0 has all the spectrum on the imaginary axis. Now the linearization around the solution $e^{iw_\lambda t}\varphi_\lambda$ gives

$$\mathsf{A}_\lambda = \begin{pmatrix} 0 & B_\lambda \\ -A_\lambda & 0 \end{pmatrix}$$

where

$$A_\lambda = -\Delta + V(x) + w_\lambda - (p+1)\lambda|\varphi_\lambda|^p$$

$$B_\lambda = -\Delta + V(x) + w_\lambda - \lambda|\varphi_\lambda|^p$$

The operator A_λ is a perturbation of A_0 therefore the question arises under what conditions A_0 is structurally stable. let$-E_0 < -E_1 \leq -E_2 \leq \cdots \leq -E_n < 0$ be the negative eigenvalues of A_0 and $0 < E_{n+1} \leq E_{n+2} \leq \cdots \leq E_{n+m} < w$ the positive eigenvalues of A_0, notice that $E = 0$ is also an eigenvalue.

Theorem 3.5: $\mathcal{A}_0$ *is structurally stable iff the following conditions hold*

(a) $E_i \neq E_j$ *for* $i = 0, 1, \ldots, n$ *and* $j = n+1, \ldots, n+m$.

(b) $E_0 < w$.

The proof follows from Theorem 2.6, condition (a) guarantees that eigenvalues of opposite signs do not collide while (b) means that there is no eigenvalue of negative sign embedded in the continuous spectrum of $\mathcal{A}_0$.

Proposition 3.6: *If the potential* $V(x)$ *is such that the conditions of Thoerem 3.5 are satisfied then the spectrum of* $\mathcal{A}_\lambda$ *for* λ *small is on the imaginary axis.*

From the above proposition one can construct examples of critical points of the non-linear equation such that the linearized equation is stable, the stability of such solutions is an open question.

Appendix

Theorem (A): *Consider an operator of the form*

$$\mathsf{A} = J\mathsf{H}$$

where $J = \begin{pmatrix} 0 & 1 \\ -1 & 0 \end{pmatrix}$, $\mathsf{H} = \begin{pmatrix} A_1 & 0 \\ 0 & A_2 \end{pmatrix}$, $A_i = -\Delta + m + V_i(x)$ $i = 1, 2$ *and* $|V_i(x)| \leq \mathsf{C}e^{-a|x|}$ $x \in \mathsf{R}^N$ *for some* C, a, m *positive constants. Then* $\sigma_{\text{ess}}(\mathsf{A}) = \sigma_{ac}(\mathsf{A}) = \pm i\sigma(-\Delta + m) = \{i\lambda : |\lambda| > m\}$.

Proof: Let $H = -\Delta + m$, $\mathsf{H}_0 = \begin{pmatrix} H & 0 \\ 0 & H \end{pmatrix}$, using Weyl's essential spectrum theorem it is easy to show that $\sigma_{\text{ess}}(\mathsf{A}) = \sigma_{\text{ess}}(J\mathsf{H}_0) = \pm i\sigma(-\Delta + m)$. Notice that $iz \in \sigma(\mathsf{A}) \Leftrightarrow \mathsf{H} + izJ$ is not invertile. Let $V = \begin{pmatrix} V_1 & 0 \\ 0 & V_1 \end{pmatrix}$ and

$$|V|^{1/2} = \begin{pmatrix} |V_1|^{1/2} & 0 \\ 0 & |V_2|^{1/2} \end{pmatrix}, \quad V^{1/2} = \begin{pmatrix} |V_1|^{1/2} sgnV_1 & 0 \\ 0 & |V_2|^{1/2} sgnV_2 \end{pmatrix}$$

then the following formula holds

$$(\mathsf{H} + izJ)^{-1} = (\mathsf{H}_0 + izJ)^{-1} - [(\mathsf{H}_0 + izJ)^{-1} V^{1/2}] \times$$
$$\times [1 + |V|^{1/2} (\mathsf{H}_0 + izJ)^{-1} V^{1/2}][|V|^{1/2} (\mathsf{H}_0 + izJ)^{-1}]$$

and

$$(\mathsf{H}_0 + izJ)^{-1} = (\mathsf{H} - izJ)\mathrm{diag}((H^2 - z^2)^{-1})$$
$$\text{where } \mathrm{diag}((H^2 - z^2)^{-1}) = \begin{pmatrix} (H^2 - z^2)^{-1} & 0 \\ 0 & (H^2 - z^2)^{-1} \end{pmatrix}$$

so that

$$|V|^{1/2} (\mathsf{H}_0 + izJ)^{-1} V^{1/2} = \begin{pmatrix} |V_1|^{1/2} H (H^2 - z^2)^{-1} V_1^{1/2}, & -iz|V_1|^{1/2} (H^2 - z^2)^{-1} V_2^{1/2} \\ iz|V_2|^{1/2} (H^2 - z^2)^{-1} V_1^{1/2}, & |V_2|^{1/2} H (H^2 - z^2)^{-1} V_2^{1/2} \end{pmatrix}$$

$$= \begin{pmatrix} |V_1|^{1/2} (H + z)^{-1} V_1^{1/2} & 0 \\ 0 & |V_2|^{1/2} (H + z)^{-1} V_2^{1/2} \end{pmatrix}$$

$$+ z \begin{pmatrix} |V_1|^{1/2} (H^2 - z^2)^{-1} V_1^{1/2}, & |V_1|^{1/2} (H^2 - z^2) V_2^{1/2} \\ |V_2|^{1/2} (H^2 - z^2)^{-1} V_1^{1/2}, & |V_2|^{1/2} (H^2 - z^2)^{-1} V_2^{1/2} \end{pmatrix}$$

Now $(H^2 - z^2)^{-1} = (H - z)^{-1}(H + z)^{-1} = (-\Delta + m - z)^{-1}(-\Delta + m + z)^{-1}$ and $(-\Delta + m \mp z)^{-1}$ can be identified with an integral operator $G_{\pm}(z; x, y)$, for example if $N = 3$ then

$$G_{\pm}(z;x,y) = \frac{\exp(i\sqrt{\pm z - m}|x-y|)}{4\pi|x-y|}$$

where $\sqrt{\pm z - m}$ is taken the one with positive imaginary part, hence $(H^2 - z^2)^{-1} \cong G_+ * G_- \cong G(z; x, y)$ where $G(z; x, y)$ is the convolution of G_+ and G_-. Because of the assumption on $V_i(x)$ $i = 1, 2$ $(1 + |V|^{1/2}(\mathsf{H}_0 + izJ)^{-1}V^{1/2})^{-1}$ exists for all $z \in \mathbf{C} - [m, +\infty)$ and has continuous boundary values as $z \to \lambda + i0$ as long as λ avoids a finite set $\mathcal{E}$ of numbers. Let $[a, b]$ be disjoint from $\mathcal{E}$ and $D = \{\vec{p} = (p_1(x), p_2(x)) \mid p_i(x) \in C_c^{\infty}(\mathbf{R}^N)\}$ the set D is dense in $L^2(\mathbf{R}^N) \times L^2(\mathbf{R}^N)$ and it is easy to show, using the previous observations, that $|\langle \vec{p}, (\mathsf{A} - z)^{-1}\vec{p}\rangle|$ is uniformly bounded on $\{\lambda + i\varepsilon \mid 0 < \varepsilon < 1;\ \lambda \in [a, b]\}$, this in turn implies using a criterion from (5) pg. 138 that the spectrum of A consists only of absolutely continuous and discrete parts. □

Remark: Notice that the critical element in the above proof is the fact that $|V_i|^{1/2}(H^2 - z^2)^{-1}V_j^{1/2}$ is an analytic family of compact operators for $z \in \mathbf{C} - ([m, +\infty) \cup [-m, -\infty))$ with continuous boundary values as $z \to \lambda + i0$, $|\lambda| > m$.

Theorem (B): *Let $\mathsf{A} = J\mathsf{H}$ where $J = \begin{pmatrix} 0 & J_2 \\ J_2 & 0 \end{pmatrix}$, $J_2 = \begin{pmatrix} 0 & 1 \\ -1 & 0 \end{pmatrix}$ and*

$$\mathsf{H} = \begin{pmatrix} A_1 & 0 \\ 0 & A_2 \end{pmatrix},$$

$$A_1 = \begin{pmatrix} -\Delta + m^2 + V_1(x), & -w \\ -w & ,\ 1 \end{pmatrix}$$

and

$$A_2 = \begin{pmatrix} -\Delta + m^2 + V_2(x), & w \\ w, & 1 \end{pmatrix}$$

Assume that $|V_i(x)| \le \mathrm{C}e^{-a|x|}$ $x \in \mathbf{R}^N$ for some C, a positive constants then $\sigma_{\mathrm{ess}}(\mathsf{A}) = \sigma_{\mathrm{ac}}(\mathsf{A}) = \pm i\sigma(-\Delta + m^2 - w^2)$.

The proof of Theorem (B) proceeds exactly as in Theorem (A) as long as one observes that

$$\begin{pmatrix} -\Delta + m^2, & w \\ w, & 1 \end{pmatrix}^{-1} = \begin{pmatrix} 1, & -w \\ -w, & -\Delta + m^2 \end{pmatrix} \operatorname{diag}(-\Delta + m^2 - w^2)$$

Finally, using the methods from (5) ch. XIII 6,7, it is easy to show the following

Theorem (C): *Let* $R = -\Delta + m + V(x)$ *where* $|V(x)| \leq \mathsf{C}e^{-a|x|}$, $x \in \mathbf{R}^N$ *for some* C, a *positive constants and let* $\psi_i(x)$ $i = 1, \ldots, n$ *a set of functions in* $C_c^\infty(\mathbf{R}^N)$. *Call* $\mathsf{P} = \{p \in L^2(\mathbf{R}^N) \mid p \perp \psi_i \ i = 1, \ldots, n\}$, $\pi =$ *orthogonal projection on* P *and* $R_1 = \pi R$: $\mathsf{P} \to \mathsf{P}$ *then*

$$\sigma_{\mathrm{ac}}(R_1) = \sigma_{\mathrm{ess}}(R_1) = \sigma_{\mathrm{ess}}(R) = \sigma(-\Delta + m).$$

References

1) V.Arnold, *"Mathematical Methods of Classical Mechanics"*, Springr-Verlag, 2nd ed., 1980.

2) M. Grillakis, J. Shatah, W. Strauss, *"Stabiity Theory of Solitary Waves in the Presence of Symmetry, I"*, J. of Fund. Anal., Vol. 74, 1987, pp. 160–197.

3) M.Grillakis, *"Linearized Instability for Nonlinear Schrödinger and Krein-Gordon Equations"*, to appear in CPAM.

4) C. Jones, *"An Instability Mechanism for RadiallySymmetric Standing Waves of a Nonlinear Schrödinger Equation"*, to appear in JDE.

5) M. Reed, B. Simon, *"Methods of Modern Mathematical Physics"*, Vol. IV, Academic Press, 1978.

6) M. Krein, *"Topics in Differential and Integral Equations and Operator Theory"*, OT7 Birkäuser, 1983, pp. 1-98.

7) W. Magnus, S. Winkler, *"Hill's Equation"*, Dover, 1979.

8) A. Weinstein, *"Nonlinear Stabilization of Quasi-Modes"*, Proc. of Symp. in Pure Math. Vol. 36 1980.

9) C. Jones and T. Küpper, *"On the infinitely many Solutions of a Semilinear Elliptic Equation"*, SIAM J. Math. Anal. Vol. 17, No. 4, July 1986.

ON NUMERICAL CHAOS IN THE NONLINEAR SCHRÖDINGER EQUATION

B.M. Herbst† Mark J. Ablowitz
Department of Mathematics and Computer Science
Clarkson University, Potsdam, NY 13676

1 Introduction

During the past twenty to twenty-five years there has been extensive interest in a class of nonlinear evolution equations which admit certain extremely stable solutions, termed *solitons.* Historically the Korteweg-de Vries (KdV) equation,

$$u_t + 6uu_x + u_{xxx} = 0, \tag{1}$$

was the first equation discovered with the solition solution and many other important properties. In 1965 Zabushky and Kruskal [1] discovered the soliton property of the KdV equation from numerical simulations. Shortly thereafter in 1967 Gardner, Greene, Kruskal and Muira [2] found that the Cauchy problem for the KdV equation corresponding to the initial values $u(x,0) = f(x)$, vanishing sufficiently rapidly as $|x| \to \infty$ could be linearized by employing methods of direct and inverse scattering. Indeed, Lax [3] showed that there was a rather general formulation by which the KdV equation could be viewed as a compatibility condition between two linear operators. He also investigated in considerable detail the interaction properties of the KdV solitons.

Subsequent studies have established the wide ranging significance and occurrence of solition solutions. In 1972 Zakharov and Shabat [4] found that the physically significant cubic nonlinear Shrödinger equation,

$$iu_t + u_{xx} + 2u^2u^* = 0, \tag{2}$$

(here $i^2 = -1$ and u^* is the complex conjugate of u), admitted solition solutions and that the Cauchy problem (for decaying data on $|x| \to \infty$) could be linearized by inverse scattering methods. In fact, Ablowitz, Kaup, Newell and Segur [5] demonstrated that these ideas applied to a class of nonlinear evolution equations, including the physically interesting sine-Gordon,

$$u_{xt} = \sin u \tag{3}$$

and modified KdV (mKdV) equation,

$$u_t + 6u^2 u_x + u_{xxx} = 0. \tag{4}$$

They termed the method of solution the Inverse Scattering Transform (IST) in analogy with the linear technique of Fourier Transforms.

Not only does the method apply to nonlinear partial differential equations, but it also applies to discrete equations. Examples include (a) the Toda lattice [6]

$$u_{ntt} = e^{-(u_n - u_{n-1})} - e^{-(u_{n+1} - u_n)} \tag{5}$$

which in a suitable continuous limit tends to the KdV equation, and (b) a differential-difference nonlinear Schrödinger equation [7]

$$iu_{nt} + (u_{n+1} + u_{n-1} - 2u_n) + u_n u_n^*(u_{n+1} + u_{n-1}) = 0, \tag{6}$$

which tends to the cubic NLS in the continuous limit. There are nonlinear partial difference equations (discrete in space and time) which are in the IST class as well [8]. There is an extensive literature on this subject and we refer the reader to the monograph [9] for a review of some of the work in this field.

It should also be mentioned that these ideas can be suitably extended in order to study certain nonlinear singular integro differential equations such as the Benjamin-Ono equation,

$$u_t + 6uu_x + H(u_{xx}) = 0, \tag{7}$$

where Hu is the Hilbert transform of u:

$$Hu(x) = \frac{1}{\pi}\mathcal{P}\int_{-\infty}^{\infty} \frac{u(\xi)}{\xi - x} d\xi, \tag{8}$$

and $\mathcal{P}$ denotes the Cauchy principal value, and various multidimensional equations, e.g. the Kadomtsev-Petviashvili equation (a 2+1 dimensional generalization of KdV)

$$(u_t + 6uu_x + u_{xxx})_x = -3\sigma^2 u_{yy}, \tag{9}$$

and the so-called Davey-Stewartson equations (a 2+1 dimensional generalization of the NLS equation),

$$\begin{aligned} iu_t + \sigma^2 u_{xx} + u_{yy} &= (\phi - |u|^2)u \\ \phi_{xx} - \sigma^2 \phi_{yy} &= \pm 2(|u|^2)_{xx}. \end{aligned} \tag{10}$$

where $\sigma^2 = \pm 1$ in both cases.

The above equations are not only of physical and mathematical interest but they also provide a fertile testing ground for analytical and numerical studies of basic nonlinear phenomena. In this paper we shall discuss certain special solutions of the NLS equation and some of its discretizations. We shall demonstrate that depending on the approximation method chosen and at intermediate levels of discretization, we find that the phenomenon of numerically induced chaos occurs. The chaos disappears with a fine enough discretization and convergence to a quasi-periodic solution is eventually obtained. There is no chaos associated with the numerical solution of (6) – an integrable version of the NLS equation. However, chaos is found for the nonintegrable difference schemes as well as Fourier spectral methods. This serves to reinforce one's intuitive feeling that a numerical scheme which represents the essential analytical features of the continuous equation is very desirable and that even higher order asymptotic accuracy (e.g. exponential accuracy for Fourier spectral schemes) may not give good approximations at intermediate levels of discretization and levels that one might intuitively think would be adequate for qualitative approximations.

2 The continuous NLS

It is convenient for our purposes to study the NLS equation in the form given by,

$$iu_t + u_{xx} + Q|u|^2 u = 0 \tag{11}$$

with periodic boundary conditions, $u(x+L,t) = u(x,t)$, and initial condition, $u(x,0) = f(x)$, Q is a real parameter and $i^2 = -1$. It is straightforward to check that this equation has a solution, $u(x,t) = a\exp(ikx - i\omega t)$, provided ω satisfies the dispersion relation, $\omega = k^2 - Q|a|^2$. It is well-known that this solution is unstable under small perturbations, see, for example, [10]. Let

$$u(x,t) = a\exp(ikx - i\omega t)(1 + \epsilon(x,t)) \tag{12}$$

where $|\epsilon| << 1$ and assume that ϵ is given by,

$$\epsilon(x,t) = \sum_{n=-\infty}^{\infty} \hat{\epsilon}_n(t)\exp(i\mu_n x), \tag{13}$$

where $\mu_n = 2\pi n/L$. If we now substitute (12) into (11), keep linear terms in ϵ and solve the resulting equations for $\hat{\epsilon}_n$, we find that

$$\hat{\epsilon}_n \propto \exp\left(-2ik\mu_n \pm \mu_n\sqrt{2Q|a|^2 - \mu_n^2}\right)t. \tag{14}$$

It follows that al modes, n, in the Fourier expansion of the perturbation are unstable for which

$$0 < \mu_n^2 < 2Q|a|^2. \tag{15}$$

Note that the number of unstable modes increase with increasing values of $Q|a|^2$. In practice we find that an increasing number of unstable modes tend to become more difficult to resolve numerically since a finer resolution is required to capture the spatial structure. In terms of the studies of so-called inertial manifolds, an increasing number of unstable modes increases the dimension of the manifold essentially requiring a higher dimensional approximation space [11].

It has been demonstrated experimentally and numerically [12] that the unstable modes will exhibit the phenomenon known as recurrence, where the initial configuration is periodically recovered. The unstable modes grow exponentially according to the linear analysis before the nonlinearity inhibits further growth; the unstable modes taking turns in dominating the spatial structure – more unstable modes leading to more complex behavior, before the initial configuration is recovered.

3 The numerical schemes

The discretizations of the NLS equation we shall discuss here are of finite difference and spectral type. The difference schemes are given by (periodic boundary conditions are used throughout; in this case given by $U_{j+N} = U_j$),

$$i\dot{U}_j + (U_{j-1} - 2U_j + U_{j+1})/h^2 + Q|U_j|^2 U_j^{(k)} = 0, k = 0, 1, \tag{16}$$

where

$$\text{(a) } U_j^{(1)} = U_j, \text{ (b) } U_j^{(2)} = \tfrac{1}{2}(U_{j-1} + U_{j+1}).$$

Note that scheme (b) is simply the integrable scheme (6) mentioned earlier. Both schemes are of second order accuracy and Hamiltonian. In case (16a) there are two constants of the motion, the L^2 norm,

$$I = \sum_{j=0}^{N-1} |U_j|^2,$$

and the Hamiltonian

$$H = -i \sum_{j=0}^{N-1} \left(|U_{j+1} - U_j|^2/h^2 - \tfrac{1}{2}Q|U_j|^4\right). \tag{17}$$

Hence, when $N = 2$ the system is integrable. The Poisson brackets are the standard ones.

The Hamiltonian structure of (16b) is given (for $h = 1$) by the Hamiltonian [13],

$$H = -i \sum_{j=0}^{N-1} \left[U_j^*(U_{j-1} + U_{j+1}) - 4Q^{-1} \ln(1 + \tfrac{1}{2}QU_jU_j^*)\right] \tag{18}$$

together with the non-standard Poisson brackets, $\{q_m, p_n\} = (1 + \frac{1}{2}Qq_np_n)\delta_{m,n}$ and $\{q_m, q_n\} = 0 = \{p_m, p_n\}$. This system has been demonstrated to be solvable by IST and there is an infinite number of conserved quantities [7]. As mentioned earlier, general solutions can be obtained on the infinite line (see [7, 9]); theoretical solutions may also be obtained on the periodic interval [14]. In fact, there are further partial difference analogues to (16b) which also have special properties and can be used as effective numerical schemes [8]. The markedly different behavior obtained from these two schemes has already been described in detail [15], allowing us to concentrate here on the differences between the integrable difference scheme, (16b) and trigonometric spectral schemes.

Assuming a uniform grid as before, the discrete Fourier transform of $U_j, j = -\frac{1}{2}N, \ldots, \frac{1}{2}N - 1$ and $U_{j+N} = U_j$, denoted by $\hat{U}_j$, is given by,

$$\hat{U}_n = \mathcal{F}_n\{\mathbf{U}\} = \frac{h}{L} \sum_{j=-\frac{1}{2}N}^{\frac{1}{2}N-1} U_j \exp(-i\mu_n x_j)$$

the inverse transform is given by

$$U_j = \mathcal{F}_j^{-1}\{\hat{\mathbf{U}}\} = \frac{L}{h}\mathcal{F}_j^*\{\hat{\mathbf{U}}^*\}.$$

and the pseudo-spectral scheme becomes

$$\dot{U}_j = -i\mathcal{F}_j^{-1}\{\mu_n^2 \mathcal{F}_n\{\mathbf{U}\}\} + iQ|U_j|^2 U_j, \tag{19}$$

or equivalently

$$\dot{\hat{U}}_n = -i\mu_n^2 \mathcal{F}_n\{\mathbf{U}\} + iQ\mathcal{F}_n\{|\mathbf{U}|^2\mathbf{U}\}. \tag{20}$$

The Hamiltonian is given by

$$H(q,p) = -i\left[\frac{L}{h} \sum_{n=-\frac{1}{2}N}^{\frac{1}{2}N-1} \mu_n^2 \hat{q}_n \hat{p}_n - \tfrac{1}{2}Q \sum_{j=-\frac{1}{2}N}^{\frac{1}{2}N-1} q_j^2 p_j^2\right],$$

where a $q_j = U_j$ and $p_j = U_j^*$ and $\hat{q}_n$ denotes the Fourier transform of q_j, etc. (Note that here we use the convention that n labels the Fourier coefficients and j the discrete function values.) The Poisson brackets are standard. A second constant of integration is given by

$$I = \sum_{j=-\frac{1}{2}N}^{\frac{1}{2}N-1} |U_j|^2$$

and it follows that the system is integrable for $N = 2$.

Both the standard finite difference, (16a), and the pseudo-spectral schemes suffer from aliasing errors due to the inherent inability of the grid to resolve waves with wavenumbers larger than $\frac{1}{2}N$. Higher wave numbers are continuously generated through the nonlinear interactions and are added onto the lower wavenumbers that can be resolved by the grid. This is easily seen by transforming the nonlinear term, $|U_j|^2 U_j$, to Fourier space,

$$\begin{aligned}\mathcal{F}_n\{|\mathbf{U}|^2\mathbf{U}\} &= \sum_{k,\ell,m} \hat{U}_k \hat{U}_\ell \hat{U}_m^* \delta_{k+\ell-m,n \bmod N} \\ &= \sum_{k+\ell-m=n} \hat{U}_k \hat{U}_\ell \hat{U}_m^* + \sum_{k+\ell-m=n\pm N} \hat{U}_k \hat{U}_\ell \hat{U}_m^* \\ & n = -\tfrac{1}{2}N, \ldots \tfrac{1}{2}N-1.\end{aligned}$$

The second term on the right-hand side represents the aliasing error. Waves with wavenumbers $n \pm N$ go under the alias of waves with wavenumber n. The aliasing errors are of particular significance for this problem. Since the low wave numbers to which they are added grow exponentially, the aliasing errors too, are exponentially amplified. There are standard techniques for removing the aliasing error from the pseudo-spectral scheme [16, 17]. In fact, removal of the aliasing error amounts to a full spectral method and it is therefore not surprising that the full spectral method performs much better on these problems than the pseudo-spectral method.

The spectral method derives from a Galerkin approximation of the partial differential equation. Writing the approximate solution as

$$U(x,t) = \sum_{n=-\frac{1}{2}N}^{\frac{1}{2}N} \hat{U}_n(t)\phi_n(x),$$

and substituting into the differential equation (11), the coefficients, $\hat{U}_n$, are calculated from

$$i\sum_j \left[\dot{\hat{U}}_j(\phi_j,\phi_n) - \hat{U}_j(\phi_j',\phi_n')\right] + Q(|U|^2U,\phi_n) = 0,$$

where

$$(\phi,\psi) = \int_0^L \phi\psi^* dx.$$

Choosing ϕ_n to be the trigonometric functions, $\phi_n(x) = \exp(i\mu_n x)$, we arrive at,

$$\dot{\hat{U}}_n = -i\mu_n^2\hat{U}_n + iQ \sum_{p+q-r=n} \hat{U}_p\hat{U}_q\hat{U}_r^*. \tag{21}$$

This is exactly the equation one obtains from removing the aliasing contribution from the nonlinear term in (20). In fact, the way we implement the full spectral method in practice, is to solve (20) with a de-aliasing procedure to get rid of the aliasing.

Again this is a Hamiltonian system with the Hamiltonian,

$$H(q,p) = -i\sum_n \left[\mu_n^2 q_n p_n - \tfrac{1}{2}Q \sum_{k,\ell,m} q_k q_\ell p_m p_n \delta_{k+\ell-m,n}\right]$$

and the standard Poisson brackets. A second constant of integration is given by [18]

$$I = \sum_n |\hat{U}_n|^2$$

implying the system (21) is integrable for $N = 2$. Although the removal of the aliasing contributions leads to a substantial improvement in the performance of the method, it is not sufficient to ensure integrability for $N \geq 4$, as our numerical experiments will indicate.

4 Numerical results

Chaotic solutions are found for the pseudo-spectral method using $Q = 4$ and the initial condition,

$$u(x,0) = 0.5(1 + 0.1\cos(\mu x)),$$

where $\mu = 2\pi/L$ and $L = 2\sqrt{2}\pi$. From (15) follows that the second mode is just on the edge of the instability region for these parameter values. The system of ODE's (19) or (20) are integrated in double precision by the Runge-Kutta-Merson routine, D02BBF, in the NAG software library with the relative tolerance specified as 10^{-6}. The results were verified with tolerances as small as 10^{-10}, to ensure that the results reported here are due to the spatial discretization.

The pseudospectral solution for $N = 16$ is shown in figure 1a, showing the time evolution of the modulus of the solution at $x = 0$. The irregular temporal behavior is evident and is confirmed by its broad-band Fourier spectrum shown in figure 1b. This is in contrast with the solution and Fourier spectrum obtained from the integrable difference scheme, (16b), shown in figures 2a and 2b for $N = 16$ and the same parameter values as in figure 1. The solution is clearly quasi-periodic and there is no sign of chaos.

Irregular temporal behavior is also observed for the full spectral scheme, (21), for $N = 8$ and the parameter values specified as before, albeit somewhat more weakly than the pseudo-spectral scheme. This is not surprising in view of our remarks concerning the role of the aliasing error. Next we choose a somewhat more difficult problem to illustrate the chaotic solutions associated with the full spectral scheme. Accordingly, we choose the problem used by Weideman [19]. The initial condition is given by,

$$u(x,0) = \begin{cases} 0.5(1 + 0.1(1 + \frac{1}{\sqrt{2}\pi}x)) & \text{if} \quad -2\sqrt{2}\pi \leq x \leq 0 \\ 0.5(1 + 0.1(1 - \frac{1}{\sqrt{2}\pi}x)) & \text{if} \quad 0 < x \leq 2\sqrt{2}\pi \end{cases} . \tag{22}$$

According to (15) the first two modes ($n = 1, 2$) fall inside the instability region. The difficulty of the problem is further enhanced by the large number of Fourier modes present in the initial condition. Figures 3a and 3b show the modulus of the spectral solution at $x = 0$ and its Fourier

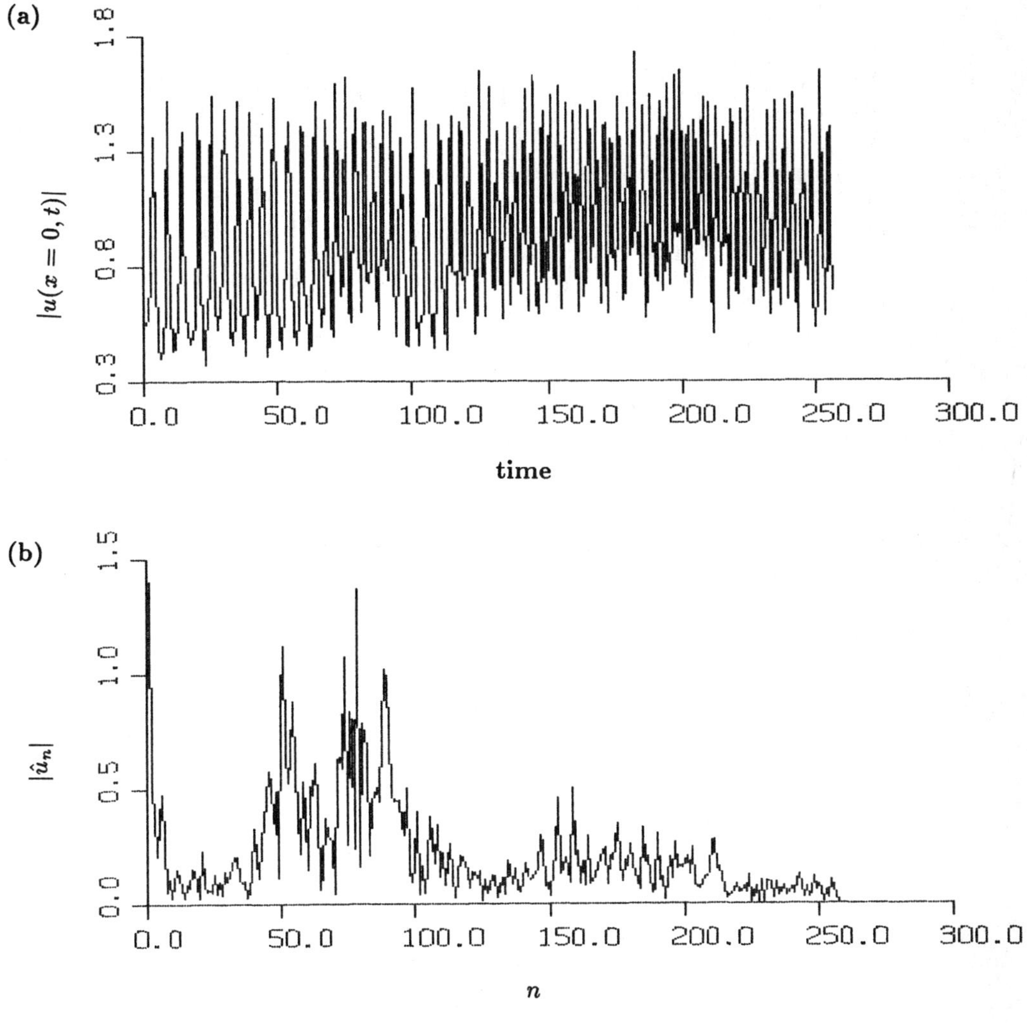

Figure 1: The pseudo-spectral solution, $N = 16$.

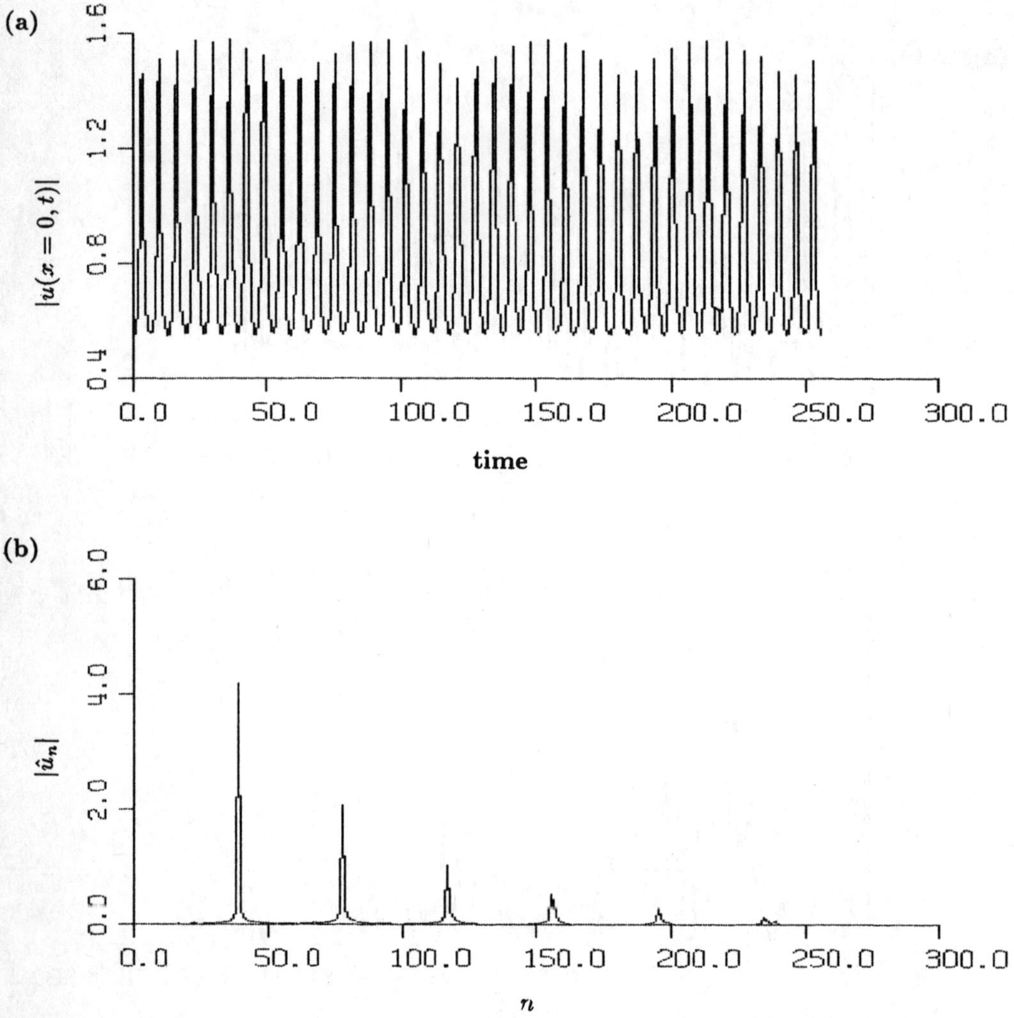

Figure 2: The integrable difference scheme, $N = 16$.

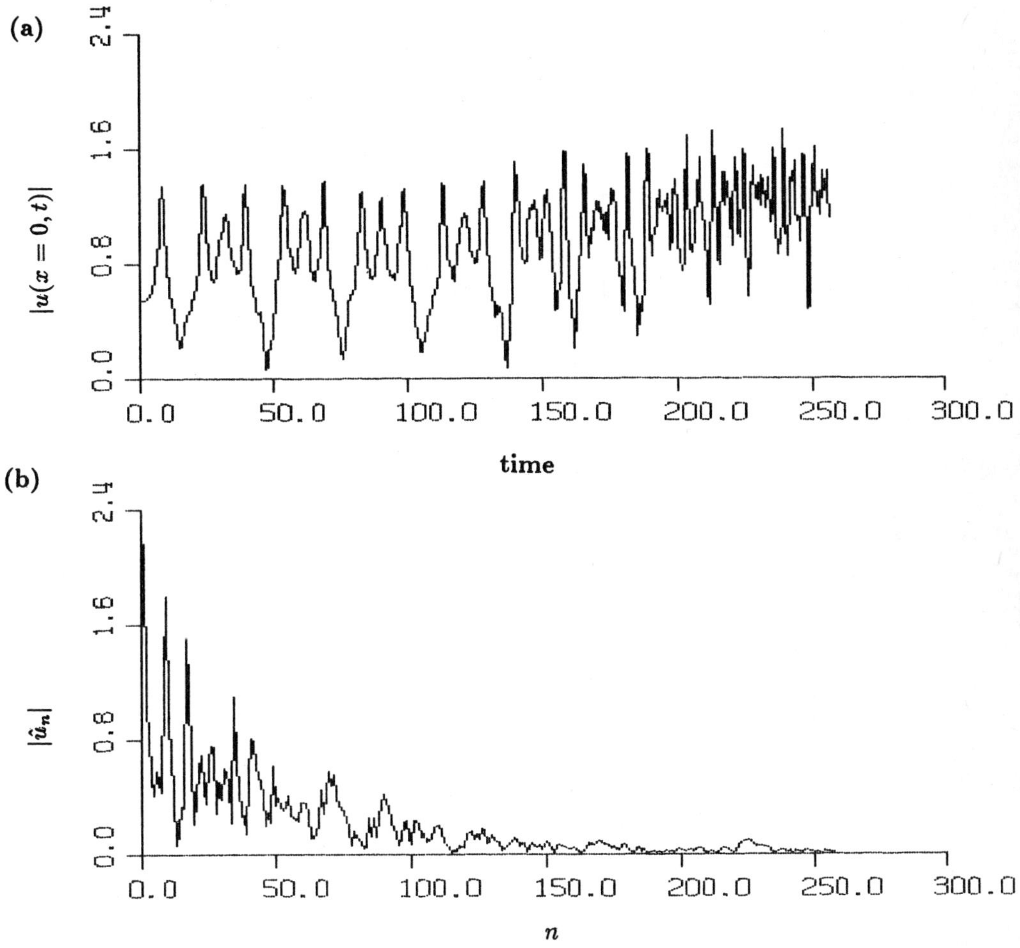

Figure 3: The full spectral solution, $N = 16$.

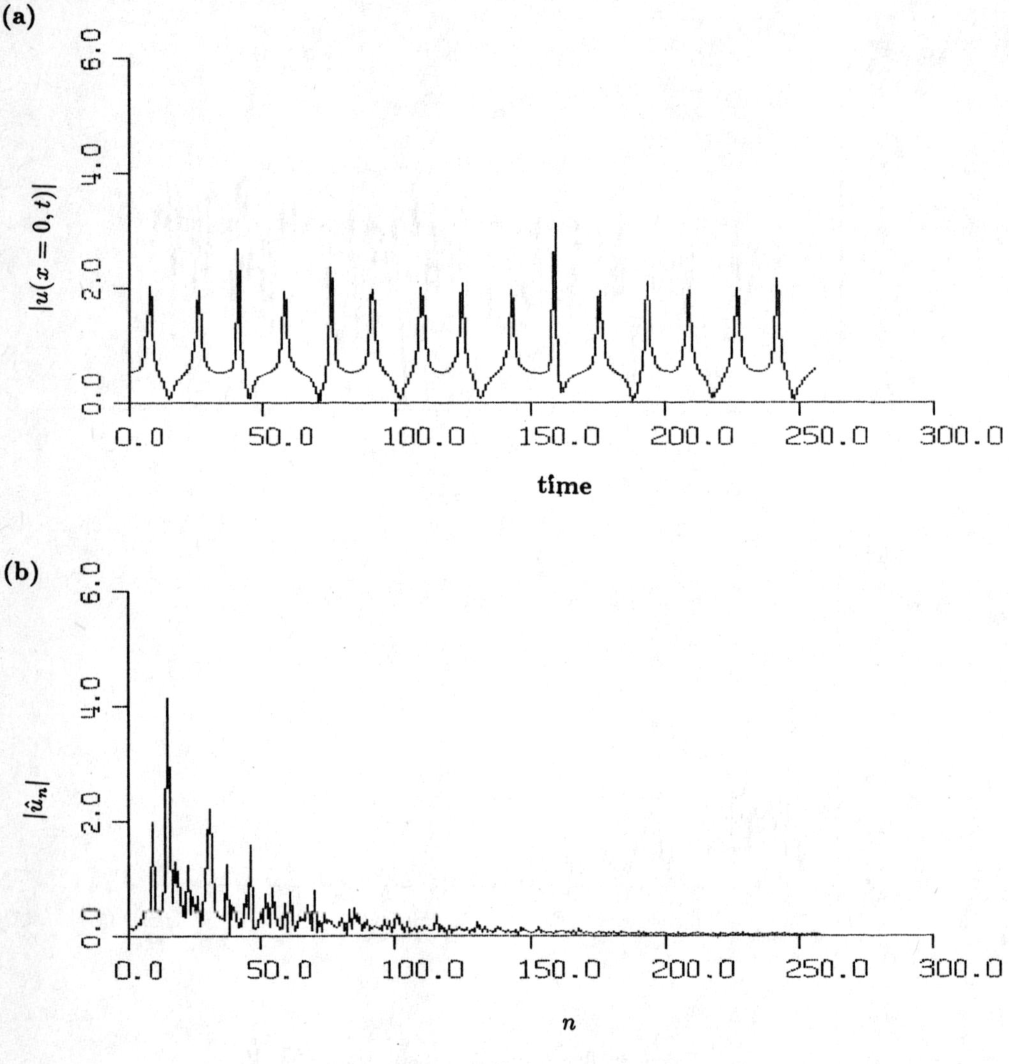

Figure 4: The integrable difference scheme, $N = 16$.

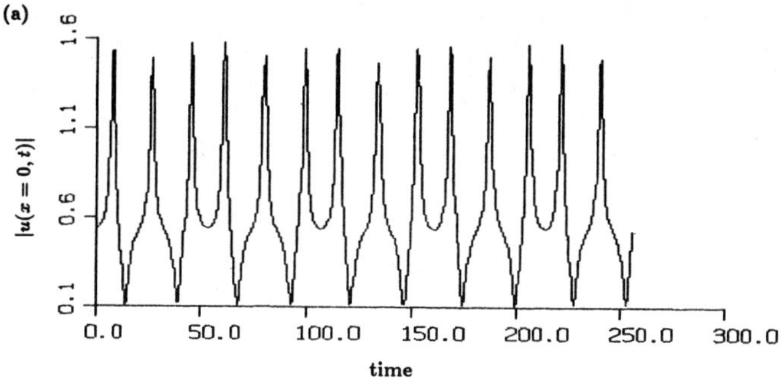

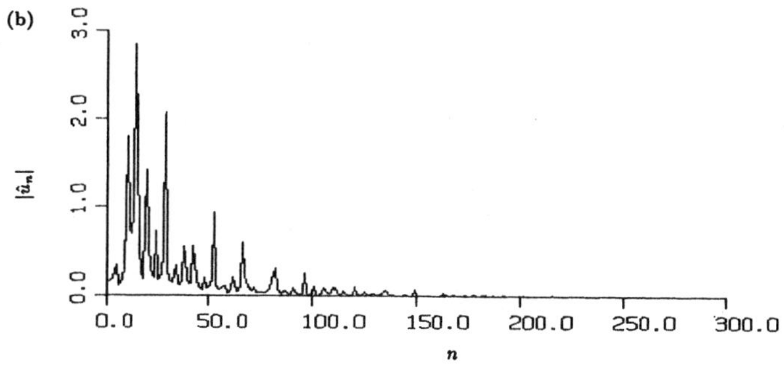

Figure 5: The integrable difference scheme, $N = 50$.

transform for $N = 16$ and figures 4a and 4b the corresponding solutions obtained from the integrable scheme, (16b), also for $N = 16$. Although the solution of the integrable scheme is not very good, it is clearly more regular than the spectral solution and already roughly assumes the qualitative structure of the true solution. The solution of the integrable scheme improves with increasing values of N and for $N = 50$ a satisfactory solution is obtained. This is shown in figure 5. The Fourier spectrum shown in figure 5b shows the large number of frequencies present in the problem. The spectral solution for $N = 50$ is shown in figure 6. Although it is a significant improvement over the solution shown in figure 3, the solution deteriorates from approximately $T = 175$ onwards. Even for $N = 64$ there are indications that the solution deteriorates for $T > 220$.

These results not only show the superiority of the integrable scheme, but it is also significant that the chaos disappears for high values of N, confirming our earlier results on the finite difference methods [15]. This is, of course, related to the convergence properties of the methods; although, it might be surprising to some that the second-order integrable scheme, (16b), performs much better than the exponentially convergent Fourier spectral scheme.

(a)

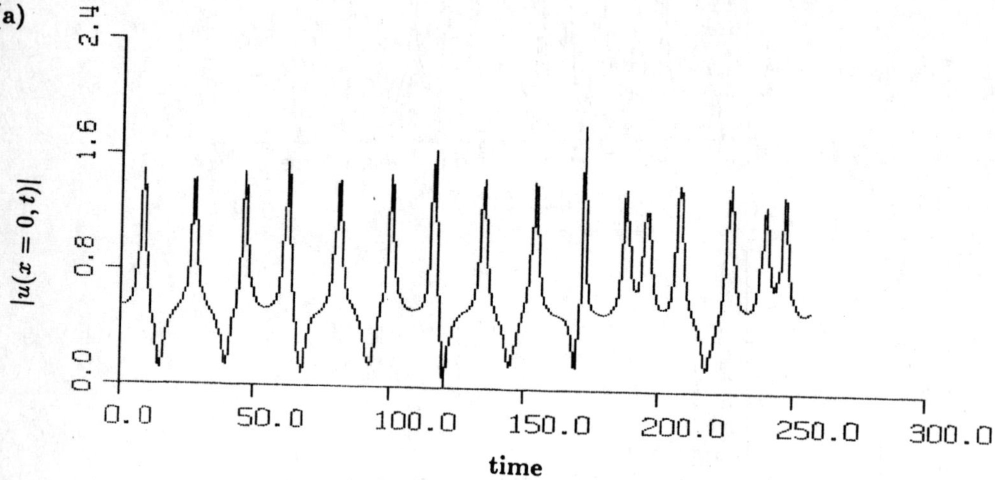

(b)

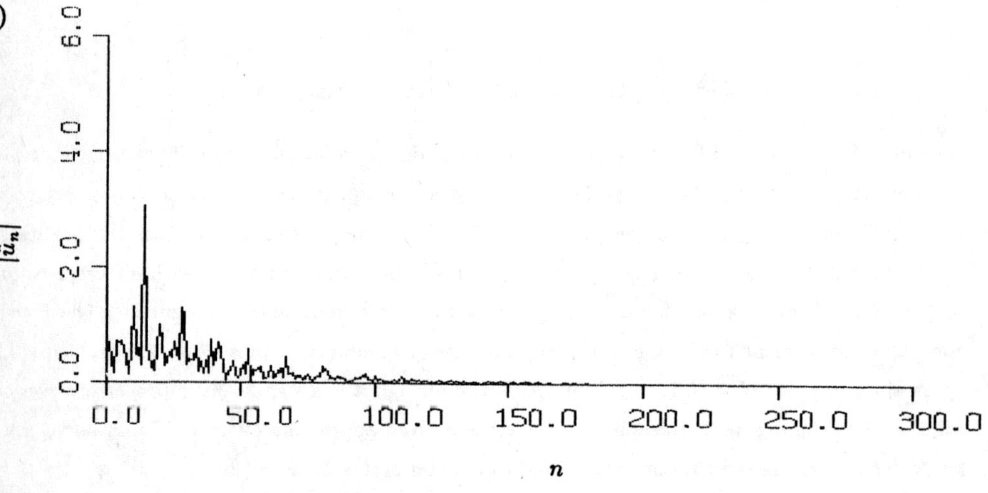

Figure 6: The full spectral solution, $N = 50$.

These results should be considered in the light of the recent work on the finite dimensionality of inertial manifolds, see, for example [11]. If $P_N u$ denote the projection of the solution of (11) onto the approximation space, e.g., Fourier spectral space, then $Q_N u$ with $Q_N = 1 - P_N$ denotes the error in the approximation. If the solutions satisfy the *cone condition*, $\|Q_N a\| < \|P_N a\|$, $a = u - \tilde{u}$ where u and $\tilde{u}$ are any two solutions of (11) for some value of N (N now denotes the dimension of the approximation space), a one-to-one relationship between the solutions of (11) and the finite dimensional approximation space can be established in a straightforward manner (see, for example, [11]). The cone condition has been established for the complex Ginzburg-Landau equation [11] and Fourier spectral approximations but, to our knowledge, not for the NLS equation. In any event, it should be fairly obvious that one can only hope to establish the cone condition, i.e., the one-to-one correspondence between the solution space and the finite dimensional approximation space, for sufficiently high values of N. Having managed that, one is assured that the qualitative behavior of the solution is reflected by the approximate solution. A further increase in the value of N will lead to a better resolution of, e.g., the spatial structure, but the intrinsic qualitative properties of the solution is captured as soon as the cone condition is satisfied. This suggests a notion of nonlinear stability, and is analogous to the situation in numerical analysis where a time step restriction is often required to ensure stability, a necessary condition for convergence. The cone condition provides a similar estimate for the spatial resolution – N needs to be sufficiently large before convergence is observed. Intermediate levels of discretization might well result in spurious numerical chaos as indicated in this paper.

Acknowledgements This work (MJA) is partially supported by the NSF, Grants No. DMS-8507658 and No. DMS-8202117, the Office of Naval Research, Grant No. N00014-76-C-0867 and the Air Force Office of Scientific Research, Grant No. AFOSR-84-0005. One of us (BMH) would like to express his appreciation for the hospitality of Clarkson University and the support of the University of the Orange Free State and colleagues in the department of Applied Mathematics.

†Permanent address: Department of Applied Mathematics, University of the Orange Free State, Bloemfontein 9300, South Africa.

References

[1] N.J. Zabusky and M.D. Kruskal. Phys. Ref. Lett., **15**, p240 (1965).

[2] C.S. Gardner, J.M. Greene, M.D. Kruskal and R.M Muira. Phys. Rev. Lett., **19**, p1095 (1967).

[3] P.D. Lax. Comm. Pure Appl. Math., **21**, p467 (1968).

[4] V.E. Zakharov and A.B. Shabat. Sov. Phys. JETP, **34**, p62 (1972).

[5] M.J. Ablowitz, D.J. Kaup, A.C. Newell and H. Segur. Stud. Appl. Math., **53**, p249 (1974).

[6] H. Flaschka. Phys. Rev. B, **9**, p1924 (1974); H. Flaschka. Prog. Theoret. Phys., **51**, p703 (1974).

[7] M.J. Ablowitz and J.F. Ladik. Stud. Appl. Math., **55**, p213 (1976).

[8] Thiab R. Taha and Mark J. Ablowitz. J. Comput. Phys., **55**, p192 (1984); J. Comput. Phys., **55**, p203 (1984); J. Comput. Phys., **55**, p231 (1984); J. Comput. Phys., **77**, p540 (1988).

[9] M.J. Ablowitz and H. Segur. Solitons and the Inverse Scattering Transform. SIAM, Philadelphia, 1981.

[10] J.T. Stuart and R.C. DiPrima. Proc. R. Soc. London, **A362**, p27 (1978).

[11] Charles R. Doering, John D. Gibbon, Darryl D. Holm and Basil Nicolaenco. Phys. Rev. Lett., **59**, p2911 (1987) and Los Alamos Report No. LA-UR 87-1546 (1987).

[12] H.C. Yuen and B.M. Lake. Phys. Fluids, **18**, p956 (1975); H.C. Yuen and W.E. Ferguson. Phys. Fluids, **21**, p1275 (1978).

[13] P.P. Kulish. Letters in Mathematical Physics, **5**, p191 (1981).

[14] N.N. Bogolyubov and A.K. Prikarpat-skii. Sov. Phys. Dokl., **27**, p113 (1982).

[15] B.M. Herbst and Mark J. Ablowitz. Numerically induced chaos in the nonlinear Schrödinger equation. Preprint.

[16] G.S. Patterson and S.A. Orszag. Phys. Fluids, **14**, p2538 (1971).

[17] Y.Salu and G. Knorr. J.Comp. Phys., **17**, p68 (1975).

[18] J.A.C Weideman and B.M. Herbst. SIAM J. Sci. Stat. Comput., **8**, p988 (1987).

[19] J.A.C. Weideman. Ph.D. Thesis. University of the OFS, Bloemfontein (1986).

SINGULAR SOLUTIONS OF THE CUBIC SCHRÖDINGER EQUATION

M.J. Landman [1], B.J. LeMesurier [2],
G.C. Papanicolaou [1], C. Sulem [3], P.L. Sulem [4]

[1] *Courant Institute of Mathematical Sciences, 251 Mercer Street, New York, NY 10012*
[2] *Department of Mathematics, University of Arizona, Tucson, AZ 85721*
[3] *CNRS, LMENS-CMA, 45 rue d'Ulm, Paris 75230 France and Department of Mathematics, Ben Gurion University of the Negev, 84105 BeerSheva, Israel*
[4] *CNRS, Observatoire de Nice, BP 139, 06003 Nice, France and School of Mathematical Sciences, Tel-Aviv University, 69978 Tel-Aviv, Israel*

Abstract

A review of recent numerical and analytical results on the nature of singular radially symmetric solutions of the Nonlinear Schrödinger Equation is presented. It is shown that in three dimensions, the solutions have locally a self-similar form with a blow-up rate $(t^* - t)^{-1/2}$. In two dimensions, the self-similarity is weakly broken and the blow up rate is found to be $((t^* - t)/\ln\ln\frac{1}{t^*-t})^{-1/2}$.

1. Introduction

The nonlinear Schrödinger equation (NLS)

$$i\partial_t\Psi + \triangle\Psi + |\Psi|^2\Psi = 0 \tag{1.1a}$$

$$\Psi(x,0) = \Psi_0(x) \tag{1.1b}$$

($x \in R^n$) arises in various physical contexts as an envelope equation in nonlinear wave motion. The invariance properties of the equation lead to the existence of the two invariants [10-12] :
the mass

$$M = \int |\Psi|^2 dx \tag{1.2}$$

and the energy

$$H = \int (|\nabla\Psi|^2 - \frac{1}{2}|\Psi|^4) dx. \tag{1.3}$$

In one dimension, the problem is integrable by inverse scattering [16]. In dimension $d \geq 2$, the solution may develop a singularity at some finite time $t = t^*$. Indeed, smooth solutions with initially finite variance $\int |x|^2|\Psi_0|^2 dx$ satisfy the identity [14]:

$$\partial_t^2 \int |x|^2|\Psi|^2 dx = 8H - 2(d-2)\int |\Psi|^4 dx. \tag{1.4}$$

For $d \geq 2$ and $H < 0$, eq.(1.4) implies that the variance vanishes in a finite time $\tilde{t}$. The inequality

$$\int |\Psi|^2 dx \leq \frac{2}{d}\Big(\int |\nabla\Psi|^2 dx\Big)^{1/2} \Big(\int x^2|\Psi|^2 dx\Big)^{1/2} \tag{1.5}$$

implies that the L^2 norm of the gradient of the solution and the maximum of the solution tend to infinity when t tends to t^* [3]. This time needs not to be $\tilde{t}$. Blowup could occur at an earlier time and it generally does. Furthermore, the condition on the initial variance does not seem necessary ; singularities were obtained numerically with data that have infinite variance and also when NLS is considered in a periodic domain [13].

In this paper, we review recent numerical and analytical results on the rate of blow-up for singular radially symmetric solutions in two (critical) and three (supercritical) dimensions obtained in refs. [5-9].

2. The dynamical rescaling

The numerical method is based on a dynamical rescaling of variables that transforms the primitive equation to a similar one that is not singular. This enables us to follow the amplification of the solution up to several orders of magnitude. Since NLS is invariant under the similarity transformation

$$\Psi(x,t) \rightarrow \frac{1}{\lambda}\Psi\Big(\frac{x}{\lambda},\frac{t}{\lambda^2}\Big) \tag{2.1}$$

where λ is a constant, it is of interest to make the change of variables

$$\xi = \frac{|x|}{L(t)}, \qquad \tau = \int_0^t \frac{1}{L(s)^2}ds, \quad , \quad u(\xi,\tau) = L(t)\Psi(x,t). \tag{2.2}$$

Eq.(1.1) then reads :

$$i\partial_\tau u + u_{\xi\xi} + \frac{d-1}{\xi}u_\xi + |u|^2 u + ia(\tau)(\xi u)_\xi = 0 \tag{2.3}$$

$$a = -L\frac{dL}{dt} = -\frac{1}{L}\frac{dL}{d\tau}. \tag{2.4}$$

The invariants (1.2) and (1.3) take the form :

$$M = L^{d-2}\int_0^\infty |u(\xi,\tau)|^2 \xi^{d-1} d\xi \tag{2.5}$$

$$H = L^{d-4}\int_0^\infty \{|u_\xi(\xi,\tau)|^2 - \frac{1}{2}|u(\xi,\tau)|^4\}\xi^{d-1} d\xi. \tag{2.6}$$

The scaling factor $L(t)$ is chosen such that a suitable norm of the rescaled solution u remains bounded. It is convenient to choose

$$L(t)^{d-1} = \frac{1}{\int |\nabla\Psi(x,t)|^2 dx} \tag{2.7}$$

As $\tau \to \infty, t \to t^*$ and $L(t) \to 0$. The behaviour of $a(\tau)$ as $\tau \to \infty$ determines the behaviour of $L(t)$ as $t \to t^*$. A similar method but without amplitude rescaling has recently been used by Kosmatov, Petrov, Shvets and Zakharov [4].

3. Singular solutions in dimension $d = 3$

There exist self similar solutions of (2.3) for which $a(\tau)$ is a strictly positive constant $\bar{a}$. They have the form

$$u(\xi,\tau) = e^{iC\tau} Q(\xi,\bar{a}). \tag{3.1}$$

After rescaling of ξ and Q by a factor $\sqrt{C}$, the profile Q satisfies the nonlinear ODE

$$Q_{\xi\xi} + \frac{d-1}{\xi} Q_\xi - Q + iK(\xi Q)_\xi + |Q|^2 Q = 0, \quad \xi > 0 \tag{3.2}$$

$$Q'(0) = 0; \quad Q(0) \quad \text{real}; \quad Q(\infty) = 0 \tag{3.3}$$

with $K = \bar{a}/C$. Furthermore, the profile Q satisfies

$$\int_0^\infty |Q_\xi|^2 \xi^{d-1} d\xi < \infty. \tag{3.4}$$

It has infinite L^2 norm

$$\int_0^\infty |Q|^2 \xi^{d-1} d\xi = \infty \tag{3.5}$$

and zero-energy

$$\int_0^\infty (|Q_\xi|^2 - \frac{1}{2}|Q|^4) \xi^{d-1} d\xi = 0. \tag{3.6}$$

The large distance behaviour of Q can easily be derived from (3.2). We get

$$Q \sim \alpha Q_1 + \beta Q_2, \qquad \xi \to \infty \tag{3.7}$$

where

$$Q_1 = |\xi|^{-1-i/K} \tag{3.8}$$

and

$$Q_2 = e^{iK\xi^2/2} |\xi|^{-d+1+i/K}. \tag{3.9}$$

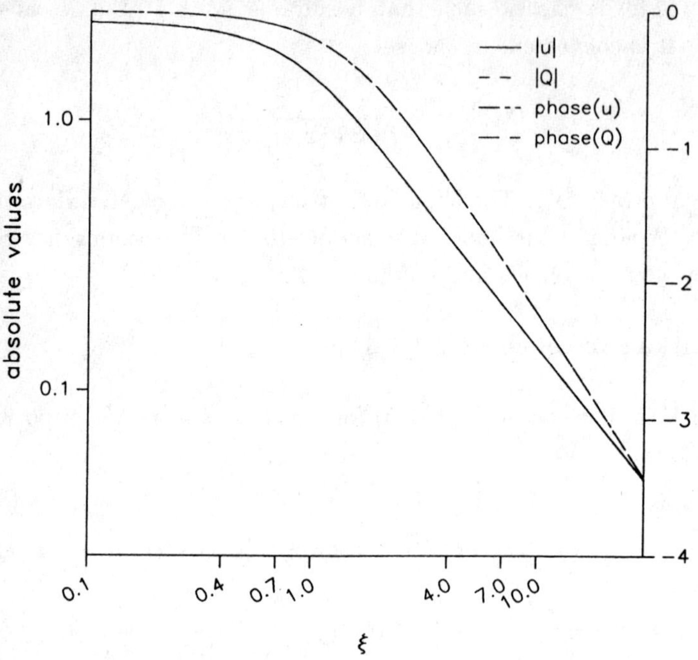

Fig.1 : $d = 3$: $\ln|u|$ $\ln|Q|$, $\arg(ue^{-i\bar{C}\tau})$, $\arg Q$ vs $\ln\xi$.

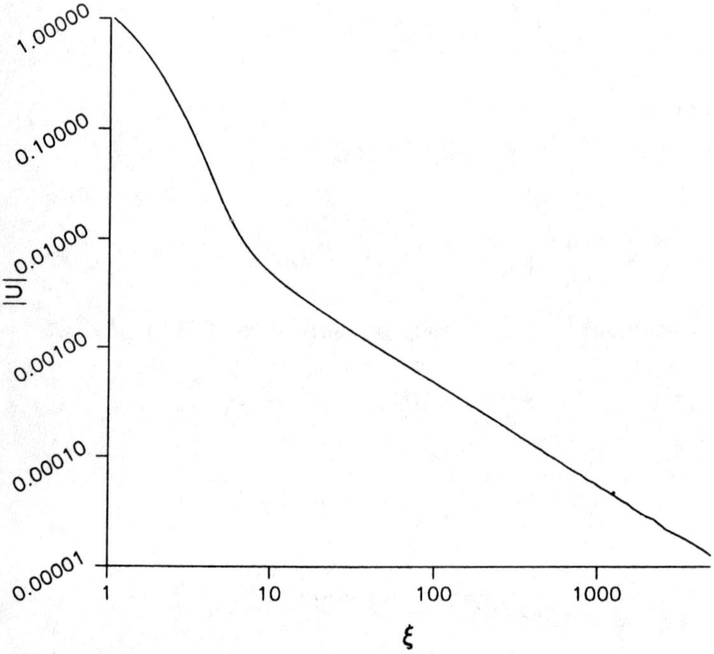

Fig.2 : $d = 2$: $\ln|u|$ vs $\ln\xi$.

Solutions satisfying (3.4) have $\beta = 0$. Eqs. (3.2)-(3.3) can be viewed as a nonlinear eigenvalue problem which has finite energy solutions only for special values of K, given the dimension $d > 2$. For example, we found by solving (3.2)-(3.3) numerically that for $d = 3$, $K = 0.917....$

The solutions thus constructed are exact self similar solutions of NLS of the form

$$\Psi(x,t) = \frac{1}{(2K(t^* - t))^{1/2}} \exp i(\frac{1}{2K} \ln \frac{t^*}{t^* - t} + \theta)\; Q(\frac{|x|}{(2K(t^* - t))^{1/2}}; K). \tag{3.10}$$

Numerical integration of eq.(2.3) for $d = 3$ indicates that singular solutions actually behave asymptotically like (3.10) [6], [9] : we observed that $a(\tau)$ tends to a positive constant $\tilde{a}$ as τ tends to infinity. The solution u has the asymptotic form (3.1): the amplitude becomes stationary; the phase is of the form $\tilde{C}\tau$, being constant in space and linear in τ. We found that $K = \tilde{a}/\tilde{C}$ has a universal value 0.917.. and the rescaled profile satisfies (3.2). Fig. 1 shows amplitudes and phases of $ue^{-i\tilde{C}\tau}$ and Q. Logarithmic scales have been used to emphasize the algebraic decay at large distance.

4. Singular solutions in dimension $d = 2$

At the critical dimension $d = 2$, there are no monotonically decaying solutions Q of (3.2)-(3.3) with $K > 0$. Indeed, writing $Q = Ae^{i\phi}$, one easily notices that regular, even solutions have a quadratic phase $\phi = K\xi^2/4$. From (3.8)-(3.9) such solutions have an amplitude displaying an oscillatory decay at infinity.

Numerical simulations of NLS in two dimensions indicate that $a(\tau)$ tends very slowly to zero as τ goes to infinity. They also show that the profile of the solution near the origin fits the positive solution R of

$$R_{\xi\xi} + \frac{1}{\xi} R_\xi - R + R^3 = 0, \tag{4.1}$$

while at large distance, it decays algebraically, the transition moving slowly to infinity with increasing τ (fig. 2) [7]. These observations suggest the construction of an asymptotic solution of NLS based on the Q - function with a positive time-dependent coefficient K tending to 0 when τ tends to infinity. Since eqs (3.2)-(3.3) have acceptable solutions only for specific values of K, given the dimension, varying K requires a variation of the space dimension. An important remark is that the function $d(K)$ satisfies

$$(\frac{d}{dK})^p (d(K) - 2)|_{K=0} = 0 \qquad \text{for all} \quad p = 0, 1, 2... \tag{4.2}$$

Consequently, an asymptotic expansion containing information to all orders in K is necessary.

We now proceed to the construction of a family of asymptotic solutions. Without restriction, we can take $C = 1$ by rescaling τ and thus identify $K(\tau)$ and $a(\tau)$. We look for a solution of

$$i\partial_\tau u + u_{\xi\xi} + \frac{d-1}{\xi}u_\xi + ia(\tau)(\xi u)_\xi + |u|^2 u = 0. \tag{4.3}$$

After the change of variables $u = e^{i\tau - ia\xi^2/4}v$, eq.(4.3) becomes :

$$i\partial_\tau v + v_{\xi\xi} + \frac{d-1}{\xi}v_\xi + (a^2 + a_\tau)\frac{\xi^2}{4}v + |v|^2 v = 0. \tag{4.4}$$

We look for a solution in the form

$$v(\xi,\tau) = P(\xi; a(\tau)) + W(\xi,\tau). \tag{4.5}$$

where $P = Qe^{ia\xi^2/4}$ satisfies

$$P_{\xi\xi} + \frac{d-1}{\xi}P_\xi - P + \frac{a^2}{4}\xi^2 P - ia\frac{d-2}{2}P + |P|^2 P = 0, \quad \xi > 0 \tag{4.6}$$

$$P'(0;a) = 0; \quad P(0;a) \quad \text{real}; \quad P(\infty;a) = 0. \tag{4.7}$$

We assume that as $\tau \to \infty$

$$a(\tau) > 0, \qquad a(\tau) \to 0 \qquad \text{and} \quad a_\tau \ll a(\tau) \quad \text{as} \quad \tau \to \infty. \tag{4.8}$$

To principal order, W satisfies

$$\begin{aligned} W_{\xi\xi} + \frac{d-1}{\xi}W_\xi - W + (a^2 + a_\tau)\frac{\xi^2}{4}W - ia\frac{(d-2)}{2}W + 2|P|^2 W + P^2\bar{W} \\ = -i(a_\tau P_a + \frac{a}{2}(d-2)P) - (a_\tau\frac{\xi^2}{4}P + \frac{d-2}{\xi}P_\xi). \end{aligned} \tag{4.9}$$

Let $P = S + iT$ with S and T real. To the leading order as $a \to 0$, S and T are solutions of

$$S_{\xi\xi} + \frac{1}{\xi}S_\xi - S + 3R^2 S = -a_\tau\frac{\xi^2}{4}R + \frac{d-2}{\xi}R_\xi \tag{4.10}$$

$$T_{\xi\xi} + \frac{1}{\xi}T_\xi - T + R^2 T = -2aa_\tau\rho - \frac{a}{2}(d-2)R \tag{4.11}$$

where

$$\rho = \frac{1}{2}P_a \tag{4.12}$$

is solution of

$$\rho_{\xi\xi} + \frac{1}{\xi}\rho_\xi - \rho + 3R^2\rho = -\frac{\xi^2}{4}R. \tag{4.13}$$

Eq.(4.10) is always solvable. In contrast, the homogeneous equation associated to (4.11) has the solution $T = R$. Thus the solvability condition for eq.(4.11) reads :

$$a_\tau = \alpha(d(a) - 2) \tag{4.14}$$

where

$$\alpha = \frac{4\int R\rho\xi d\xi}{\int R^2\xi d\xi} = \frac{1}{2}\frac{\int R^2\xi^3 d\xi}{\int R^2\xi d\xi} > 0. \tag{4.15}$$

The next step is to obtain the behaviour of $d(a)$ when a tends to 0. Once $d(a)$ is known, eq(4.14) is an ODE that determines $a(\tau)$ which in turn, by (2.4) determines the blow up rate $L(t)$ for singular solutions of the NLS equation in critical dimension $d = 2$.

We write eq(3.2) in terms of amplitude A and phase ϕ of Q :

$$A_{\xi\xi} + \frac{1}{\xi}A_\xi - A + A^3 - \psi(\psi + a\xi)A = 0 \tag{4.16a}$$

$$(2\psi + a\xi)A_\xi + (\psi_\xi + \frac{1}{\xi}\psi_\xi + a) = 0 \tag{4.16b}$$

where $\psi = \phi_\xi$. Eq.(4.16b) rewrites

$$\frac{d}{d\xi}\{\xi(\psi + a\frac{\xi}{2})A^2\} = -(d-2)A^2\psi. \tag{4.17}$$

Thus,

$$\psi + a\frac{\xi}{2} = -\frac{d-2}{\xi A^2}\int_0^\xi A^2\psi d\xi \tag{4.18}$$

Eqs.(4.16a) and (4.18) show that when a tends to 0 and d tends to 2 with fixed ξ, A reduces to the R function and ϕ to the quadratic phase $-a\xi^2/4$ predicted by Zakharov and Synakh [17]. From the asymptotic behaviour of Q for large ξ, we have $A \sim \mu/\xi$ and $\psi \sim -1/a\xi$. This implies that the integral $\int_0^\xi A^2\psi d\xi$ converges. For large ξ, the integral range can be extending to infinity. The main contribution comes when ξ is of order one, leading to

$$\int_0^\xi A^2\psi d\xi \sim -aI^2/2 \tag{4.19}$$

as $a \to 0$, where

$$I^2 = \int_0^\xi R^2\xi d\xi. \tag{4.20}$$

This enables us to compute the coefficient μ by equating the unbounded terms in (4.18) and thus

$$\mu \sim (d-2)^{1/2}I \quad \text{as} \quad a \to 0. \tag{4.21}$$

Substituting (4.18) in the amplitude equation (3.16a), we have

$$A_{\xi\xi}+\frac{1}{\xi}A_{\xi}-A+A^{3}$$
$$+\{\frac{a^{2}\xi^{2}}{4}-\frac{(d-2)^{2}}{\xi^{2}A^{4}}[\int_{0}^{\xi}\psi A^{2}d\xi]^{2}\}A=0. \tag{4.22}$$

For $a\xi \ll 1$, the terms in the brackets are negligible and $A(\xi) \sim R(\xi)$ which at large ξ behaves like $\xi^{-1/2}e^{-\xi}$. For $a\xi \gg 1$,

$$A(\xi) \sim \frac{(d-2)^{1/2}I}{\xi} \quad \text{and} \int_{0}^{\xi}\psi A^{2}d\xi \sim -\frac{a}{2}I^{2}. \tag{4.23}$$

At the edge of the inner region $\xi \sim 1/a$,the integral should prevent $\xi^{2}a^{2}/4$ from dominating in the bracket of (4.22). Otherwise, A will oscillate and will not remain positive. Thus, at the edge of the inner region,

$$\frac{d-2}{\xi A^{2}}\int_{0}^{\xi}\psi A^{2}d\xi \tag{4.24}$$

should also be of order unity. Using the fact that in this region $A^{2} \sim \xi^{-1}e^{-2\xi}$, we get

$$d-2 \sim a^{-1}e^{-\lambda/a} \tag{4.25}$$

where λ is a constant. A WKB argument due to Zakharov [15] gives $\lambda = \pi$. We notice that this relation ensures patching of the inner and outer expansions of A at $\xi = 0(1/a)$.

We are now able to obtain the rate of blow-up of the family of singular solutions we have constructed. Putting together (4.14) and (4.25), we have

$$a_{\tau} \sim -\frac{1}{a}e^{-\lambda/a}. \tag{4.26}$$

The asymptotic solution of (4.26) as τ goes to infinity is

$$a(\tau) \sim \frac{\lambda}{\ln \tau}. \tag{4.27}$$

Eq.(2.4) then determines the asymptotic form of the scaling factor $L(t)$ as $t \to t^{*}$. We find

$$L(t) \sim \sqrt{\frac{2\lambda(t^{*}-t)}{\ln\ln\frac{1}{t^{*}-t}}}. \tag{4.28}$$

The asymptotic behaviour of the family of singular solutions we have constructed is then

$$\Psi(\mathbf{x},t) \sim \frac{1}{L(t)}e^{i\tau(t)}Q(\frac{|\mathbf{x}|}{L(t)};a(t)) \tag{4.29}$$

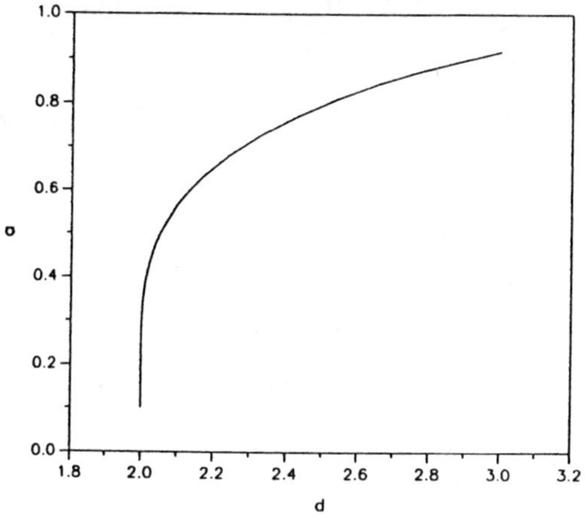

Fig.3 : Numerical computation of a vs d for the eigenvalue problem (3.2)-(3.3).

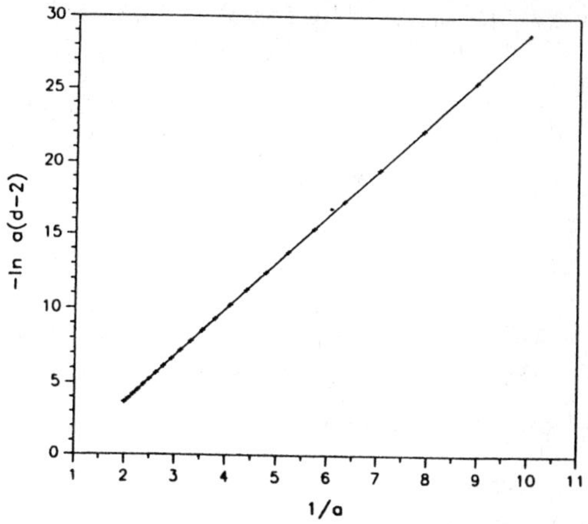

Fig.4 : Plot of $\ln a(d-2)$ vs $1/a$.

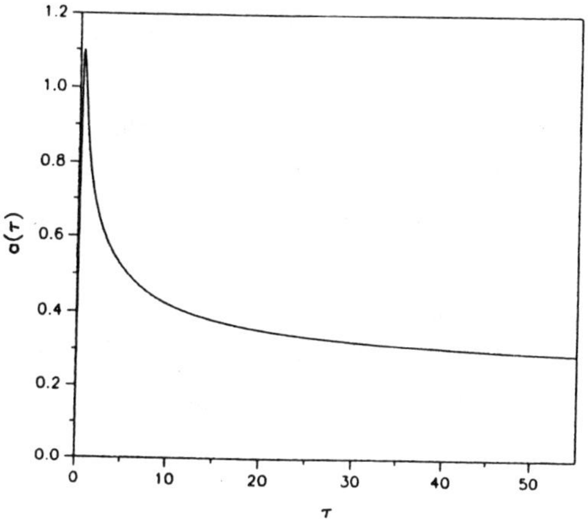

Fig.5 : The function $a(\tau)$ plotted vs τ computed in the evolution problem (2.3).

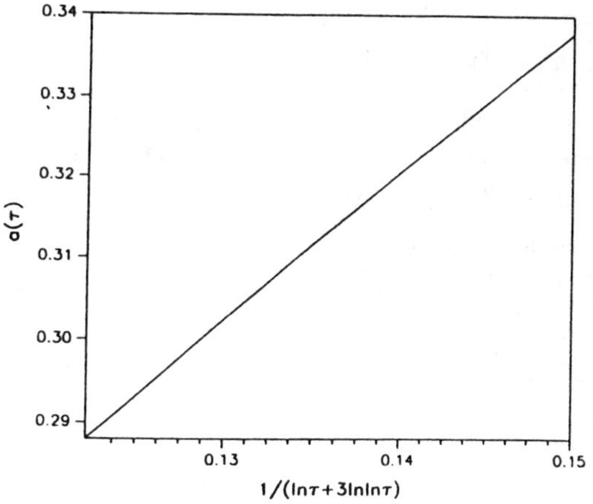

Fig.6 : The function $a(\tau)$ vs $1/(\ln\tau + 3\ln\ln\tau)$ from the same data as Fig.5 excluding the transient.

where $L(t)$ is given by (4.28),

$$\tau(t) \sim \frac{1}{2\lambda} \ln \frac{1}{t^* - t} \ln \ln \frac{1}{t^* - t} \tag{4.30}$$

and

$$a(t) \sim [\lambda \ln \ln \frac{1}{t^* - t}]^{-1}. \tag{4.31}$$

The rate of blow-up (4.28) of the scaling factor was obtained using a more physical argument by Fraiman [2]. The paper contains an unfortunate misprint on the formula for the blowing rate (ln instead of ln ln.)

The asymptotic result (4.25) is supported by numerical integration of the nonlinear eigenvalue problem (3.2)-(3.3) for $d(a)$ and $Q(\xi; a)$. It was solved by a continuation method using the software package AUTO [1]. As the parameter a is varied, the family of solutions with the appropriate value of $d(a)$ is determined. Fig.3 is a plot of a versus d. The curve appears to approach $d = 2$ with all the derivatives becoming infinite as $a \to 0$. In fig.4, we plotted $-\ln a(d-2)$ versus $1/a$. We see an excellent linear least-squares fit over the range $0.1 \leq a \leq 0.4$ with the slope $\lambda = 3.14$. We also checked the behaviour of A and ψ (figs 3-4 of [9]). We saw that it is the same behaviour as that observed for the profile in the simulation of the evolution problem (fig.2).

In order to validate the asymptotic analysis, we now proceed to the comparison of the analytic construction of singular solutions and the numerical simulations of the evolution problem based on the dynamical rescaling. Fig.5 shows the evolution of $a(\tau)$ obtained numerically with the initial condition $\Psi_0(x) = 4e^{-x^2}$. In fig.6, $a(\tau)$ is plotted versus $1/(\ln \tau + 3 \ln \ln \tau)$ and we see a linear scaling with precision of few percent. Its slope however is still far from 3.14, the reason being that the asymptotic regime in which (4.27) holds has not yet been reached. Nevertheless, the functional form of the asymptotic behaviour has been captured in the simulations .

5. Conclusion

A method of dynamical rescaling of variables enables us to study the focussing singularities of the isotropic NLS equation with cubic nonlinearity. In three dimensions (supercritical case), these calculations strongly corroborate the asymptotic self-similarity of the singular solutions. In addition, a unique form of the limiting rescale profile is observed.

In two-dimensions (critical case), the self-similarity is weakly broken and the construction of singular solutions requires the resolution of a singularly perturbed nonlinear eigenvalue problem. The rate of blow up then reveals a *loglog* correction in time to the square root singularity found in three dimensions.

The present study concentrates on isotropic solutions. An important question is whether anisotropic initial conditions for the NLS equation focus into the isotropic singularities studied here.

Acknowledgements

We thank V.E. Zakharov for pointing Fraiman's paper to our attention .

References

1. D.A. Aronson, E.J. Doedel, H.G. Othmer, Physica **25D** (1987) 20
2. G.M. Fraiman, Zh. Eksp. Teor. Fiz. **88**, 1985, 390; Sov. Phys. JETP **61** (1986) 228
3. R.T. Glassey, J. Math. Phys. **18** (1977) 1974
4. N.E. Kosmatov, I.V. Petrov, V.F. Shvets, V.E. Zakharov, *Large Amplitude Simulation of Wave Collapse in Nonlinear Schrödinger equations*, Preprint, 1988, Academy of Sciences of the USSR, Space Research Institute
5. M.J. Landman, G. Papanicolaou, C. Sulem, P.L. Sulem, Phys. Rev. **A 38** (1988) 3837
6. B. LeMesurier, G. Papanicolaou, C. Sulem, P.L. Sulem, in *Directions in Partial Differential Equations*, M.G. Crandall, P.H. Rabinovitz and R.E. Turner, eds. Academic Press (1987) 157
7. B. LeMesurier, G. Papanicolaou, C. Sulem, P.L. Sulem, Physica **31D** (1988) 78
8. B. LeMesurier, G. Papanicolaou, C. Sulem, P.L. Sulem, Physica **32D** (1988) 210
9. D.W. McLaughlin, G. Papanicolaou, C. Sulem, P.L. Sulem, Phys. Rev. **A 34** (1986) 1200
10. I. Rasmussen, K. Rypdal, Phys. Scr. **33** (1986) 481
11. I. Rasmussen, K. Rypdal, Phys. Scr. **33** (1986) 498
12. K. Rypdal, I. Rasmussen, K. Thomses, Physica **16D** (1985) 339
13. P.L. Sulem, C. Sulem, A. Patera, Comm. Pure Appl. Math. **37** (1984) 755
14. V.E. Zakharov, Zh. Eksp. Teor. Fiz. **62** (1972) 1745; Sov. Phys. JETP **35** (1972) 908
15. V.E. Zakharov, Private communication
16. V.E. Zakharov, A.B. Shabat, Zh. Eksp. Teor. Fiz. **61**, 1971, 118; Sov. Phys. JETP **34** (1972) 62
17. V.E. Zakharov, V.S. Synakh, Zh. Eksp. Teor. Fiz. **68** (1975) 940; Sov. Phys. JETP **41**, 1976, 465

HAMILTONIAN DEFORMATIONS AND OPTICAL FIBERS

Curtis R. Menyuk
Department of Electrical Engineering
University of Maryland
Baltimore, MD 21228

ABSTRACT

In integrable, nonlinear systems an arbitrarily shaped initial pulse is known to break up into several solitons and a dispersive wave component. Similar behavior is often observed when substantial Hamiltonian deformations which destroy the system's integrability are present, as long as the Hamiltonian deformations have no explicit dependence on space or time. By contrast, this behavior is usually destroyed by non-Hamiltonian deformations even when they are quite small. Hence, it is usually sufficient to know a deformation's character to immediately determine its effect on solitons. Application of this result to optical fiber communication is discussed.

I. INTEGRABLE EQUATIONS

It may seem odd at first that anything which sounds as esoteric as Hamiltonian deformations could have something useful to tell us about optical fibers. We believe, however, that our results are a nice example of how a physical/mathematical principle when properly understood can lead to important insights into the operation of real-world devices.

Many, if not most, physical systems exhibit turbulent or chaotic behavior in at least some regimes. Such systems are appropriately modeled by equations like the Navier-Stokes equation which has turbulent solutions at high Reynolds numbers and is used to study fluids. In many important cases, however, the physical systems exhibit nice, coherent behavior over a wide range of parameters. That is particularly the case in devices which are useful for something, as opposed to systems which are handed to us by nature, since one usually wants the device to behave in a nice, predictable manner.

Nonlinear field equations which always exhibit coherent behavior include the sine-Gordon equation

$$u_{xt} = \sin u, \tag{1}$$

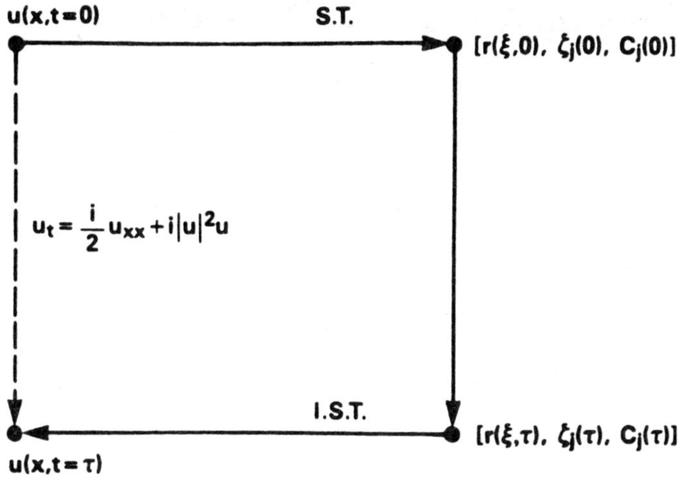

FIGURE 1. Schematic illustration of the way in which the spectral transform and its inverse can be used to solve the nonlinear Schrödinger equation.

which has been used to model self-induced transparency [1], the Korteweg-de Vries equation

$$u_t - 6uu_x + u_{xxx} = 0, \tag{2}$$

which has been used to model water waves in shallow channels [2] and ion-acoustic waves in plasmas [3], and the nonlinear Schrödinger equation

$$iu_t + \frac{1}{2}u_{xx} + |u|^2u = 0, \tag{3}$$

which has been used to model Langmuir waves in plasmas [4] and light pulses in optical fibers [5]. We will be discussing this last application in far more detail at a later point.

These equations are often referred to as "integrable." That is to say, they have a number of special properties which the vast majority of field equations do not have. Among the most important of these properties is a spectral transformation which can be considered to be a nonlinear Fourier transform. The spectral transform can be used to solve these special equations just like the usual Fourier transform can be used to solve linear field equations. The transformation procedure is shown schematically in Fig. 1 for the nonlinear Schrödinger equation, assuming that the initial data $u(x,t=0)$ falls off sufficiently rapidly as $x \to \pm\infty$ [6]. The spectral transformation yields $[r(\xi,0), \zeta_j(0), C_j(0)]$. The quantity $r(\xi,0)$ depends continuously on the variable ξ and is directly analogous to the usual Fourier transform, although it is not identical. Physically, it corresponds to a dispersive wave component whose amplitude vanishes as $t \to \infty$. In addition, the spectral transform yields a number $N \geq 0$ of discrete pairs (ζ_j, C_j) which have no analogy in the usual Fourier transform. These pairs correspond to solitons, nonlinear wave packets which propagate without dispersing.

Knowing the solution at $t = 0$, it is possible to immediately write down the solution at $t = \tau$. It is [6]

$$\begin{aligned} r(\xi,\tau) &= r(\xi,0)\exp(2i\xi^2\tau), \\ \zeta_j(\tau) &= \zeta_j(0), \\ C_j(\tau) &= C_j(0)\exp(2i\zeta_j^2\tau). \end{aligned} \tag{4}$$

One can use the inverse spectral transform to determine $u(\xi,\tau)$. The significance of the spectral transform is that it allows us to determine $u(\xi,\tau)$ in three steps, shown as solid lines in Fig. 1, no matter what the size of τ. If one were to use the direct route shown as a dashed line in Fig. 1, one would in general need to cut the time axis into a number of pieces proportional to τ and determine the solution iteratively. Thus, there exists some time τ beyond which the indirect approach always wins.

The issue which concerns us is that there are many systems which behave integrably in at least one important respect. Initial data breaks up into a dispersive wave component and a number of solitons, or, more precisely, solitary waves. We make no distinction between solitons and solitary waves in keeping with the standard practice of experimental physicists who use the word "soliton" even though their systems are never strictly integrable, and, hence, they never observe strict mathematical solitons.

It is natural to suppose that these systems which appear to behave integrably can be well-modelled by one of the integrable equations. If the actual system were to be perturbed away from the integrable system by an amount of order ϵ, then one might expect that the integrable behavior would only appear for a time of order ϵ^{-1}. On a longer time scale, solitons would be destroyed. This expectation is borne out in practice when the perturbations are dissipative or have an explicit space or time dependence; however, when the perturbations are Hamiltonian and independent of space and time, that is no longer the case. Indeed, the systems appear to act integrably on arbitrarily long time scales. Moreover, they continue to act integrably when the Hamiltonian deviations are so large that they can no longer be referred to as perturbations, but must be considered substantial deformations. Why are systems so rugged under the influence of Hamiltonian deformations, and what are the implications for practical devices like optical fibers? We will be addressing these issues in the following sections.

While this sort of behavior can be seen in a large number of real physical systems, we mention here two numerical examples closely related to the nonlinear Schrödinger equation

$$iu_t + \frac{1}{2}u_{xx} + |u|^2u = -\{1 - |u|^2 - \exp(-|u|^2)\}u, \tag{5}$$

which has been used to model Langmuir waves in plasmas [7] and

$$iu_t + \frac{1}{2}u_{xx} + |u|^2u = i\beta u_{xxx}, \tag{6}$$

which has been used to study light pulses in optical fibers near the zero-dispersion point [8,9]. The deformations in both Eqs. (5) and (6) are Hamiltonian. We emphasize numerical results because in simulated systems the effect of dissipation can be completely

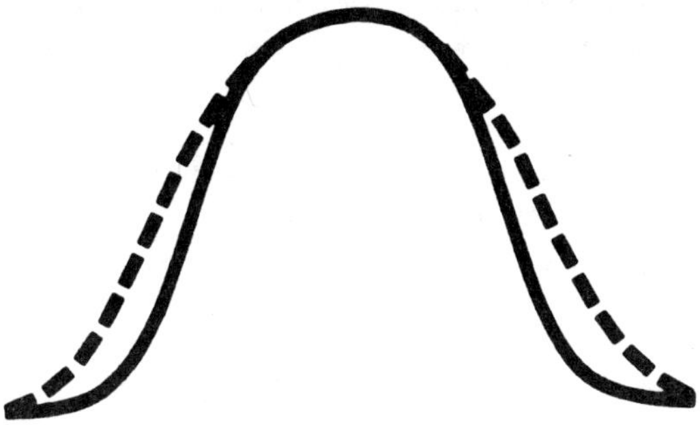

FIGURE 2. Schematic illustration of shape renormalization. Under the influence of a Hamiltonian perturbation, a soliton's shape will change from that shown as a solid line to that shown as a dashed line.

eliminated which can never be the case in real systems. In numerical solutions of Eq. (5), initial data are seen to break into a soliton and dispersive waves when $|u|$ is as large as 2, so that the term on the right is making a large contribution. Similar results are found when Eq. (6) is solved with β arbitrarily large. Clearly, then, the right-hand side can be a large deformation indeed! It should be noted that while solitons continue to exist, their shapes and frequency shifts are observed to change from what is predicted by the nonlinear Schrödinger equation, as shown schematically in Fig. 2.

II. HAMILTONIAN SYSTEMS

In order to demonstrate that Eqs. (5) and (6) are Hamiltonian, it is sufficient to show that they can be derived from a Hamiltonian functional. In the case of Eq. (6), this functional is

$$H = -\frac{i}{2}\int_{-\infty}^{\infty} dx \left[u_x u_x^* - |u|^4 + i\beta(u_x u_{xx}^* - u_{xx} u_x^*)\right]. \tag{7}$$

Letting $q \equiv u$ and $p \equiv u^*$, one can show

$$\dot{q} = \frac{\delta H}{\delta p}, \qquad \dot{p} = -\frac{\delta H}{\delta q}, \tag{8}$$

as is appropriate for Hamiltonian systems, where the derivatives $\delta x/\delta y$ are functional derivatives. A similar result can be obtained for Eq. (5). Such systems are often referred to as infinite-dimensional Hamiltonian systems because each point in x can be considered a separate degree-of-freedom. When one states that a Hamiltonian system is integrable, one generally means that a canonical transformation exists which yields a Hamiltonian independent of the new coordinates, depending only on the new momenta. This point of view seems different from that of the previous section where we said that the nonlinear Schrödinger equation could be solved by making a spectral transformation;

in fact, these two points of view are equivalent. The spectral transformation turns out to be a canonical transformation which yields a Hamiltonian only depending on the momenta.

Before demonstrating this point explicitly, it is useful to turn to a simpler example to explain how these canonical transformations work. They are important because when integrable field equations with Hamiltonian perturbations are considered, it is possible to find an infinite series of canonical transformations which eliminates order-by-order the dependence on the coordinates. This result explains qualitatively why integrable behavior is rugged under the influence of Hamiltonian deformations. (At least when the deformations are small!)

The example we will consider is a simple, finite-dimensional system

$$H = \sum_i \frac{\omega_i}{2}(p_i^2 + q_i^2). \tag{9}$$

The canonical transformation $(p_i, q_i) \rightarrow (P_i, Q_i)$, where

$$p_i = (2P_i)^{1/2} \cos Q_i, \qquad q_i = (2P_i)^{1/2} \sin Q_i, \tag{10}$$

reduces the Hamiltonian to the desired form

$$H = \sum_i \omega_i P_i, \tag{11}$$

which depends only on the momenta. As a consequence, the momenta are constant in time, while the coordinates vary linearly. Writing the equations of motion,

$$\dot{P}_i = 0, \qquad \dot{Q}_i = \omega_i, \tag{12}$$

we obtain,

$$P_i = P_{i,0}, \qquad Q_i = Q_{i,0} + \omega_i t, \tag{13}$$

where $P_{i,0}$ and $Q_{i,0}$ are constants of integration. In similar fashion, if we make the transformation $u \rightarrow [P(\xi), Q(\xi), P_j, Q_j]$, where, in terms of the spectral data,

$$\begin{aligned} P(\xi) &= \frac{i}{\pi} \ln[1 + |r(\xi)|^2], \qquad Q(\xi) = \arg r(\xi), \\ P_j &= 2i\zeta_j, \qquad Q_j = -\ln C_j, \end{aligned} \tag{14}$$

we find that the transformed Hamiltonian becomes [6]

$$H = \int_{-\infty}^{\infty} d\xi \, [2\xi^2 P(\xi)] + \frac{i}{6} \sum_j P_j^3, \tag{15}$$

which only depends on the momenta. Hence, just as in the previous case, the momenta are constant in time while the coordinates vary linearly.

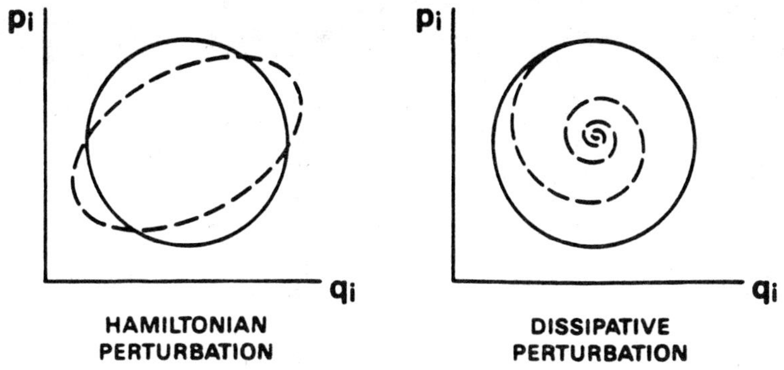

FIGURE 3. Effect of Hamiltonian and non-Hamiltonian perturbations. A Hamiltonian perturbation slightly deforms the trajectory, but it remains neutrally stable. While the trajectory does not generally close on itself in the original coordinates, it does in appropriate perturbed coordinates. A dissipative perturbation, no matter how small, leads to a spiral trajectory which ultimately falls into the origin.

Suppose now that we perturb the finite-dimensional system by adding cubic terms to the Hamiltonian,

$$H = \sum_i \frac{\omega_i}{2}(p_i^2 + q_i^2) + ap_1^3 + bp_1^2 q_1 + \cdots \tag{16}$$

In the limit where p_i and q_i are small, this perturbation only makes a small contribution to the Hamiltonian. As long as all the ω_i are incommensurable, it is possible to find a canonical transformation, using the Lie transform method or the Poincaré-von Zeipel method, which eliminates the cubic terms at the expense of introducing fourth and higher order terms [10], *i.e.*, there exists a transformation $[p_i, q_i] \rightarrow [\tilde{p}_i, \tilde{q}_i]$, such that the Hamiltonian becomes

$$H = \sum_i \frac{\omega_i}{2}(\tilde{p}_i^2 + \tilde{q}_i^2) + \tilde{a}\tilde{p}_1^4 + \cdots \tag{17}$$

The fourth order terms can then be eliminated by making another, analogous transformation, and we can continue in this fashion order-by-order. Physically, this series of transformations is possible because when q_i and p_i are sufficiently small, the effect of the cubic perturbations, for the vast majority of initial conditions, is to deform the orbit of the pair (p_i, q_i) without destroying its neutral stability. By contrast, a dissipative perturbation, no matter how small, will lead to a fundamental change in orbit topology as shown qualitatively in Fig. 3.

A similar series of transformations exists for Hamiltonian perturbations of integrable field equations. We recall that the original transformation $u \rightarrow [P(\xi), Q(\xi), P_i, Q_i]$ yields quantities which evolve linearly in time when u_t is given by the nonlinear Schrödinger equation. That is no longer the case once the equations are perturbed.

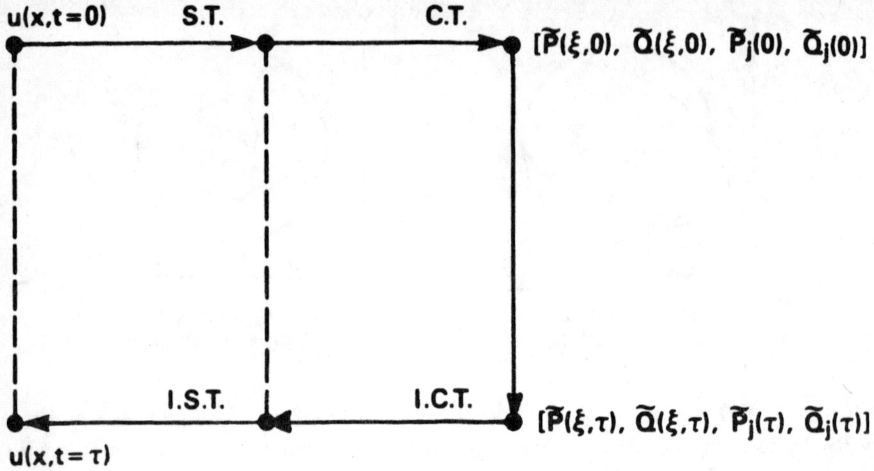

FIGURE 4. Schematic illustration of the integration procedure for the perturbed nonlinear Schrödinger equation when the perturbations are Hamiltonian. One first makes a spectral transformation followed by a series of canonical transformations to arrive at variables which evolve linearly in time. Having calculated the new variables at the time τ, one reverses the original sequence of transformations to determine u.

However, at any given order, the canonical transformations yield a new set of quantities $[\tilde{P}(\xi), \tilde{Q}(\xi), \tilde{P}_i, \tilde{Q}_i]$ which evolve linearly in time *through the order to which we are working, i.e.*,

$$\begin{aligned} \tilde{P}(\xi) &= \tilde{P}_0(\xi), \qquad \tilde{Q}(\xi) = \tilde{Q}_0(\xi) + \Omega(\xi)t, \\ \tilde{P}_j &= \tilde{P}_{j,0}, \qquad \tilde{Q}_j = \tilde{Q}_{j,0} + \Omega_j t, \end{aligned} \tag{18}$$

where $\tilde{P}_0(\xi)$, $\tilde{Q}_0(\xi)$, $\tilde{P}_{j,0}$, $\tilde{Q}_{j,0}$, $\Omega(\xi)$, and Ω_j are all constant in time. Hence, just as in the integrable case, it is possible to integrate the equation in a fixed number of steps, as shown schematically in Fig. 4, independent of the length of time τ over which one wishes to determine the solution. Why then are these perturbed equations not also considered integrable? The reason is that in general this series of transformations is only convergent for special choices of the initial conditions; otherwise, the series is merely asymptotic, and only a finite number of transformations can be usefully made.

At every order of the transformation, one finds that the topology of the solution is unchanged; it still consists of a number of solitons which do not change in time when well-separated and a dispersive wave component [11,12]. Hamiltonian perturbations lead to no fundamental changes in the structure of the solution, in contrast to dissipative perturbations.

It should be noted that the results just described have only been demonstrated in detail when the underlying, integrable system is the Korteweg-de Vries equation [11,12], although the nature of the derivation makes it seem clear that similar results will hold when the underlying system is the nonlinear Schrödinger equation or any of a set of similar field equations.

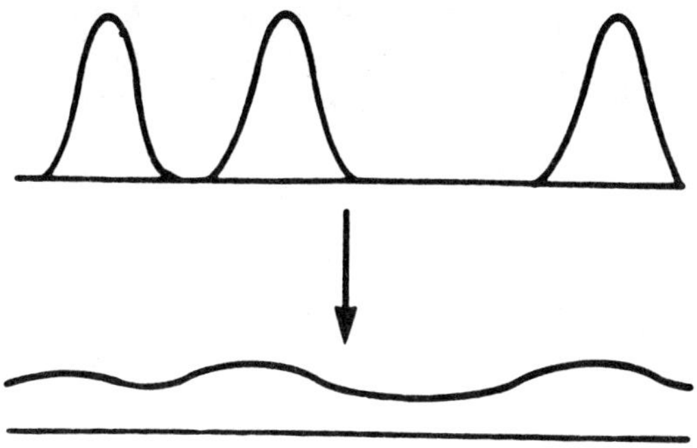

FIGURE 5. The effect of dispersion is illustrated schematically. As pulses propagate along the fiber, they broaden, eventually overlapping.

III. OPTICAL FIBERS

Optical fibers consist of a glass core surrounded by a glass cladding; the index of refraction in the core is slightly higher than in the cladding, implying that waves will propagate. Essentially, they are trapped by total internal reflection [13].

If the core is sufficiently small, $\simeq 8\ \mu$m in diameter or less, then only a single mode, the HE_{11} mode, propagates, eliminating intermodal dispersion. Nonetheless, single mode dispersion remains a serious problem limiting the (bit rate) $\times$ (propagation length) values which can be attained in modern-day systems. The way in which dispersion limits the bit rate is shown in Fig. 5. A train of pulses is launched in the fiber. When a pulse is present in a given time slot, it is counted as a 1-bit, and, when it is absent, it is counted as a 0-bit. As the pulses propagate along the fiber, the dispersion causes spreading; after a long length, it is impossible for the detection system to tell whether there is a 1-bit or a 0-bit in any given slot.

For any given length, there is an optimum pulse size which yields the maximum bit rate possible. If the pulse is too narrow initially, then it has a large bandwidth and spreads very quickly due to the dispersion. If the pulse is too large initially, then it stays too large. It is conventional to measure pulse widths in units of time. If we write the initial pulse width as τ_0, then the final pulse width after going through a fiber of length L is

$$\tau = \tau_0 + \frac{\lambda^3}{c^2} n \left| \frac{d^2 n}{d\lambda^2} \right| \frac{L}{\tau_0}, \tag{19}$$

where λ is the light's wavelength, n is the index of refraction in the fiber, and c is the speed of light in a vacuum. The minimum fiber loss rate is 0.2 db/km when $\lambda = 1.55\ \mu$m, from which we infer a 20 km propagation length before the signal loss becomes severe [5,13]. From Eq. (19), we then infer a maximum bit rate of 5 Gbit/sec. While this figure is quite large, the bit rates in communication systems have been rising roughly exponentially as a function of time over the last two centuries, with a break to a faster rise after 1950, as shown in Fig. 6. Unless this curve magically bends over in the near future, it is clear that this bit rate will soon be achieved.

Some years ago, Hasegawa and Tappert [14] suggested using the Kerr nonlinearity to compensate for the dispersion. They showed that in the wavelength range $\lambda >$ 1.3 μm, the so-called anomalous dispersion regime, light pulses are well-described by the nonlinear Schrödinger equation and that dispersionless propagation is therefore possible. This idea has since been tested experimentally and found to be feasible [15,16].

Significant deformations of the nonlinear Schrödinger equation, both Hamiltonian and non-Hamiltonian, can exist in real fibers. Hamiltonian deformations include cubic dispersion, birefringence, and finite radial effects. Non-Hamiltonian deformations include attenuation and Raman or Brillouin scattering. From the results of the previous sections, we may infer the following: Hamiltonian deformations, even large deformations, will have no adverse effect on the solitons; their shapes may be slightly different from what the nonlinear Schrödinger equation predicts, but they will still exist and propagate. By contrast, non-Hamiltonian deformations are very destructive and must be dealt with in some way. The power of this result is that it is not necessary to do any detailed analysis; it is only necessary to determine the nature of the deformation, and one immediately knows whether it is likely to cause trouble.

As an example, we may consider the behavior of pulses which are injected at the zero dispersion point [8,9], $\lambda \simeq 1.3$ μm. At this point, the usual quadratic dispersion goes to zero, unveiling the effect of the cubic dispersion. Using appropriately normalized variables, one then finds

$$iu_\xi - iu_{sss} + |u|^2 u = 0, \tag{20}$$

where s represents the length along the fiber and τ the time variation in the group velocity frame. Note that we have reversed the roles of space and time from the "standard" roles seen in Eqs. (1–3); we do so because the pulses in fibers are initially specified for all time at a given point in space, rather than the reverse. We can obtain Eq. (20) from Eq. (6) by letting $\xi = t$, $s = x/\beta^{1/3}$, and by letting $\beta \to \infty$.

It is of great practical interest to operate as close to the zero dispersion point as possible. Since the dispersion is minimal at this point, the power needed to generate a soliton is also minimal. The power requirement can be reduced to the point where a single laser diode can generate the pulses—a very desirable result indeed! Workers previous to Wai, *et al.* [9] had assumed that pulses launched at the zero dispersion point would simply break apart rather than generate a soliton. Having noted however that Eq. (20) can be generated by a Hamiltonian deformation of the nonlinear Schrödinger equation, albeit infinite, and motivated by the results which have been presented in the previous two sections, Wai, *et al.* decided to examine this question more closely. Shown in Fig. 7 is the evolution of an initially Gaussian pulse injected at the zero dispersion

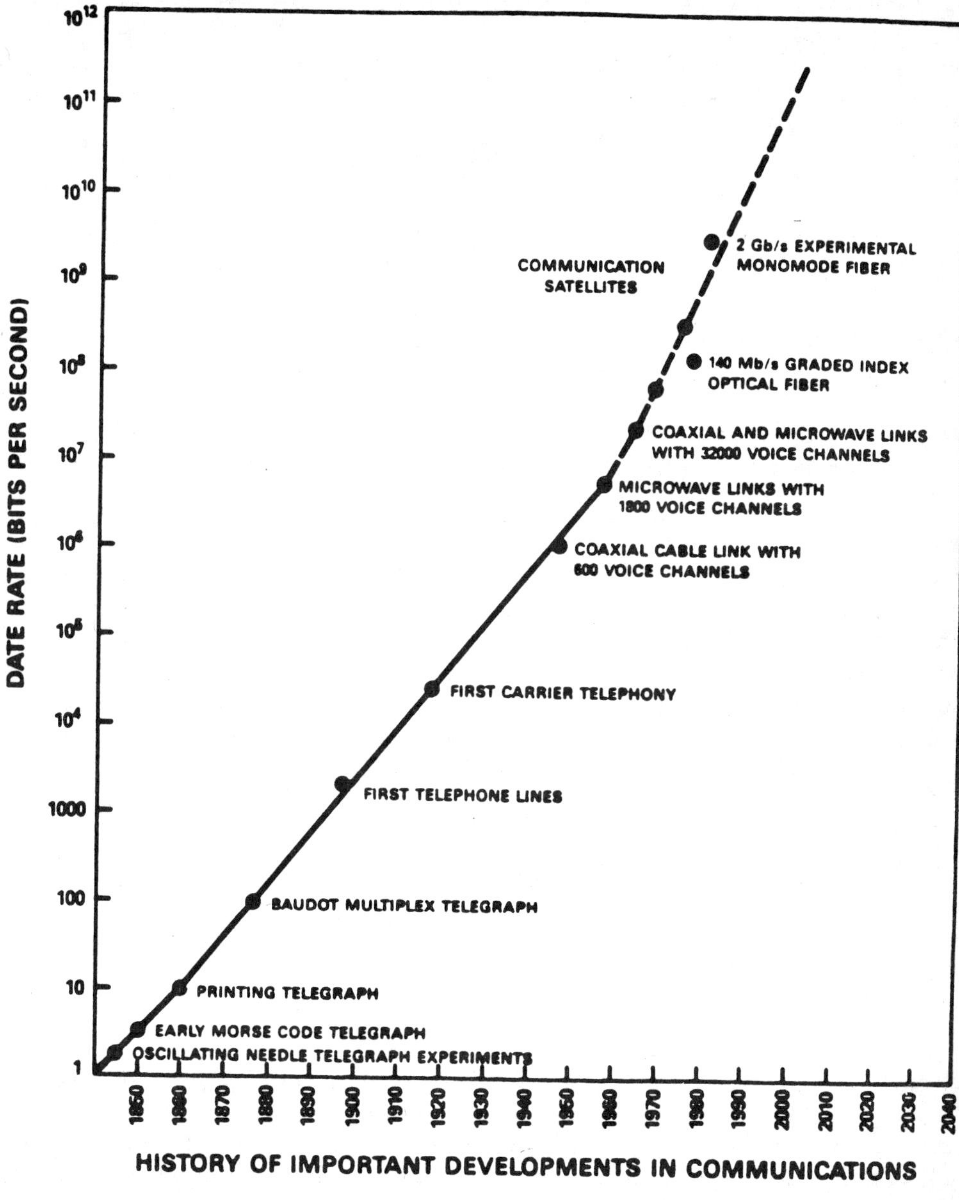

FIGURE 6. Bit rates in communication systems over the last two centuries. Note the break in the curve which occured about thirty years ago, roughly coincident with the invention of the laser.

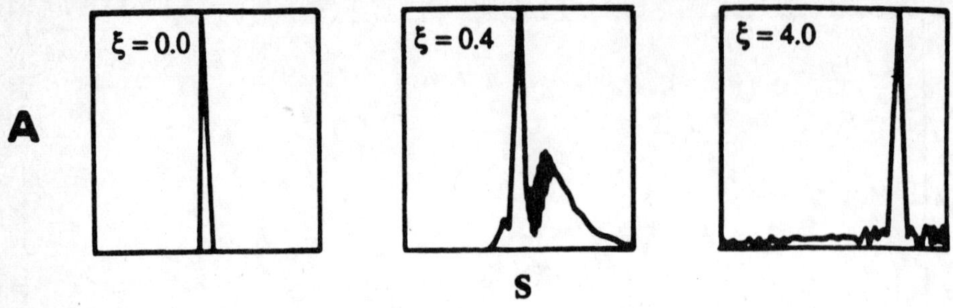

FIGURE 7. Evolution of an initially Gaussian pulse as it propagates along an optical fiber.

point. At large distances, a soliton can clearly be seen emerging! To verify the existence of a soliton, we have looked for stationary solutions of the form

$$u(\xi, s) = u(s - \xi/v)\exp(i\Omega\xi), \tag{21}$$

which converts Eq. (20) from a partial differential equation into an ordinary differential equation. They have found these stationary solutions and checked that they satisfy the original partial differential equation. They have also found that the center frequency of the solitons is shifted down from the zero dispersion point into the anomalous dispersion regime. From a physical standpoint, we might say that the initial pulse has adjusted its frequency in order to minimize the effect of the deformation. Hamiltonian deformations are benign because of the ability which pulses have to adjust to them.

Similarly motivated, Menyuk [17,18] has studied pulse propagation in linearly birefringent fibers. The basic equations in normalized units are

$$\begin{aligned} iu_\xi + i\delta u_s + \frac{1}{2}u_{ss} + (|u|^2 + \frac{2}{3}|v|^2)u &= 0, \\ iv_\xi - i\delta v_s + \frac{1}{2}v_{ss} + (\frac{2}{3}|u|^2 + |v|^2)v &= 0. \end{aligned} \tag{22}$$

The quantity δ is a measure of the birefringence strength, while u and v are complex envelopes of the two fundamental polarizations. At the same central freqency, the two fundamental polarizations have slightly different group velocities which is the origin of the terms proportional to δ. If the coefficient 2/3 in the cross-coupling terms is replaced with 1, then Eq. (22) becomes a version of Manakov's equation [19] and is integrable. The Hamiltonian for Eq. (22) may be written in the form

$$\begin{aligned} H = \frac{i}{2}\int_{-\infty}^{\infty} \Big[& i\delta(u_s u^* - u u_s^* - v_s v^* + v v_s^*) - (|u_s|^2 + |v_s|^2) \\ & + |u|^4 + \frac{4}{3}|u|^2|v|^2 + |v|^4 \Big]\, ds, \end{aligned} \tag{23}$$

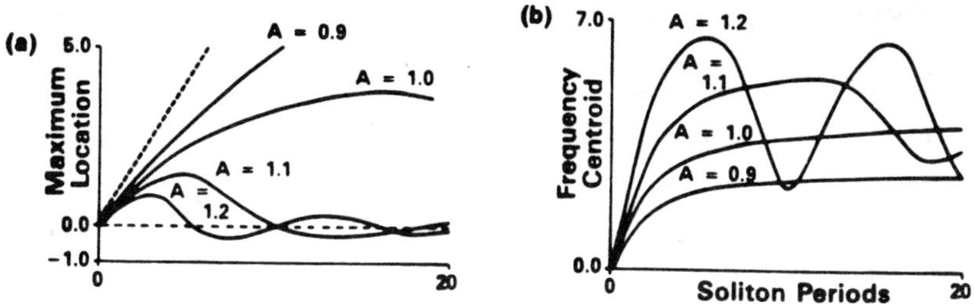

FIGURE 8. The maximum and frequency centroid of the u-polarization is shown as a function of distance at several values of the initial amplitude A.

where (u, u^*) and (v, v^*) are the canonically conjugate pairs. Thus, Eq. (22) is a Hamiltonian deformation of an integrable system, and we expect it to have soliton solutions. One such solution is

$$\begin{aligned} u &= \left(\frac{3}{5}\right)^{1/2} \exp\left[-i\delta s + \frac{i}{2}(1+\delta^2)\xi\right] \operatorname{sech} s, \\ v &= \left(\frac{3}{5}\right)^{1/2} \exp\left[i\delta s + \frac{i}{2}(1+\delta^2)\xi\right] \operatorname{sech} s. \end{aligned} \tag{24}$$

More to the point, we would expect fairly arbitrary initial conditions to produce such solitons, and that is exactly what occurs as long as δ is not too large.

From Eq. (24), it follows that the two polarizations have different central frequencies. The shift in central frequency is just enough to ensure that the two polarizations move at the same group velocity. When the two polarizations are initially at the same central frequency, then the nonlinearity will move at least part of the energy in each polarization to the same central frequency. The two polarizations then self-trap and form a soliton. This effect is shown explicitly in Fig. 8. Our initial conditions are

$$u(\xi = 0, s) = v(\xi = 0, s) = (A/\sqrt{2}) \operatorname{sech} s. \tag{25}$$

The location of the u-maximum is shown as A increases, as well as the frequency centroid of the u-polarization. Below a threshold in A, no soliton forms and the u-maximum increases forever. Above the threshold a soliton forms, the u-maximum oscillates, and the frequency centroid also oscillates. In Fig. 8, we have set $\delta = 0.5$. Qualitatively similar results occur at different values of δ. Distance is measured in soliton periods. One soliton period corresponds to $\xi = \pi/2$.

IV. CONCLUSION

Solitons persist in the face of large Hamiltonian deformations which have no explicit dependence on space or time while non-Hamiltonian deformations usually destroy them. This result has important technical implications for light propagation in optical fibers. By simply determining whether a deformation is Hamiltonian or non-Hamiltonian, we can immediately tell whether or not it is likely to cause trouble. Since this result is quite general, it is of importance not only in fibers, but in many other physical systems as well.

REFERENCES

[1] S. L. McCall and E. L. Hahn, "Self-induced transparency by pulsed coherent light," Phys. Rev. Lett. **18**, 908–911 (1967).

[2] D. J. Korteweg and G. De Vries, "On the change of form of long waves advancing in a rectangular canal, and on a new type of long stationary waves," Philos. Mag. Ser. 5 **39**, 422–443 (1895).

[3] H. Washimi and T. Taniuti, "Propagation of ion acoustic solitary waves of small amplitude," Phys. Rev. Lett. **17**, 996–998 (1966).

[4] V. E. Zakharov, "Collapse of Langmuir waves," Sov. Phys. JETP **35**, 908–914 (1972).

[5] A. Hasegawa and Y. Kodama, "Signal transmission by optical solitons in a monomode fiber," Proc. IEEE **69**, 1145–1150 (1981).

[6] M. J. Ablowitz and H. Segur, *Solitons and the Inverse Scattering Transform* (SIAM, Philadelphia, 1981). See Chapter 1.

[7] M. D'Evelyn and G. J. Morales, "Properties of large amplitude Langmuir solitons," Phys. Fluids **21**, 1997–2008 (1978).

[8] P. K. A. Wai, C. R. Menyuk, Y. C. Lee, and H. H. Chen, "Nonlinear pulse propagation in the neighborhood of the zero dispersion wavelength of monomode optical fibers," Optics Lett. **11**, 464–468 (1986).

[9] P. K. A. Wai, C. R. Menyuk, Y. C. Lee, and H. H. Chen, "Soliton at the zero-group-dispersion wavelength of single-mode fiber," Optics Lett. **12**, 628–630 (1987).

[10] A. J. Lichtenberg and M. A. Lieberman, *Regular and Stochastic Motion* (Springer-Verlag, New York, 1983). See Chapter 2.

[11] C. R. Menyuk, "Origin of solitons in the 'real' world," Phys. Rev. A **33**, 4367–4374 (1986).

[12] C. R. Menyuk, "Application of Lie methods to autonomous Hamiltonian perturbations: Second order calculation," in *Nonlinear Evolutions*, edited by J. P. P. Léon (World Scientific Publ., Singapore, 1988), pp. 571–592.

[13] G. K. Keiser, *Optical Fiber Communication* (McGraw-Hill, New York, 1983). See Chapter 3.

[14] A. Hasegawa and F. Tappert, "Transmission of stationary nonlinear optical pulses with birefringent fibers," Appl. Phys. Lett. **23**, 142–144 (1973).

[15] L. F. Mollenauer, R. H. Stolen, and J. P. Gordon, "Experimental observation of picosecond pulse narrowing and solitons in optical fibers," Phys. Rev. Lett. **45**, 1045–1048 (1980).

[16] L. F. Mollenauer and R. H. Stolen, "Solitons in optical fibers," Laser Focus **18**(4), 193–198 (1982).

[17] C. R. Menyuk, "Stability of solitons in birefringent optical fibers. I: Equal propagation amplitudes," Optics Lett. **12**, 614–616 (1987).

[18] C. R. Menyuk, "Stability of solitons in birefringent optical fibers. II: Arbitrary amplitudes," JOSA B **5**, 392–402 (1988).

[19] S. V. Manakov, "On the theory of two-dimensional stationary self-focussing of electromagnetic waves," Sov. Phys. JETP **38**, 248–253 (1974) [Zh. Eksp. Teor. Fiz. **65**, 505–516 (1973)].

THE JOSEPHSON FLUXON MECHANICS

G. Reinisch and J.C. Fernandez
Observatoire de Nice
Boîte Postale No. 139
F-06003 Nice Cedex, France

ABSTRACT

We present recent results concerning sine-Gordon kink soliton dynamics, and emphasize their physical meaning in the frame of the physics of long Josephson junctions.

This paper emphasizes some theoretical and experimental results concerning the sine-Gordon (S.G.) kink mechanics in the frame of possible laboratory experiments using Long Josephson Junction (LJJ) devices. It will discard any intermediate calculations which is already detailed in previous papers, and will focus on the physics involved in the results.

Let us first remind some important preliminary properties concerning the S.G. equation, the LJJ and their kink soliton solutions [1 – 4]. In its dimensionless form, the S.G. equation reads :

$$\phi_{TT} - \phi_{XX} + \sin\phi = 0 \tag{1}$$

and is immediately deduced either from Lagrange's principle related to the following Lagrange density:

$$\Lambda = \frac{1}{2}\phi_T^2 - \frac{1}{2}\phi_X^2 - 1 + \cos\phi \quad , \tag{2}$$

or from Hamilton's principle applied to the following hamiltonian density (deduced from Λ through the Legendre transformation):

$$h = \frac{1}{2}\phi_T^2 + \frac{1}{2}\phi_X^2 + 1 - \cos\phi \quad . \tag{3}$$

Equation (1) was first introduced by Seeger and Kochendorfer, in solid state physics [5 – 6], then by Perring and Skyrme [7], in order to propose a one-dimensional theory of a scalar field modelling a classical particle. The two next important steps of its history were the emphasis of its pedagogical power by use of the very simple chain of coupled pendula, made by Scott [8], and the solution of the related inverse scattering transform problem obtained by Ablowitz, Kaup, Newel and Segur [9]. This sketch of the S.G. saga is clearly both subjective and drastically simplified. It must be regarded to more as a sequence of a few important steps than as an extensive historical account. For the present purpose, we emphasize an additional important feature of the S.G. history, which is the link with LJJ's [10].

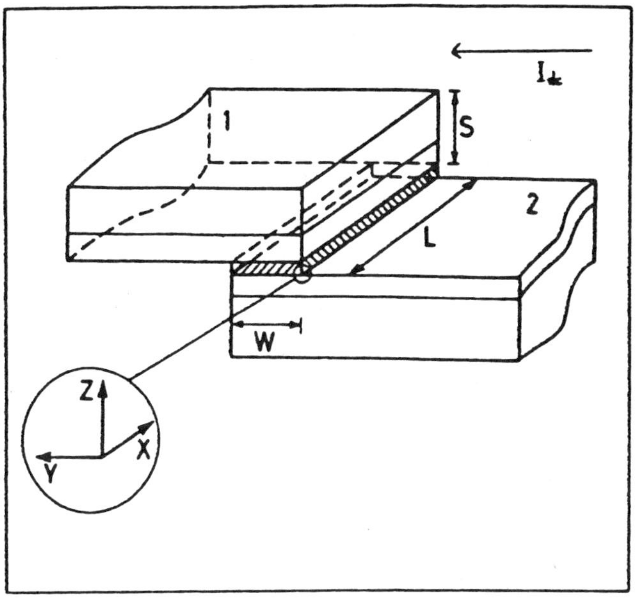

Fig. 1: Typical geometry of a Josephson junction (here an overlap junction). Typical material is $Sn - Sn_xO_y - Sn$. Typical dimensions are $S \sim 4000 \mathring{A}$; $t^* \sim 10 - 40 \mathring{A}$; $L \sim 1mm$; $L/\lambda_0 \sim 5 - 40 \mathring{A}$; $W/\lambda_j 0 \sim 20 \mu m$. The temperature is around $4°K$, $\bar{c} \sim$ few % of c and $\omega_0 \sim 10^{11} Hz$. After S. Pagano.

A LJJ, consisting in two "long" (a few hundred of μm) superconductive layers separated by a very thin oxyde barrier (thickness of order of 20 Å), and overlapping over a narrow strip (whose width w is about 30 μm) according to figure **1**, has its dynamical state described in first approximation by the following partial differential equation (P.D.E.) [1 – 4].

$$\phi_{tt} - \frac{1}{\mathcal{L}C} \phi_{xx} + \frac{2e\, J_o}{\hbar\, C} \sin \phi = - \frac{1}{\mathcal{R}C} \phi_t + \frac{2e}{\hbar C} I(x,t) \quad , \qquad (4)$$

where $\mathcal{L}$, C, J_o, I are respectively the LJJ inductance, capacitance, the maximum Josephson current and the external bias current per unit of LJJ length. For dimension reasons, $\mathcal{R}$ must then be the product of a resistance (the quasi-particle resistance) by the unit of LJJ length. Equivalently one may say that $\mathcal{R}\mathcal{C}$ is the product of the total LJJ resistance by the total LJJ capacitance. The phase $\phi(x,t)$, which measures the local phase difference between the two macroscopic wave functions describing each superconductive layer, is related to the local Josephson current i crossing the insulating layer (per unit of LJJ length) through the first Josephson equation :

$$i_J = J_o \ \sin\phi \qquad , \tag{5}$$

while the second Josephson equation gives the voltage $V(x,t)$ across the insulating barrier as a function of the time derivative of $\phi\,(x,t)$:

$$V\,(x,t) = \frac{\hbar}{2e}\,\phi_t(x,t) \tag{6}$$

($\hbar$ and e are respectively the reduced Planck constant and the electron charge). Then the total current (per unit of LJJ length) through the oxyde layer reads:

$$i_z = i_J + \frac{\hbar}{2e\mathcal{R}}\phi_t + \frac{\hbar C}{2e}\phi_{tt} \qquad . \tag{7}$$

The second term of the r.h.s. of equation (7) represents dissipative effects due to quasi-particle tunnelling, while the third term is related to the energy capacitively stored in the barrier, which yields a displacement current.

Defining as follows the LJJ "plasma" frequency (which is of order of a few hundreths of GHz)

$$\omega_o = \left[\frac{2e\ J_o}{\hbar\ C}\right]^{1/2} \qquad , \tag{8}$$

the LJJ Swihart[11] velocity (of order of a few tenths of the light velocity in vacuum)

$$\bar{c} = [\mathcal{L}\ C]^{-1/2} \qquad , \tag{9}$$

the LJJ penetration depth (of order of some fifty microns)

$$\lambda_o = \bar{c}\ \omega_o^{-1} \qquad , \tag{10}$$

the LJJ reduced time T and space variable X

$$T = \omega_o t \quad ; \quad X = \lambda_o^{-1}\,x \qquad , \tag{11}$$

the LJJ quasiparticle loss factor (related to the so-called McCumber number β_c through $\alpha = \beta_c^{-1/2}$, with β_c typically of order 10^4)

$$\alpha = \left[\frac{\hbar}{2e\, J_o\, \mathcal{R}^2\, C}\right]^{-1/2} = Q^{-1} \quad , \tag{12}$$

where Q is the quality factor:

$$Q = \omega_o\, \mathcal{R} C \quad , \tag{13}$$

and the normalized bias current

$$\gamma(X,T) = J_o^{-1}\, I(\lambda_0\, X\, ,\, T/\omega_0) \quad , \tag{14}$$

we finally obtain the following PDE :

$$\phi_{TT} - \phi_{XX} + \sin\phi = -\alpha\,\phi_T + \gamma(X,T) \quad , \tag{15}$$

where the *l.h.s.* is the dimensionless S.G. operator given by equation (1). Note that, in these reduced units, sin ϕ measures the reduced Josephson current (cf. equation 5) :

$$j_J = J_o^{-1}\, i_J = \sin\phi \quad , \tag{16}$$

while ϕ_T measures the reduced voltage v :

$$\phi_T = V_o^{-1}\, V = v \quad , \tag{17}$$

where

$$V_o = \frac{\hbar\,\omega_o}{2e} \quad . \tag{18}$$

Let us emphasize that equation (7),together with the following expression of the surface current density vector field propagating along the interface between each superconductive layer and the oxyde barrier, within the London penetration depth :

$$\vec{i}_S = \lambda_0^2 (J_0/w) \vec{\nabla}\phi \quad , \tag{19}$$

where $\vec{\nabla} = (\partial/\partial x, \partial/\partial y)$ and w is the LJJ width, and by use of the following equation of continuity :

$$\frac{i}{w} = \vec{\nabla} \bullet \vec{i}_S \quad , \tag{20}$$

yields the two-dimensional perturbed S.G. equation:

$$i_z = \lambda_0^2 J_0 (\phi_{xx} + \phi_{yy}) \quad . \tag{21}$$

Then, assuming $\phi_{total}(x,y,t) = \phi(x,t) + g(y)$, where $g \ll 1$, one recovers the 1+1 dimensional PDE (4) [12].

Equation (1) propagates solitons (kink – or antikink) described by the following formula :

$$\phi(X,T) = 4\tan^{-1}\, exp[-\kappa(u)\sigma(X - uT)] \quad , \tag{22}$$

where

$$\kappa = [1 - u^2]^{-1/2} \tag{23}$$

is the Lorentz factor related to the covariance of the original PDE (1) with respect to Lorentz transformations, and describes the Lorentz contraction of the soliton (22) propagating with the velocity u. The additional factor σ accounts for kinks ($\sigma = -1$) or antikinks ($\sigma = +1$).

The (anti)kink energy is given by:

$$E_S = \int_{-\infty}^{\infty} dx\ h[\phi_S(X,T); \frac{\partial}{\partial T}\phi_S(X,T); \frac{\partial}{\partial X}\phi_S(X,T)] = 8\kappa \tag{24}$$

(cf. equation (3) .

What is such a soliton in the LJJ framework ? Consider, for sake of simplicity, a static soliton : $u = 0$. Since ϕ varies from zero to 2π in the case of, say, a kink (22,23), such a LJJ soliton is a Josephson current distribution (along the X dimension) proportional to sin ϕ according to equation 16). It may therefore be modelled by two opposite "tunnelling" Josephson currents "located" at $X = X_{1,2}$, where $X_{1,2}$ respectively correspond to the values of the phase difference ϕ equal to $\pi/2$ and $3\pi/2$. According to formulas (22,23) , we have $| X_2 - X_1 | \sim 1$.

Therefore, definitions (11) show that the two opposite Josephson "tunnelling" currents are centered about two points which are separated by the Josephson penetration depth λ_o. Then two adapted surface currents close the circuit. Hence a soliton in a LJJ is a **closed current density loop** which is **extremely large** compared to its width, basically of the order of twice the London penetration depth (i.e. a few thousands of Å). This current loop generates a quantum of magnetic flux, whose value is equal to $h/2e = 2.07\ 10^{-15}$ Weber. For this reason, the (anti)kink is often called in LJJ physics a (anti)fluxon.

The energy which is carried along by such a propagating fluxon reads:

$$E_{fluxon} = E_0\, E_S \quad , \tag{25}$$

where E_S is given by equation (24) and E_0 is:

$$E_0 = \frac{\hbar}{2e} J_0\, \lambda_0 \quad . \tag{26}$$

1 THE TWO BASIC PERTURBATIONS RELATED TO THE LJJ : THE DRIVER AND THE DAMPING

1.1 The reflective boundary conditions for a finite damped line

Assume that the (reduced) length of the system is **finite** :

$$-\frac{\ell}{2} \leq X \leq \frac{\ell}{2} \qquad ,\ell = \lambda_0^{-1} L \quad , \tag{27}$$

and that the following boundary conditions (b.c.) are adopted :

$$\phi_X \left(\pm\frac{\ell}{2}\right) = 0 \tag{28}$$

(open - or free - b.c.). Then these reflective boundary conditions create two virtual antisoliton states on the extended line $-3\ell/2 \leq X \leq 3\ell/2$, according to the following receipt:
one kink located at Y such that : $-\frac{1}{2}\ell \leq Y \leq \frac{1}{2}\ell$ creates two virtual antikinks located at $\tilde{Y}_1$ and $\tilde{Y}_2$ such that :

$$\frac{1}{2}[Y + \tilde{Y}_1] = -\frac{1}{2}\ell \; ; \frac{1}{2}[Y + \tilde{Y}_2] = +\frac{1}{2}\ell \quad , \tag{29}$$

and we have the equivalent formulation for an antikink located at Y. Hence the particular boundary condition (27, 28) are equivalent to a soliton-antisoliton collision process at both ends of the LJJ (see figures **2** and **3**). This very simple physical analogy allows to proceed from a one-soliton dynamical problem on the finite line to a 3-soliton problem on the infinite line, and hence to use the corresponding results concerning the dissipative case [**13** – **17**].

Therefore, when a kink (say) collides with either end of the LJJ, it changes into an antikink, experiences both an **kinetic** energy loss, of order of :

$$\Delta\, E_K = -2\,\pi^2\,\alpha \quad , \tag{30}$$

and a so-called "phase shift", which means that **after** the collision, the reflected soliton – now an antikink – is **in advance** with respect to the position obtained by the extrapolation from the impacting kink velocity, according to :

$$\delta = 2[1 - u^2]^{1/2}\, ln \mid u \mid \tag{31}$$

(see figure **3**).

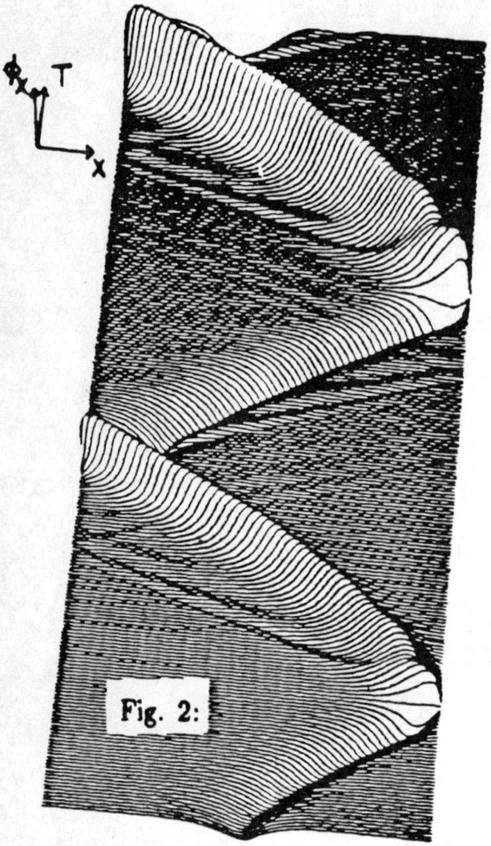

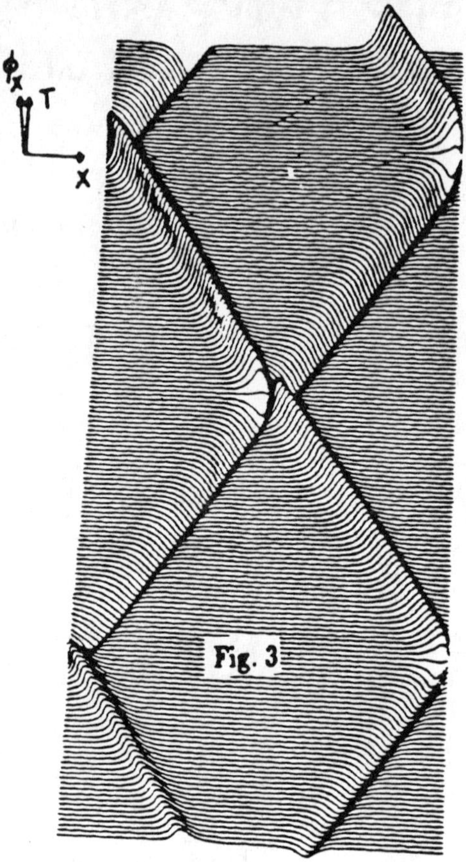

Fig. 2: The shuttling dynamics of a driven damped S.G. soliton $\ell = 20$. $\gamma = 0.1$. $\alpha = 0.1$. $0 \leq t \leq 150$. Note the transient regime before the soliton reaches its limit velocity u_∞, together with its corresponding phonon emission.

Fig. 3: Two unperturbed S.G. solitons colliding symetrically with either LJJ end. Note that the reflection is rigorously identical with the two-soliton collision process which occurs at the center of the line. Note also the spatial advance experienced by each colliding soliton after the shock, which is described by the phase-shift equation (31). Here $\ell = 40$; $\alpha = \gamma = 0$; $u = 0.6$; $0 \leq t \leq 100$.

Actually, formula (31) simply describes the phase-shift experienced by a kink (even in the unperturbed case $\alpha = \gamma = 0$) when it collides with an antikink, as a consequence of the kink-antikink attraction force [18].

Let us introduce the simplest collective-coordinate $Y(T)$ by assuming that the dynamical behaviour of the solitary-wave solution of equation (15) may be describe by a single "collective" degree of freedom derived from equation (22), according to:

$$\phi(X,T) = 4\ \tan^{-1}\ exp\left[-\sigma\ \frac{X - Y(T)}{(1-\dot{Y}^2)^{1/2}}\right] \quad . \tag{32}$$

Then we have the two following reflection boundary conditions respectively related to the change of velocity and the change of position experienced by the pointlike particle after its reflection, according to equations (30) and (31) [19]:

$$\dot{Y}_+ = -\dot{Y}_-\left[1 - \frac{\pi^2\alpha(1-\dot{Y}^2)^{3/2}}{4\dot{Y}^2}\right] \quad . \tag{33}$$

$$Y_+ = sign(Y_-)\left[\frac{1}{2}\ell + 2(1-\dot{Y}^2)^{1/2} ln|\dot{Y}|\right] \quad , \tag{34}$$

where "plus" and "minus" respectively means : "just after" and "just before" the reflection.

In presence of a weak damping $\alpha << 1$, a slowly impacting soliton ($|\dot{Y}| << 1$) is killed at the reflection at either end of the LJJ when :

$$\dot{Y}_{min}^{\ 2} = \frac{1}{2}\pi^2\alpha \rightarrow |\dot{Y}_{min}| = \pi\sqrt{\frac{\alpha}{2}} \quad . \tag{35}$$

This means that the soliton kinetic energy, whose value is $4\dot{Y}^2$ in the dimensioness units (11) (cf. equation (24)) is now equal to the energy loss (30). Therefore if the impacting kink velocity $\dot{Y}_-$ is less than the threshold value defined by equation (35), the soliton will annihilate into a decaying soliton-antisoliton bound state (see figure 4).

1.2 The constant driver $\gamma << 1$.

Consider equation (15) with $\alpha = 0$, and $\gamma = constant << 1$. Assume that the initial condition is given by formula (22), with $u = 0$ and, say, $\sigma = +1$ (static antikink located at $X = 0$). Then, on the "far wings" of the antikink, i.e. for $|X| >> 1$, the PDE (15) may be linearized about 0 (2π), with $\phi_{XX} \equiv 0$ (provided the driver is small enough), and one is left with a weakly driven harmonic oscillator, which can be reduced to the undriven harmonic oscillator by assuming :

$$\phi = \tilde{\phi} + \sin^{-1}\gamma \sim \tilde{\phi} + \gamma \quad . \tag{36}$$

This means that if the (anti)kink now connects the two new fundamental energy states - coresponding to $\phi(-\infty) = 2\pi + \sin^{-1}\gamma$ and $\phi(+\infty) = \sin^{-1}\gamma$, instead of (respectively) 2π and 0 for the "unperturbed" antikink (22) -, there is no effect of the driving field γ on the linear part of the field $\phi(x,t)$. In this case, one may consider that the effect of the driver is restricted to the nonliner part of the field, when $\phi \sim \pi$, i.e. about the soliton "center". Assuming that the field $\tilde{\phi}$ may be described by the ansatz equation (32). we obtain in the case of a constant driver the so-called "newtonian S.G. soliton dynamics" by use of several simple perturbation -and/or hamiltonian- methods [20 – 27]. In terms of the collective-coordinate $Y(T)$, we have indeed:

$$\ddot{Y} = \frac{\pi\,\sigma}{4}(1 - \dot{Y}^2)^{3/2}\,\gamma \quad . \tag{37}$$

Note that, if the initial condition is the original soliton given by equation (22), its far wings $\phi \sim 0(2\pi)$ will oscillate as an X-independant flat state with an amplitude 2 $\sin^{-1}\,\gamma$ and a frequency equal to unity about the value $< \phi_\infty >= \sin^{-1}\,\gamma$.

The existence and stability of this (spatially) asymptotic oscillating flat state (with or without damping) can be established with classical mathematical analysis, most simply in the frame of the Nonlinear Schrödinger (NLS) approximation [28]. A hysteresis diagram depicting the solution of the algebraic equation which yields the time evolution of the (complex) amplitude of the flat state is obtained, and its stability is stressed. As the ampitude of the flat state increases, the most unstable mode changes from a flat ($k = 0$) state, to a $k_1 = 2\pi/\ell$ state (where ℓ is the reduced LJJ length), and then on, to a $k_2 = 2(2\pi/\ell)$ state, and so on...

These waves will interact in a complicated way with the nonlinear part of the soliton, approximated by equation (32), and the dynamics will no longer be newtonian : instead of a constant initial acceleration given by $\ddot{Y} \sim \frac{1}{4}\,\pi\,\sigma\,\gamma$, as in equation (37), we obtain $\ddot{Y} = \frac{1}{8}\,\pi\,\sigma\,\gamma\,t^2$: the solitary wave starts more slowly [20]. The above interaction between linear and nonlinear waves of the PDE (15) may then be described in terms of coherent or (random-phase-approximation) incoherent coupling processes [26 – 27].

Therefore, the nonlocal range of the driver γ ensures the length ℓ of the system to become a (sometimes) important parameter for the dynamical description of the system [28 – 30].

1.3 The effect of damping : $\alpha << 1$.

When a weak damping is present, the above nonlocal effect of the driver, i.e. the possible generation of linear "plasma" waves $\omega^2 = 1 + k^2$ perturbating the newtonian dynamics (37) , becomes a transient effect over a time of order α^{-1} (see figure **2**). The asymptotic state for ϕ is described by equations (36) and (32), applied to $\tilde{\phi}$. We have: [22 – 24]:

$$\ddot{Y} = \frac{\pi\,\sigma}{4}(1 - \dot{Y}^2)^{3/2}\,\gamma - \alpha(1 - \dot{Y}^2)\dot{Y} \quad , \tag{38}$$

(compare with equation (37). In particular, when the (anti)kink now reaches the following (absolute value of the) velocity :

$$u_\infty = \left[1 + \frac{16\,\alpha^2}{\pi^2\,\gamma^2}\right]^{-1/2} \quad , \tag{39}$$

we have $\ddot{Y} = 0$. Actually, velocity (39) is the limit velocity which is asymptotically reached after the (complicated) transient regime described above, since it corresponds to a **stable** stationnary dynamical state, where $\dot{Y} = \ddot{Y} = 0$ (see figure **2**).

1.4 The translation of the above results in terms of LJJ physics.

Let us consider the reflective boundary conditions (28). In terms of LJJ physics, they represent open circuit ends since ϕ_x is proportional to the current, according to equation (19). Equivalently, according to Maxwell equations, they mean that there is no external magnetic field $H_\perp$ along the transverse (y) direction. If such a field would be present, they would change into the following boundary conditions :

$$\phi_x(\pm\frac{L}{2}) = \frac{2\,e\,d}{\hbar\,c}\;H_\perp \quad . \tag{40}$$

and the simple equivalence between soliton reflection and collision would no more exist.

Note that,considering the 2-dimensional perturbed S.G. equation (21), similar b.c. as in equation (40) yield:

$$\phi_y(\pm\frac{w}{2}) = \frac{2\,e\,d}{\hbar\,c}\;H_{\|}(\pm\frac{w}{2}) \quad , \tag{41}$$

where $H_{\|}$ is the magnetic field component along the x-direction. Even when no external magnetic field is present, such a component always exists,due to the current I flowing in the junction (see figure (1) .Then,extrapolating linearly ϕ_y between the two values given by equation (41) leads to the function $g(y)$ which allows the reduction of equation (21) to equation (4).

Now, consider that the soliton is shuttling in the LJJ, according to the reflection conditions (28). If the junction length is large enough to allow the velocity of the fluxon to relax rapidly to the limit velocity (39) after each reflection (see figure **2**), then the current-voltage characteristics curve (or $I - \bar{\bar{V}}$ characteristics) takes the well-known "Zero-field-step" (ZFS) profile [1 – 4], according to the following formulas, immediately deduced from equations (17), (22) and (39):

$$\bar{V} = \frac{V_0}{L}\int_{-\frac{L}{2}}^{\frac{L}{2}} dx\;\phi_T = -\frac{\lambda_0\,V_0}{L}\int_{-L/2\lambda_0}^{L/2\lambda_0} dX\;u\;\phi_X \;\simeq\; 2\pi\,\lambda_0\,V_0\,u_\infty\,\sigma\,L^{-1} \quad , \tag{42}$$

or

$$\bar{V} = 2\,\pi(\frac{\hbar}{2e})\;\bar{c}(\sigma\,u_\infty)L^{-1} = \frac{\pi\,\hbar\,c}{e\,L}\;|\,u_\infty\,|\;\frac{\bar{c}}{c} \quad , \tag{43}$$

$$\bar{\bar{V}} = Lim_{T_0=\infty}\,\frac{1}{T_0}\int_0^{T_0} dT\,\bar{V} = \bar{V} \quad . \tag{44}$$

$\bar{V}$ means a spatial average of the voltage V over the LJJ length; $\bar{\bar{V}}$ is the time average of $\bar{V}$ over a time interval $T_0 \gg 1$, and is beleaved to represent the actual measured voltage. The absolute value of u_∞ appears because, when γ is assumed, say, positive,a kink ($\sigma = -1$) has a negative limit velocity ($-u_\infty$), according to equation (37), while the antikink ($\sigma = +1$), which is obtained after the kink reflection, relaxes toward the positive velocity, $+u_\infty$. Introducing the constant $\Phi_0 = \pi\, \hbar c/e = 6.2\ 10^-5 volts.cm$, where c is the velocity of light in vacuum—- note that $\Phi_0 = c(magnetic\,flux\,quantum)$—-, we obtain the following ZFS characteristics (cf. equation (39)) :

$$\gamma(\bar{V}) = \frac{16\alpha}{\pi} \frac{\bar{V}}{\left[\frac{\Phi_0^2\, \bar{c}^2}{L^2\, c^2} - \bar{V}^2\right]^{1/2}} = \frac{16\alpha}{\pi} \frac{\vartheta}{\left[1 - \vartheta^2\right]^{1/2}} \quad , \tag{45}$$

where:

$$\vartheta = \frac{L\, c\, \bar{V}}{\Phi_0\, \bar{c}} = \frac{1}{2\pi} \left[\frac{L}{\lambda_0}\right] \left[\frac{\bar{V}}{V_0}\right] \; . \tag{46}$$

Finally, let us mention the important complementary experimental detection of the fluxon shuttling regime, in terms of electromagnetic emission. With the reflective boundary conditions (28)), each time a fluxon is reflected at either end of the LJJ into an antifluxon, there is an emission of a short and weak (power $\sim 10^{-10}\ W$) electromagnetic pulse, due to the very rapid $4\,\pi$- change of the phase $\phi(\pm\frac{l}{2})$, during the reflection (i.e. collision) time. One may consider that this emission is related to the energy loss (30) which occurs during the kink reflection.

Therefore, the voltage $V(\pm\frac{l}{2})$ appears as a periodic time series of spikes separated by the period $2\,L/\bar{c}\,u_\infty$, each spike having an amplitude equal to $4\pi\ V_0/\delta$ where δ is the collision time, of order $\lambda_0/\bar{c}\ u_\infty$ .Hence, the frequency of the emitted radiation is

$$\nu_{em} = \frac{\bar{c}\, u_\infty}{2L} = \frac{e\,\bar{V}}{h} = \frac{1}{2}\ \nu_J \tag{47}$$

(cf. equation (43)), where ν_J is the Josephson frequency (equal to $2e\,\bar{V}/h$), while its spectrum appears as a large collection of very narrow spectral features consisting in harmonics of the fundamental (47).

On the other hand,the power emitted may be estimated, according to equations (24, 25, 26, 30), as follows:

$$P \sim \frac{E_0\, \Delta\, E_K}{(2L/\bar{c})} = \frac{\pi\, \alpha}{2} \left(\frac{\lambda_0}{L}\right) \left(\frac{\bar{c}}{c}\right)\ J_0\ \Phi_0 \quad . \tag{48}$$

Taking, as an example, [31], $J_0 = [270\, A.cm^{-2}].[15\ 10^{-4} cm]$, $L = 35\,\lambda_0$, $\bar{c} = 0.017\,c$, $\alpha = 0.038$, we have: $\nu_{em} = 3.6\,GHz$ and $P = 7\ 10^{-10} W$. This value of P is somewhat optimistic with respect to the upper threshold of the emitted power from a single LJJ,which is estimated to be about $200 picowatts$ [32]. The reason is, of course, that a significant part of power P is dissipated inside the LJJ,as linear phonon modes. The importance of this part strongly depends on the external load, at the LJJ end, and is not taken into account in the above calculation.

1.5 The existence of phase-locked fluxon limit cycles : the case of an inhomogeneous external a.c. driving field.

Assume now that the external field γ is space and time dependent, according to [**33**]:

$$\gamma(X,T) = \varepsilon \cos KX \cos \Omega T \quad . \tag{49}$$

In terms of LJJ physics,this means that we assume a standing-wave pattern for the electric field propagating in the superconducting transmission line (4). The external field γ reads,in the "true" units:

$$\gamma(x,t) = \varepsilon \cos(\lambda_0^{-1} K x) \cos(\omega_0 \Omega t) \quad , \tag{50}$$

and the phase velocity of the $\pm K$ electric waves is (cf. equation (10)):

$$u_\varphi = \frac{\omega_0 \, \lambda_0 \, \Omega}{K} = \bar{c}\left(\frac{\Omega}{K}\right) \quad . \tag{51}$$

Let us choose the following matching conditions:

$$u_\varphi = \bar{c} \quad , \tag{52}$$

$$\lambda_0^{-1} K = \pi \, L^{-1} \quad . \tag{53}$$

Then,we have:

$$\Omega = K = \pi \, / \, l \quad , \tag{54}$$

with $l = \lambda_0^{-1} L$. Note that the constant driver may be considered as a special case where $\Omega = K = 0$. Hence,the ZFS regime has the structure of a fluxon limit cycle which is phase-locked to the $\Omega = 0$ frequency, with an infinite degeneracy concerning the parameter l (see figure **2**). The aim of this part is to show that choosing a $\Omega \neq 0$ frequency according to equations (54) breaks-up this degeneracy and leads to the so-called limit "C-cycles" which are phase-locked to the frequency Ω of the external a.c. field γ.

We consider for instance the following set of (reduced) parameters:

$$\alpha = 0.01 \, , \, \varepsilon = 0.2 \, , \, \Omega = K = 0.06 \, , \, l = \pi \, K^{-1} = 52.4 \quad , \tag{55}$$

and perform numerical simulations of the PDE [(15), (49)], using the method of characteristics. In a real experimental situation, it is impossible to control the initial data corresponding to the position and velocity of the fluxon when the a.c. field is switched on. Nor it is possible to control the parity of this initial fluxon. Actually, the experimental way to reach a zero d.c. bias driving of a fluxon in presence of an a.c. bias current is to slowly decrease the value of the d.c. bias along the ZFS, once the a.c. field (50) is switched on. Therefore, we perform a series of numerical simulations which are supposed to cover a reasonably extensive range of such arbitrary initial data . We choose for each numerical experiment an initial antikink, according to the following series of initial data (see equations (22) and (23):

$$\phi(X,0) = 4\tan^{-1}\exp[-\kappa\,\sigma\,(X - X_0)] \quad , \tag{56}$$

$$\phi_T(X,0) = -u\,\phi_X(X,0)\ ; \quad \kappa = (1-u^2)^{-\frac{1}{2}} \tag{57}$$

$$u = -0.9 + n(0.2)\,;\; 0 \leq n \leq 9\ ; \quad X_0 = 14.5\ ;\; \sigma = +1 \quad . \tag{58}$$

The PDE (15) is numerically integrated up to $T = 20000$ in all cases. Defining respectively the fluxon position $Y(t)$ and the velocity $\dot{Y}(t)$ according to the following formulas :

$$\phi_X(Y,T) = \max_{-\frac{1}{2}l \leq X \leq +\frac{1}{2}l} [|\phi_X(X,T)|] \ , \ \dot{Y} = -\frac{\phi_T}{\phi_X}(X = Y, T) \quad , \tag{59}$$

, then displaying the fluxon dynamics in its phase space $\{Y, \dot{Y}\}$ (taking into consideration the parity σ),we obtain for all numerical simulations (56-58) asymptotic results which clearly show the presence of two attractive limit "C-cycles" phase-locked to the external a.c. field (see figure **5**). Each of these two C-cycles may be obtained from the other by the symmetry of the system $Y \to -Y$; $\dot{Y} \to -\dot{Y}$; $\sigma \to -\sigma$. Moreover, each such C-cycle is actually described twice within a period $2\pi/\Omega$ of the external field, half a period as a (say) antikink, and then as a kink. Therefore, if we only consider the (anti)soliton position and velocity, in the phase space $\{Y, \dot{Y}\}$ without taking into account the parity of the fluxon, the corresponding limit C-cycle displays a phase-locking of the system to the external frequency Ω with a ratio equal to 2. Then, the a.c. driven and damped (anti)fluxon is bouncing against a *single* LJJ end at each half-period of the external field.

Therefore, the emission of the electromagnetic pulses by the (anti)fluxon at each reflection , described by equations (47), (48), occurs at the frequencies $2(n+1)\Omega$. This property should lead to an unambigous experimental evidence of such an a.c. driven and damped (anti)fluxon phase-locked regime.

Note that, since the time averaged voltage $\bar{\bar{V}}$ (cf. definition (44)) is zero for such a "C-cycle" fluxon dynamics (cf. equations (42), (43) and the related comments), the vertical step in the $I - \bar{\bar{V}}$ curve corresponding to such phase-locked regimes is located about the origin ($I = 0; \bar{\bar{V}} = 0$), and cannot be experimentally observed.

A trivial theoretical approach - which nevertheless gives excellent results, and, hence, emphasizes the appealing pointlike particle description of the S.G. kink dynamics - consists in substituting the ansatz (32), together with the b.c. (33) and (34), in the original PDE (15) and (49). Then the projection of the resulting equation onto the soliton Goldstone mode as detailed in [34] and [35] leads to the following equation of motion, which accounts quite well for the above PDE

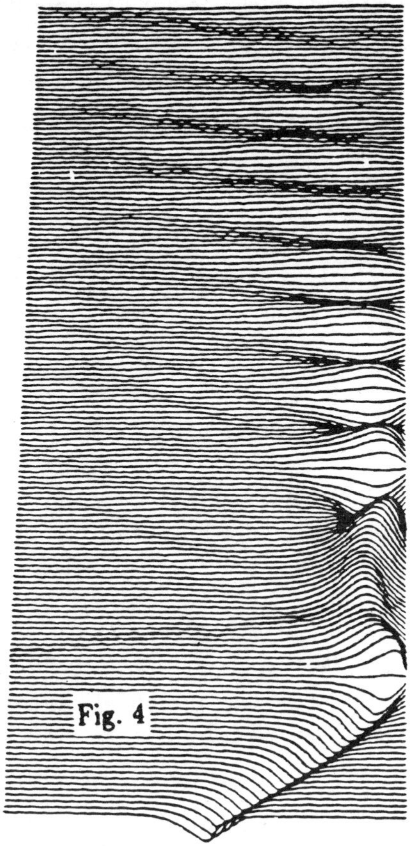

Fig. 4

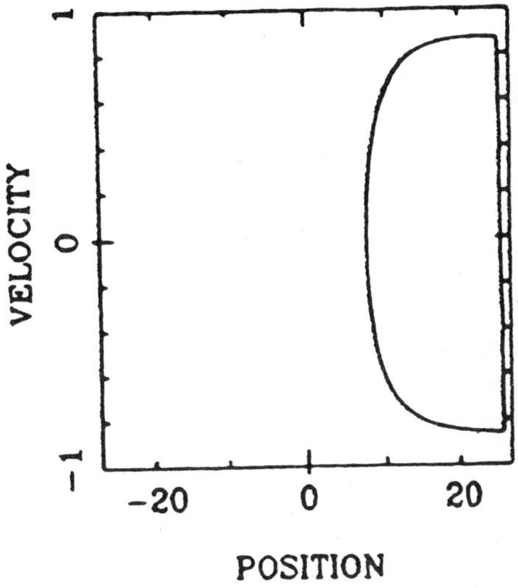

Fig. 5

Fig. 4 An antikink with initial velocity $u = 0.8$ colliding at the right end of a damped LJJ: $\alpha = 0.05$. All other perturbation parameters are zero. The drawing in perspective (for t) displays the soliton ϕ_X annihilating at its reflection.

Fig. 5 The right-sided limit C-cycle obtained from the initial conditions $n = 1$ to $n = 9$, after a transient time of order 4000,displayed in the $[Y, \dot{Y}]$*phase-space*.

numerical results concerning the phase-locked "C-cycle" regimes (including the transients):

$$\ddot{Y} = \frac{\pi\sigma}{4}(1 - \dot{Y}^2)^{3/2}\,[\varepsilon \cos KY \;\cos \Omega T] - \alpha(1 - \dot{Y}^2)\dot{Y} \tag{60}$$

(compare with equation (38)). The discrepancy between the PDE and the ODE results is less than 3/100.

Finally, note the following symetry which keeps the equation of motion (60) invariant:

$$Y \to \bar{Y} = bY\,;\; t \to \bar{t} = bt\,;\; L \to \bar{L} = bL\,; \tag{61}$$

$$\Omega \to \bar{\Omega} = b^{-1}\Omega\,;\; \alpha \to \bar{\alpha} = b^{-1}\alpha\,;\; \varepsilon \to \bar{\varepsilon} = b^{-1}\varepsilon \quad . \tag{62}$$

In these transformations, the velocity $\dot{Y}$ remains unchanged, and if an asymptotic "C-cycle" is obtained for, say, the range of parameters (55)-(58), the existence of other similar limit cycles can be deduced from the above rescaling, provided $\bar{\varepsilon} \ll 1$. The particular case $b = -1$ is equivalent to the symetry $Y \to -Y; \dot{Y} \to -\dot{Y}; \sigma \to -\sigma$ in the ODE (60), which was already mentionned above.

2 THE MAXIMUM JOSEPHSON CURRENT POTENTIAL WELL.

2.1 A S.G. model for classical kink mechanics.

Introducing a long-range (compared to λ_0 : cf. equation (10)) modulation of the maximum Josephson current density, according to the following (reduced) equation :

$$\phi_{TT} - \phi_{XX} + [1 + \frac{1}{4} W(X)] \sin \phi = 0 \quad , \tag{63}$$

leads to the best model of point-like S.G. kink classical newtonian mechanics, where the spatial modulation $W(X)$ plays the role of the potential, and the (anti) kink (32) the role of the point-like particle, whose position is defined by the single degree of freedom $(Y(T)$ [36]. In particular, it is important to note that : i) the small parameter in equation (63) is **not** the amplitude of the well W, but its gradiant. Hence we assume :

$$\varepsilon \sim \lambda_0 / \mathcal{L}_{\prime} \ll 1 \quad , \tag{64}$$

where $\mathcal{L}_{\prime}$ is a characteristic length of $W(x)$. Note that inequality (64) simply means that the reduced characteristic length of the potential well $W(X)$ is great compared to unity (which is the soliton dimensionless width).

ii) Typical values of the potential amplitude may exceed 20 [34]. In such conditions, the only linear waves which may be excited are the standing (Schrödinger-like) waves obtained by linearization of the PDE (63) about the soliton [37]. Theses waves do not depend any more on the length of the whole system, as was the case for the "phonons" described in §**1.2**. Therefore. the restrictions to the concept of a pointlike mechanics made in this §**1.2** disappear.

To the contrary, the effect of the standing waves, far from perturbating the classical kink dynamics, can be shown [38] to lead to the following exact newtonian dynamics of the pointlike kink, when this latter, described by the PDE (63), is assumed trapped in the low-energy part of an harmonic potential well $W(X) = \beta X^2$ ($\beta << 1$; see figure **6**) :

$$8 \ddot{Y} = - \frac{d}{dX} W(X) \,|_{X=Y} \; + 0(\beta^2) \quad . \tag{65}$$

where we assume (nonrelativistic kink dynamics):

$$W(Y) << 8 \quad . \tag{66}$$

The spectrum of the small (linear) oscillations about a given static kink located at the bottom of the harmonic potential well displays the discrete structure of a series of sharp peaks which, with the exception of the fundamental, describe internal oscillations of the kink [37]. On the other hand, the fundamental frequency of this spectrum :

$$\Omega_N = \frac{1}{2}\sqrt{\beta} \quad , \tag{67}$$

is the newtonian frequency of the oscillations (65) about $Y = 0$ of the kink-like particle, the (reduced) mass of which is 8 (cf. equation (24)).

Actually,it is the eigenfrequency related to the perturbed Goldstone (translation) eigenmode of the **stationary** kink-like solitary wave defined by the following ODE, which is derived from the PDE (63) by assuming $\phi_{TT} \equiv 0$[38] :

$$-\phi_{XX} + [1 + \frac{1}{4} W(X)] \sin\phi = 0 \quad , \tag{68}$$

$$\phi_X(\pm\infty) = 0 \quad . \tag{69}$$

Therefore, the link between the collective-coordinate $Y(T)$ introduced in formula (32) and the above spectral Sturm-Liouville description of the dynamical states of the confined kink is achieved through formulas (67 -69). Equation (68) defines the relevant profile which actually describes the extended particle field bouncing back and forth in the bottom of the (harmonic) potential well,at the exact newtonian frequency (67),up to order $o(\beta^2)$.

Considering now the kink internal oscillation spectrum, a second collective-coordinate $k(T)$ taking these oscillations into account may be introduced by use of the following ansatz (cf. formula (22)) :

$$\phi(X,T) = 4\tan^{-1} \; exp[-\sigma\, k(T)(X - Y(T)] \quad . \tag{70}$$

Of course, two single degrees of freedom, $Y(T)$ and $k(T)$, cannot display the same amount of information as a whole spectrum consisting in an infinite series of discretes features which range from the fundamental (67) up to (high) frequency levels corresponding to top parts of the well. Nevertheless, the agreement between numerical experiments performed on the original PDE (63) and the description obtained by use of the ansatz (70) is surprisingly good [34].

Let us now consider the damped and driven PDE (63):

$$\phi_{TT} - \phi_{XX} + [1 + \frac{1}{4}W(X)]\sin\phi = -\alpha\,\phi_T + \gamma(T) \tag{71}$$

(cf. equation (15). By direct substitution of the ansatz (70) into (71), and then projecting both members of the resulting equation onto the soliton Goldstone translation mode $\partial\phi/\partial Y$ and "Goldstone" internal mode $\partial\Phi/\partial k$ [35], we obtain the two ODE's governing respectively the time evolution of the collective-coordinate $Y(T)$ and $k(T)$ [34].

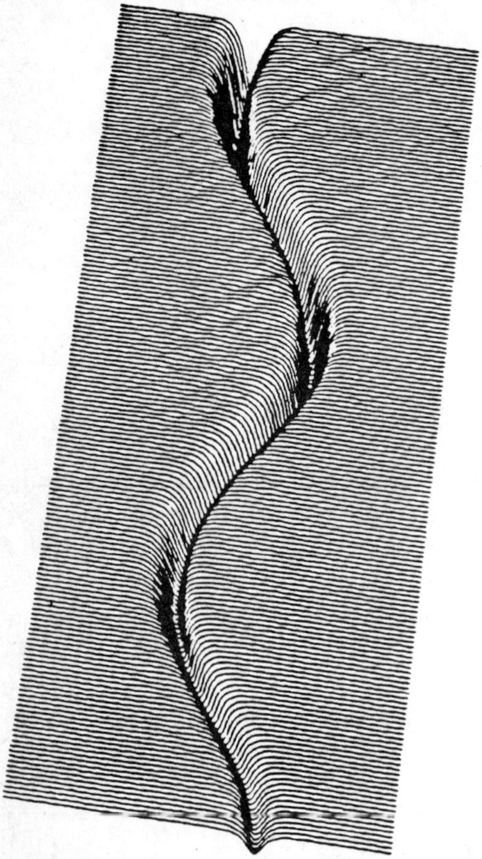

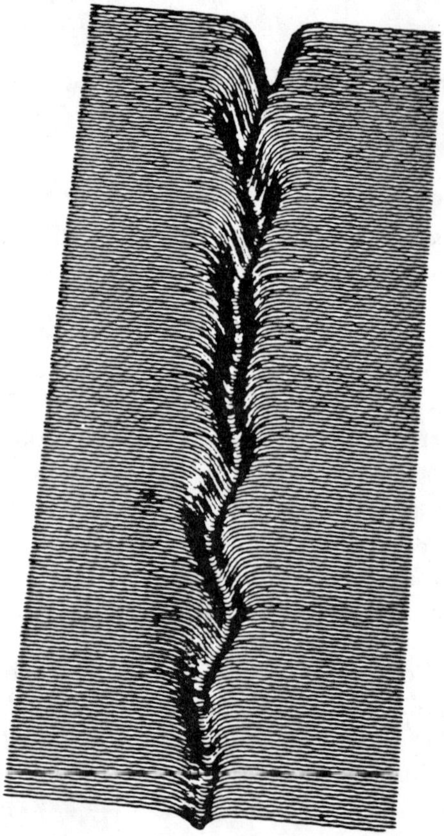

Fig. 6 The oscillation of an antikink about the bottom of an harmonic potential well, where $\beta = 0.09$. The antikink (displayed as ϕ_X versus X in t-perspective) was initially located at a distance equal to 6 with respect to the bottom of the well.

Fig. 7 An initial antikink (displayed as ϕ_X versus X in t-perspective) initially located at the bottom of a harmonic potential well, where $\beta = 0.09$. The a.c. external driving ($\varepsilon = 0.01; \Omega = 0.1$) is off-resonance . There is no damping.

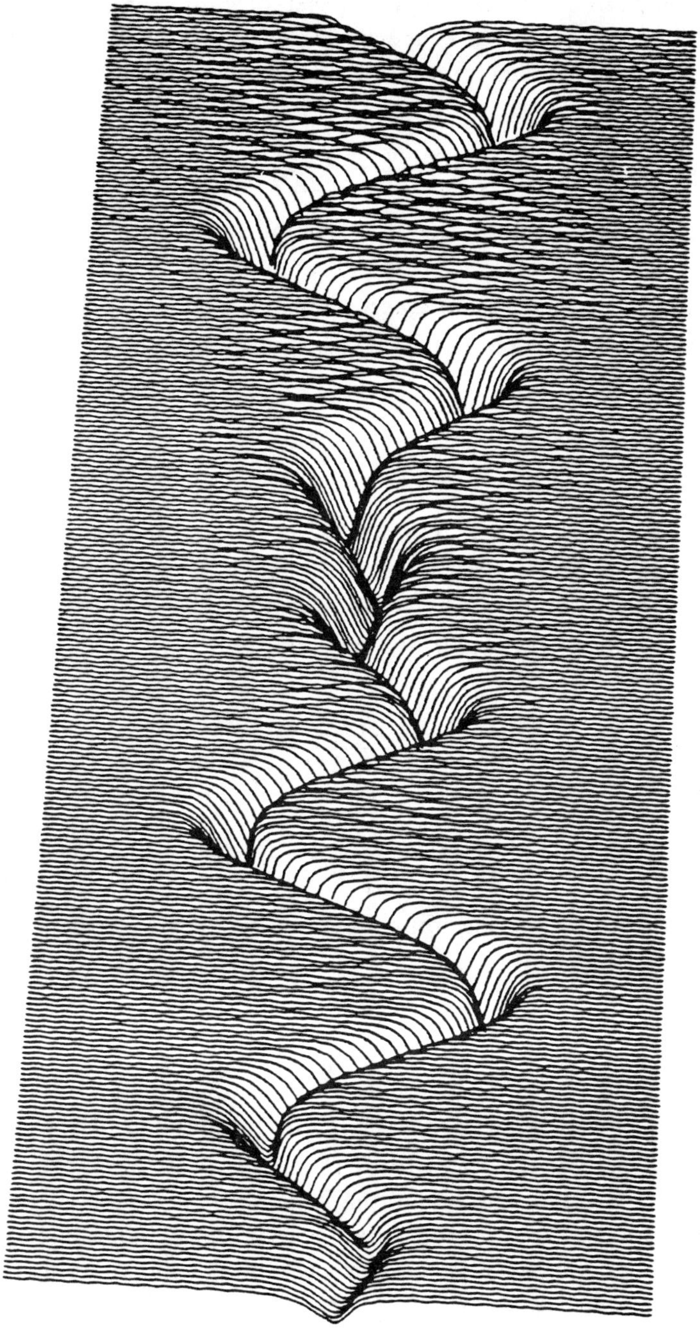

Fig. 8 The same situation as in figure **7'**,but with a resonant external driver: $\varepsilon = 0.01; \Omega = 0.127$.

This system, consisting in two ODE's, can even be simplified further by taking into account the high value of the frequency of the oscillations of $k(T)$. We finally obtain a WKB-type of the kink profile modulation, according to the following equation [34] :

$$k(T) = k[Y, \dot{Y}] = \left[\frac{1 + \frac{1}{4}\, W(Y)}{1 - \dot{Y}^2}\right]^{\frac{1}{2}} \quad . \tag{72}$$

This equation simply means that the slow spatial variation of the factor $[1 + \frac{1}{4}W(X)]$ – see inequality (64) – in the l.h.s. of equation (63) allows a local renormalisation of time and space,according to: $\tilde{T} \sim [1 + \frac{1}{4}W(X)]^{\frac{1}{2}}\, T$ and $\tilde{X} \sim [1 + \frac{1}{4}W(X)]^{\frac{1}{2}}\, X$. The single ODE left, which describes the soliton dynamics (see figures **7** and **8**), reads :

$$\ddot{Y} = -\frac{1}{8\,k^2}\,\frac{d\,W(Y)}{d\,Y} + \frac{1 - \dot{Y}^2}{8\,k}\,[2\,\pi\,\sigma\,\gamma(T) - 8\,\alpha\,k\,\dot{Y}] \quad , \tag{73}$$

where k is given by equation (72). The first term of the r.h.s. of equation (73) describes the trapped kink dynamics, whose low-energy approximation is given by equation (65). The second term refers to equation (60). Due to the presence of the potential well W in the system, the boundary conditions (33 and 34) must be corrected as follows :

$$\dot{Y}_+ = [-\dot{Y}_-]\,\left[1 - \frac{\alpha\pi^2[1 - \dot{Y}^2]^{3/2}}{4[1 + \frac{1}{4}\,W(\pm\frac{\ell}{2})]\;\dot{Y}^2}\right] \tag{74}$$

$$Y_+ = sign(Y_-)\left[\frac{\ell}{2} + 2\left[\frac{1 - \dot{Y}^2}{1 + \frac{1}{4}\,W(\pm\frac{\ell}{2})}\right]^{\frac{1}{2}}\, ln \mid \dot{Y} \mid\right] \quad , \tag{75}$$

, while the threshold value for the impacting kink velocity at either end of the system becomes (cf. (35)) [39]:

$$\dot{Y} = \pi\left[\frac{\alpha}{2\,[1 + \frac{1}{4}\,W(\pm\frac{\ell}{2})]}\right]^{\frac{1}{2}} \quad . \tag{76}$$

The LJJ technology allowing a smooth spatial modulation of the maximum Josephson current density is a present research subject in its own [40]. It can be achieved either by a spatial modulation of the CdS barrier parameters, using light-sensitive junctions, or by N^+ ion implantation techniques. This latter method seems most promising, as it recently allowed a potential well extending over 7 λ_0 about the LJJ center, with a dip of 50% of the edge value of the critical current [40].

2.2 The periodic driving of damped,confined fluxons: existence of limit cycles phase-locked to Ω or $\frac{1}{2}\Omega$.

Assume :

$$\gamma(T) = \varepsilon \cos(\Omega T) + \chi \quad , \tag{77}$$

with

$$\varepsilon << 1 \quad ; \chi << 1 \quad . \tag{78}$$

Hence the PDE (71) now reads :

$$\phi_{TT} - \Phi_{XX} + [1 + \frac{1}{4} W(X)] \sin \phi = -\alpha\phi_T + \varepsilon \cos(\Omega T) + \chi \quad , \tag{79}$$

where we now consider the following family of potential wells :

$$W(X) = a[1 - sech\ b\ X] \quad . \tag{80}$$

with

$$\ell^{-1} << b << 1 \quad , \tag{81}$$

according to inequality (64). The kink equation of motion is given by equations (72-73). It is a second order ODE which describes (as did equation (60)) the dynamics of a periodically driven and damped nonlinear oscillator (note that the b.c.'s (74, 75) introduce an additional nonlinearity). Its three-dimensional configuration space may therefore exhibit resonances and chaotic regions. The occurence of such regimes crucially depend on the value of the parameters χ, ε, Ω, a, b, α, ℓ which may be considered each as a control parameter. Let us first choose Ω as the control parameter. Figure **9** displays a typical sequence of limit (asymptotic) cycles for increasing external frequency Ω, which are phase-locked either to $1/2\ \Omega$ (fig. **a**, **d**, **e**, **f**), or to Ω (fig. **b**, **c**). They correspond respectively to resonances in which the frequency of the cycle and the frequency of the external field are in the simple commensurate ratio 1/2 and 1. The signature of such phase-locked cycles on the current-voltage characteristics is very simple : since the (reduced) voltage is equal to ϕ_T (cf. 17), a limit cycle which is phase-locked to $\frac{1}{2}\ \Omega$ as displayed on figure **10** leads to the voltage

$$\bar{\bar{V}}_{1/2} = \bar{V}_0\ v_{1/2} = \bar{V}_0 \frac{\Delta\phi}{\Delta T} \approx \bar{V}_0 \frac{4\pi}{4\pi/\Omega} = \Omega \bar{V}_0 \quad , \tag{82}$$

while a limit cycle phase-locked to Ω yields :

$$\bar{\bar{V}}_1 = 2\Omega\ \bar{V}_0 \quad , \tag{83}$$

where (cf. equations (8), (18)) :

$$\bar{V}_0 = \frac{\hbar}{2eL} \int_{-\frac{L}{2}}^{\frac{L}{2}} dx\ \omega_0(x) \quad . \tag{84}$$

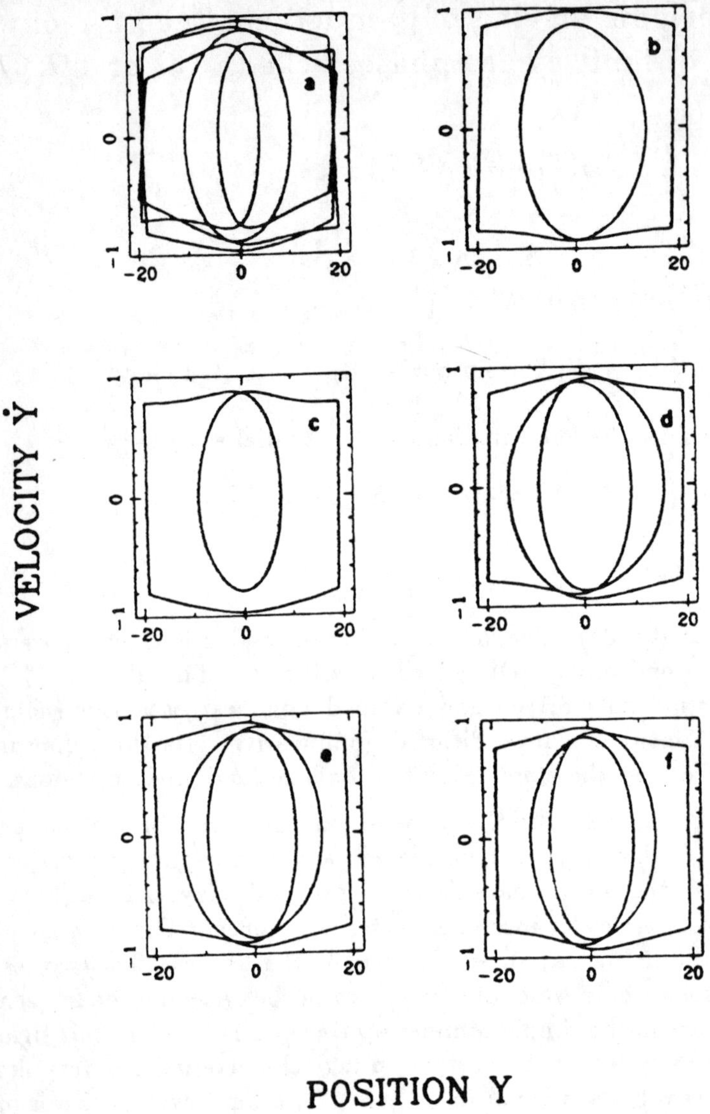

Fig. 9 A sequence of limit cycles for fixed bias field χ and increasing external frequency Ω, which are phase-locked either ti $\Omega/2$ (**a**,**d**, **e**,**f**), or to Ω (**b**,**c**). Note the peculiar sophisticated cycle displayed on figure **7a**. The parameters are: $L = 40$; $a = 20$; $b = 0.095$; $\varepsilon = 0.1$; $\alpha = 0.01$; $\chi = 0.02$. Figures **7a-f** respectively correspond to the values of Ω equal to: 0.03 ; 0.035 ; 0.04 ; 0.045 ; 0.047 ; 0.049 ; 0.051. All cycles were obtained by the numerical simulation of the ODE system between time values 17000 and 20000 (in reduced T-units). These trajectories were checked by direct numerical simulations of the PDE system. We obtained equivalent results.

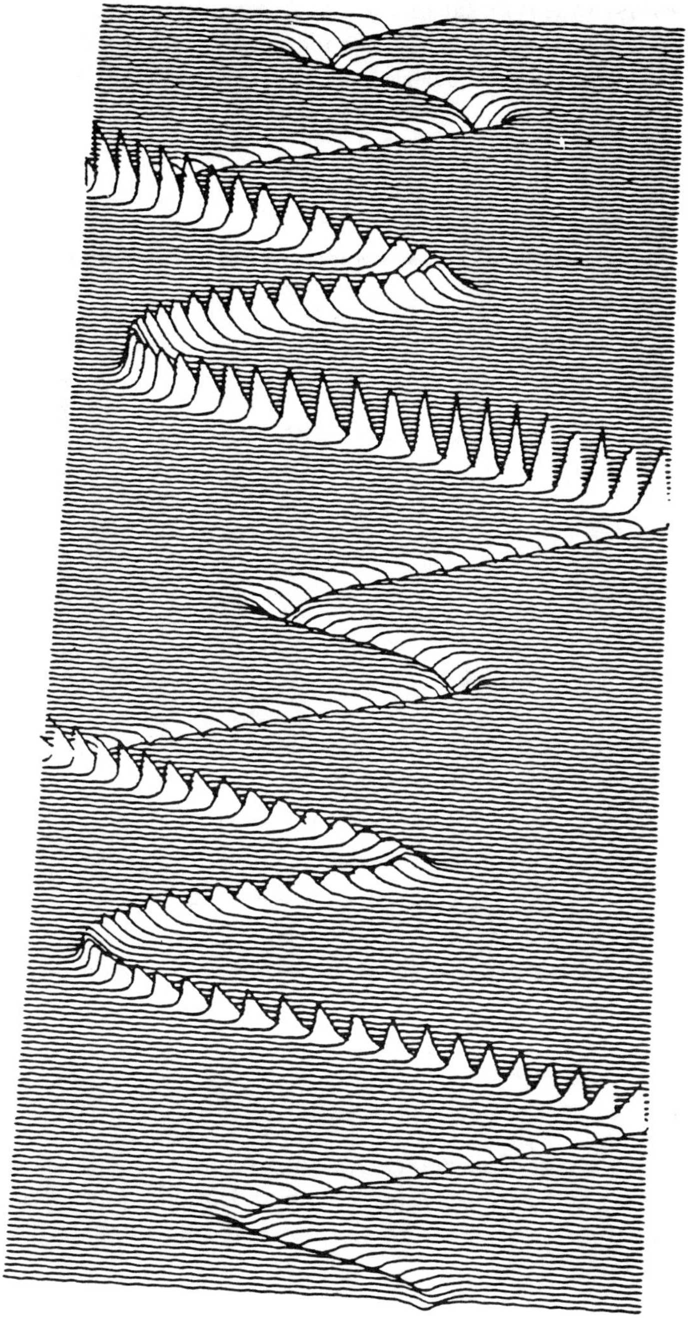

Fig. 10 The limit cycle corresponding to figure **9e**, displayed over a time interval slightly greater than 4 external periods, as ϕ_X versus X in T-perspective.

Since these values are independant of the parameter χ (the d.c. bias) as long as the phase-locking holds, the signature of such phase – locked fluxon regimes in the $I - \bar{\bar{V}}$ curve is simply a vertical step at the voltage value (82) or (83) : see figure 11.

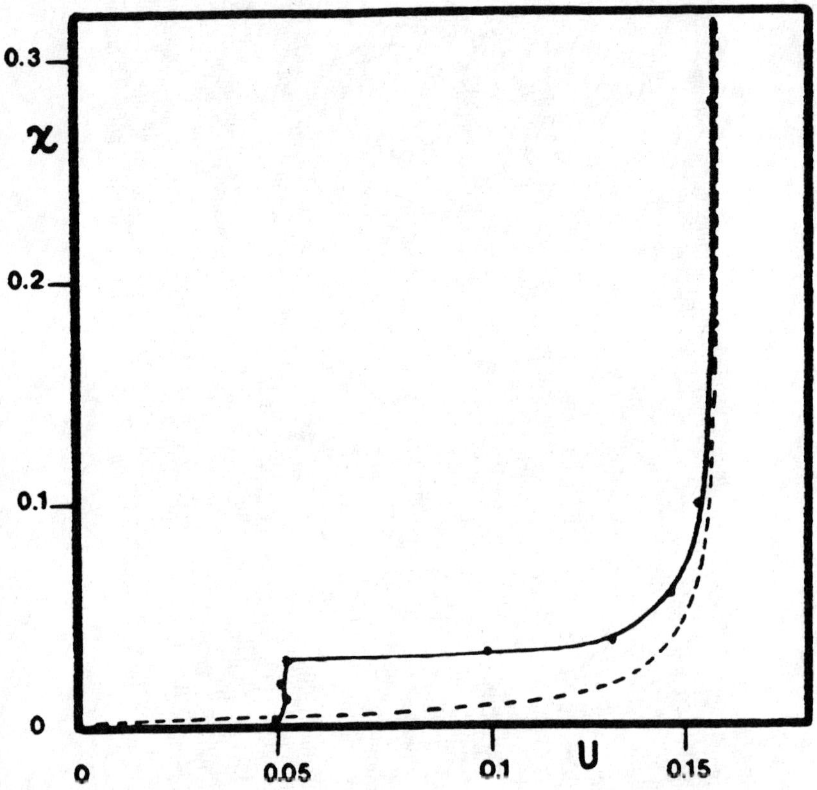

Fig. 11 Current-Voltage-Characteristics corresponding to the same choice of parameters as in figure 9, except $\Omega = 0.055$. The phase-locking of the fluxon trajectory to Ω leads to the vertical step cutting-off the ZFS at $\Omega = 0.05$.

2.3 The period-doubling route to chaos : coexistence of temporal chaos with spatial (soliton) coherence [41] – [42].

-

In the preceeding section, the control parameter was the external frequency Ω. Let us now consider the external d.c. bias amplitude χ as the new control parameter (cf. definition (77)). Figure **12** displays a sequence of limit cycles obtained from the numerical solution of the ODE system (72-75) for fixed external field frequency Ω and increasing bias values χ. We have checked that each part of the trajectory between two reflections was described within one period of the external field. Therefore,this sequence of phase diagrams clearly displays a period-doubling route. As far as the bifurcations values could reasonably be well determined,they agreeded within 10/100 with Feigenbaum's accumulation formula [43]. Since we started with a $(Y = 0, \dot{Y} = 0)$ initial condition, we indifferently obtained "left-asymetric" (figures **12 b,e,f**) or "right-asymetric" cyles (figures **12 a,c,d**), according to the symetry (62),where $b = -1$.

These trajectories were checked by direct numerical simulations of the PDE system (79-81), which exhibited a similar sequence of period-doubled limit cycles as those displayed on figure **12**, the only difference being essentially quantitative: the values of the control parameter χ for the sequence of bifurcations are slightly shifted with respect to the bifurcation values obtained by use of the ODE system (72-75) .

The last stage of this period-doubling route leads to a fully chaotic state, when $\chi \sim 0.03$. Its study then meets a major difficulty related to the "soliton death threshold" (76). Indeed, if one wishes to use the direct PDE system (79-81) in order to study this chaotic state, the soliton sooner or later has its velocity below threshold (76), and disappears. To perform such a study , we make a numerical integration of the ODE system (72- 75) **without** the threshold condition (76). Then we infer from the equivalence of both ODE and PDE descriptions as long as the trajectory remains a sub-harmonic limit cycle the relevance of the ODE description in the final chaotic state . We obtain a strange- attractor phase portrait by displaying Poincare sections of the soliton position Y and velocity $\dot{Y}$ over more than 700 periods of the a.c. external field (77) : see figure **13** and ref. **41**.

We also note that, the "death condition" being relaxed, the system is actually reduced to a sort of (relativistic and with quasi-reflective boundaries) damped and driven Duffing oscillator which has many interesting asymptotic regimes and exhibits various routes to chaos [42]. The latter is characterized by the calculation of positive Lyapunov exponents and by the coding of strange motions by the long unstable periods.

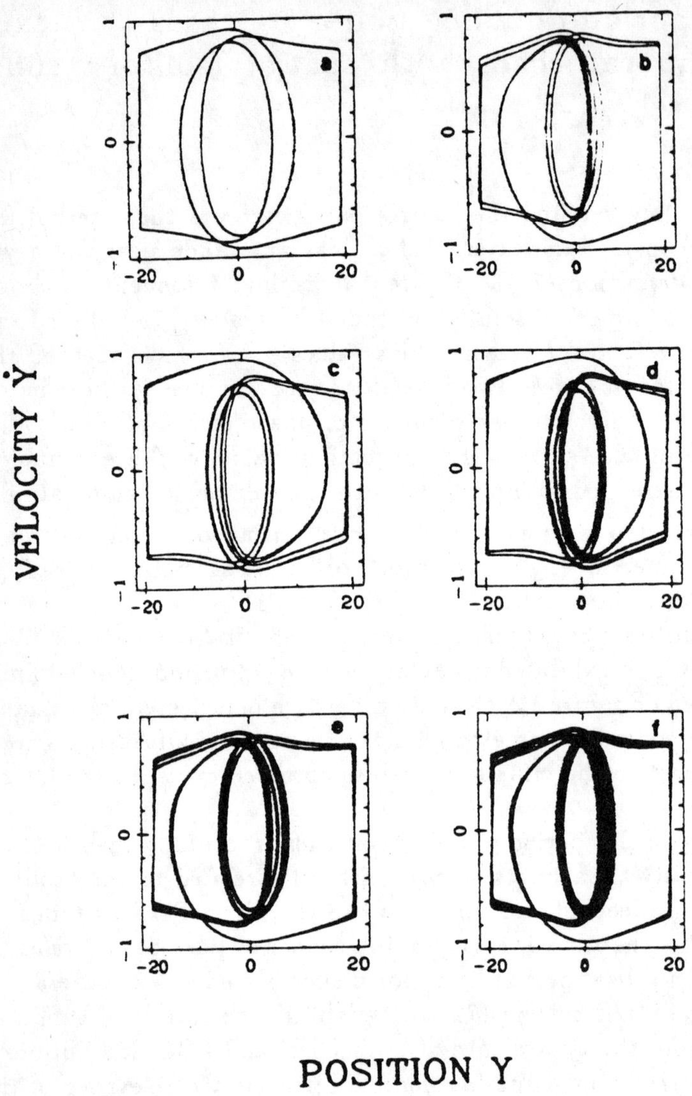

Fig. 12 A sequence of limit cycles obtained from the numerical solution of the ODE system for fixed external field frequency Ω and increasing bias values χ, clearly displaying a period-doubling route to chaos. Figures **a-f** respectively correspond to values of χ equal to: 0.023 ; 0.027 ; 0.0271 ; 0.0272 ; 0.0273 ; 0.0274. All other parameters are the same as for Figure **11**. These trajectories were checked by direct numerical simulations of the PDE system.

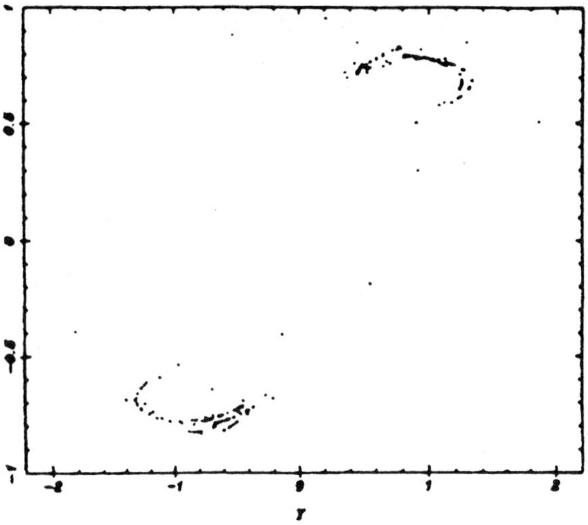

Fig. 13 The strange-attractor folded structure obtained for the ODE oscillator , displayed in the projection of Poincare sections in the $[Y, \dot{Y}]$ plane . We have $\chi = 0.028$, all other parameters being as for Figure **12**.

References

[1] A. Barone and G. Paterno, "Physics and applications of the Josephson effect", John Wiley, New York, (1982) and references therein.

[2] Pedersen N.F., "Solitons in long Josephson junctions", In Advances in Superconductivity, Ed. B. Deaver and J. Ruvalds,Plenum, NY,149, (1982).

[3] Pedersen N.F., "Solitons in Josephson transmission lines", In Solitons,Ed. by S.E. Trullinger,V.E. Zakharov and V.L. Pokrovsky,Elsevier Sc. Pub. 469, (1986).

[4] Pagano S., "Nonlinear dynamics in long Josephson junctions", Phd Thesis, Lyngby, (1987) and references therein.

[5] Seeger A. and Kochendorfer A., "Theorie des deplacements dans des lignes d'atomes a une dimension. II Deplacements arbitrairement disposes et accelerees", Ztschr. f. Phys. 130, 321, (1951).

[6] Seeger A. and Kochendorfer A., "Theorie des deplacements dans des lignes d'atomes a une dimension. III Deplacements, mouvements propres et leur interaction", Ztschr. f. Phys. 134, 173, (1953).

[7] Perring J.K., Skyrme T.H.R., Nucl. Phys. 31, 550, (1962).

[8] Scott A.C., "A nonlinear Klein Gordon equation", American J. of Phys., 37, 52, (1969).

[9] Ablowitz M.J., Kaup D.J., Newell A.C. and Segur H., "Method for solving the sine Gordon equation", Phys. Rev. Lett. 30, 1262, (1973).

[10] Josephson B.D., "Supercurrents trough barriers", Adv. Phys. 14, 419, (1965).

[11] Swihart J.C., "Field solution for a thin film superconducting strip transmission line", J. Appl. Phys. 32, 461, (1961).

[12] Eilbeck J.C., Lomdahl P.S., Olsen O.H. and Samuelsen M.R., "Comparison between one-dimensional and two-dimensional models for Josephson junctions of overlap type", J. Appl. Phys. 57, 861, (1985).

[13] Pedersen N.F. and Welner D., "Comparison between experiment and perturbation theory for solitons in Josephson junctions", Phys. Rev. B29, 2551, (1984).

[14] Pedersen N.F., Samuelsen M.R. and Welner D., "Soliton annihilation in the perturbed sine Gordon system", Phys. Rev. B30, 4057, (1984).

[15] Bodin P., Pedersen N.F., Samuelsen M.R. and Welner D., "Analytical results for the fluxon-antifluxon annihilation process in Josephson junctions", IEEE Trans. Magn. MAG-21, 636, (1985).

[16] Welner D., "Boundary effects in sine Gordon systems", Phd Thesis, Lyngby, Denmark, (1985).

[17] Olsen O.H., Pedersen N.F., Samuelsen M.R., Svensmark H. and Welner D., "Perturbation treatment of boundary conditions for fluxon motion in long Josephson junctions", Phys. Rev. B33, 168, (1986).

[18] Rubinstein J., "Sine Gordon equation", J. Math. Phys.. 11, 258, (1970).

[19] Reinisch G., Fernandez J.C.,Flytzanis N., Taki M., and Pnevmatikos S. , "Phase-lock of a weakly biased inhomogeneous long Josephson junction to an external microwave source", to appear in Phys. Rev. A, (1988).

[20] Fernandez J.C., Gambaudo J.M., Gauthier S. and Reinisch G., "Sine Gordon soliton do not behave like newtonian particles", Phys. Rev. Letters 46, 753, (1981).

[21] Olsen O.H. and Samuelsen M.R., "Sine Gordon soliton dynamics", Phys. Rev. Lett. 48, 1569, (1982).

[22] Olsen O.H. and Samuelsen M.R., "Sine Gordon 2 pi kink dynamics in the presence of small perturbations", Phys. Rev B28, 210, (1983).

[23] McLaughlin D.W. and Scott A.C., "Perturbation analysis of fluxon dynamics", Phys. Rev A18, 1652, (1978).

[24] Levring O.A., Samuelsen M.R. and Olsen O.H., "Exact and numerical solutions to the perturbed sine-Gordon equation", Physica D11, 349, (1984).

[25] Fernandez J.C. and Reinisch G., Phys. Rev. Letters, 48, 1570, (1982).

[26] Fernandez J.C. and Reinisch G., "Specific sine Gordon soliton dynamics in the presence of external driving forces", Phys. Rev. B24, 835, (1981).

[27] Reinisch G. and Fernandez J.C., "Wave mechanics of sine Gordon solitons", Phys. Rev. B25, 7352, (1982).

[28] Bishop A.R., Forest M.G., McLaughlin D.W., Overman E.A., "A quasi periodic route to chaos in a near integrable pde", Physica 23D, 293, (1986).

[29] Bennett D., Bishop A.R. and Trullinger S.E., "Coherence and chaos in the driven,damped sine-Gordon chain", Z. Phys. ,Condensed Matter B47, 265, (1982).

[30] Bishop A.R., Fesser K., Lomdahl P.S. and Trullinger S.E., "Influence of solitons in the initial state on chaos in the driven damped sine Gordon system", Physica 7D, 259, (1983).

[31] Pedersen N.F., Welner D. "Comparison between experiment and perturbation theory for solitons in Josephson junctions" Phys. Rev.B,29, 2551, (1984).

[32] Pagano S.: Comunication in the E.E.C. Conference:"Nonlinear Stimulated Effects in Josephson Devices", Capri, Sept. 1988.

[33] Fernandez J.C., Grauer R., Reinisch G., "Phase-locked dynamical regimes to an external microwave field in a long, unbiased Josephson junction" submitted, (1988).

[34] Fernandez J.C., Goupil M.J., Legrand O. and Reinisch G., "Relativistic dynamics of sine-Gordon solitons trapped in confining potentials", Phys. Rev. B34, 6207, (1986).

[35] Legrand O., "Kink-antikink dissociation and annihilation: A collective-coordinate description", Phys. Rev. A36, 5068, (1987).

[36] Leon J. J.P., Reinisch G. and Fernandez J.C., "sine-Gordon model for classical kink mechanics", Phys. Rev. B27, 5817, (1983).

[37] Fernandez J.C., Leon J.J.P. and Reinisch G., "Discrete resonances of sine-Gordon solitons trapped in a harmonic potential well", Phys. Letters 114A, 161, (1986).

[38] Fernandez J.C., Passot T., Politano H., Reinisch G. and Taki M., "Sturm-Liouville description of sine-Gordon soliton dynamic", Phys. Rev. B37, 7342, (1988).

[39] Reinisch G., Fernandez J.C., Flytzanis N., Taki M. and Pnevmatikos S., "Phase-lock of a weakly biased inhomogeneous long Josephson junction to an external microwave field", to appear in Phys. Rev. B (nov. 1988)

[40] Camerlinguo C., Russo M. : Comunication in the E.E.C. Conference:"Nonlinear Stimulated Effects in Josephson Devices", Capri, Sept. 1988.

[41] Fernandez J.C., Goupil M.J. and Reinisch G., in preparation (1988).

[42] Fournier J.D.,Goupil M.J., and Smith L., work in progress, private comunication,(1988)

[43] Berge P., Pomeau Y., and Vidal C., "L'ordre dans le chaos", Hermann, 1984.

REAL LATTICES MODELLED BY THE NONLINEAR SCHRÖDINGER EQUATION AND ITS GENERALIZATIONS.(*), (**)

Michel Remoissenet. Laboratoire O.R.C,
Université de Bourgogne, 21000 Dijon France.

Abstract.

We present the analysis of two dimerized lattices : a bi-inductance electrical network with macroscopic wave modes , an antiferromagnetic chain whith microscopic spin waves. Using the multiple scale technique of reductive perturbation we show that the original discrete equations of motion can be reduced to a Nonlinear Schrödinger equation with complex coefficients for the first system and two coupled Nonlinear Schrödinger equations for the second system. The possible solutions of these equations are discussed in relation with our numerical simulations and real experiments.

1.INTRODUCTION.

In the modern theory of nonlinear dispersive waves one of the fundamental equation is the Nonlinear Schrödinger (NLS) equation. It is a generic equation for describing the modulation of a wave in a nonlinear medium [1-3] . The NLS equation arises in a wide variety of fields in physics of continuous systems, such as fluids [4] , plasmas [5] , optical fibers [6] . It has also found applications in discrete systems, such as atomic lattices [7] , magnetic chains [8,9] and electrical networks [10] . In these systems the solutions of the NLS equation describe the slow space time evolution of the envelope of a career wave with fast oscillations.

Among these discrete systems, the dimerized lattices or "bi-lattices" with two different elements per unit cell, are interesting to study because different kinds [11,12] of waves modes coexist and can interact. Here we present the analysis of two different " bi-lattices " :

- a bi-inductance electrical network where the characteristic scale of the waves is macroscopic,
- an antiferromagnetic chain where the characteristic scale of the spin waves or excitations is microscopic.

We show that in the small amplitude limit, these systems can be modelled by generalizations of the NLS equation, the possible solutions of which are discussed in relation with numerical simulations an real experiments.

2. BIINDUCTANCE ELECTRICAL NETWORK.

2.1. Characteristic equations.

We consider a circuit where each unit cell counts of four elements : two linear inductors L_1 and L_2 in the series branch and two identical nonlinear capacitances in the shunt branches. A set of the fundamental equations for the transmission line is obtained .

$$L_1 \frac{dI_{2n}}{dt} = V_{2n-1} - V_{2n} \tag{2.1a}$$

$$L_2 \frac{dI_{2n+1}}{dt} = V_{2n} - V_{2n+1} \tag{2.1b}$$

$$\frac{dQ_{2n}}{dt} = I_{2n} - I_{2n+1} \tag{2.1c}$$

$$\frac{dQ_{2n+1}}{dt} = I_{2n+1} - I_{2n+2} \tag{2.1d}$$

Here V_{2n} denotes the AC voltage across the 2nth capacitor. I_{2n} the circuit in the induction L_2, $Q_2(V_{2n})$ which represents the charge stored in the 2nth capacitor is a nonlinear function of V_{2n} From eqs (2.1) one obtains a set of coupled nonlinear equations where n = 1,2 2... N.

$$\frac{d^2Q_{2n}}{dt^2} = \frac{1}{L_2}(V_{2n+1} - V_{2n}) - \frac{1}{L_1}(V_{2n} - V_{2n+1}) \tag{2.2a}$$

$$\frac{d^2Q_{2n+1}}{dt^2} = \frac{1}{L_1}(V_{2n} - V_{2n+1}) - \frac{1}{L_2}(V_{2n+1} - V_{2n+2}) \tag{2.2b}$$

In the limit of low voltages $Q(V_{2n})$ can be approximated by :

$$Q(V_{2n}) = C_o\,(V_{2n} - aV^2_{2n}) \tag{2.3}$$

where the nonlinear parameter a is a constant and C_0 is the constant linear capacitance. Denoting by W_n the voltage of the even cells (L_2) and by V_n the voltage of the odd cells (L_1) from eqs (2.2) and (2.3) we get :

$$(W_n - aW^2_n)_{tt} = \beta\,(V_n - W_{n+1}) - \alpha\,(W_n - V_{n+1}) \tag{2.4a}$$

$$(V_n - aV^2_n)_{tt} = \beta\,(W_n - V_{n+1}) - \alpha\,(V_{n+1} - W_{n+1}) \tag{2.4b}$$

where $\alpha = 1/L_1C_0$ and $\beta = 1/L_2C_0$.

2.2. Modulated waves

We now look for modulated waves in the semi-discrete approximation [7] i.e. we shall use the continuum approximation for the envelope but not for the carrier wave. So we use the multiple scales technique of reductive perturbation. In this method the variables x and t are extended to many independant variables each with a different scale :

$$X_n = \varepsilon^n x \quad , \quad t_n = \varepsilon^n t \ , \tag{2.5}$$

with $\varepsilon<<1$ and $n = 0,1, 2, \ldots\ldots..$

A space or time derivative should be replaced by

$$\frac{d}{dx} = \frac{\partial}{\partial x} + \varepsilon\frac{\partial}{\partial x_1} + \varepsilon^2\frac{\partial}{\partial x_2} + \ldots\ldots. \ , \quad \frac{d}{dt} = \frac{\partial}{\partial t} + \varepsilon\frac{\partial}{\partial t_1} + \varepsilon^2\frac{\partial}{\partial t_2} + \ldots\ldots\ldots \tag{2.6}$$

A general oscillating solution for the voltage $V_n(t)$ of the odd cells can be written in the form of an asymptotic series :

$$V_n(t) = \sum_{e}^{\infty} \varepsilon^e V_e(n,t) = \sum_{em}^{\infty} \varepsilon^e V_{em}(n,t)e^{im\theta(n,t)} + C.C \tag{2.7}$$

and a similar expression for $W_n(t)$.

The method is to substitute (2.5), (2.6) and (2.7) in (2.4) and equate powers of ε for the same harmonic. To $O(\varepsilon^3)$ we find that $V_n(t)$ is

$$V_n(t) = \varepsilon\, V_{11}(n,t)e^{i\theta(n,t)} + \varepsilon^2[\, V_{02}(n,t) + V_{22}(n,t)e^{2i\theta(n,t)}\,] + C.C \tag{2.8}$$

with a similar expression for $W_n(t)$. Here $\theta = 2kn-\omega t$ is the phase and C.C. denotes the complex conjugate. Expressions (2.8) for $V_n(t)$ are $W_n(t)$ are substituted in eqs (2.4). In the resulting equations there are terms like $V_{n\pm1}$ or $W_{n\pm1}$ which are expanded in a continuum approximation around V(x,t) and W(x,t) with $2n \rightarrow x$. Therefore the fast changes of the phase θ are correctly taken into account by taking differences for the discrete variable n. We obtain two equations for V(x,t) and W(x,t) which can be decoupled using the following ansatz [12] :

$$\begin{aligned} W(x,t) = {} & \sigma e^{ik}\,[\, V_{11} + b_1\varepsilon V_{11,x} + \frac{b_2}{2}\varepsilon^2 V_{11,xx} + c_1\varepsilon^2 V_{22}V^*_{11}\,]\, e^{i\theta} + C.C \\ & + \sigma_o[\, V_{02} + d_1\varepsilon V_{02,x} + \frac{d_2}{2}\varepsilon^2 V_{22,xx}\,] \\ & + \sigma_2\varepsilon^2 e^{ik}\,[\, V_{22} + c_1\varepsilon V_{22,x} + \frac{c_2}{2}\varepsilon^2 V_{22,xx} + c_g\varepsilon V_{11}V^*_{11,x}\,]\, e^{2i\theta} + C.C. \end{aligned} \tag{2.9}$$

where we keep up to 2 nd derivations of V and the appropriate nonlinear terms to balance dispersion and nonlinearity. The coefficients of the "ansatz" (2.9.) are determined from the two equation for V(x,t) which must be identical.To O(ε) we get a linear dispersion relation of the form

$$\omega^2 = (\alpha+\beta) \pm [(\alpha+\beta)^2 - 4\alpha\beta\sin^2 k]^{1/2} \tag{2.10}$$

Here the + (-) sign correspond to the High Frequency H F (Low Frequency LF) branch of the dispersion relation. Using the transformations $s = x - v_g t$ where $v_g = \partial\omega/\partial k$ is the group velocity and $T = \varepsilon t$, after lengthy calculations to, $O(\varepsilon^3)$ we obtain a Nonlinear Schrödinger with complex coefficients.

$$iV_{11,T} + PV_{11,ss} + Q\,|V_{11}|^2 V_{11} = 0 \tag{2.11}$$

Here $P = P_r + iQ_i$ and $Q = Q_r + iQ_i$ where P_r, P_i, Q_r and Q_i depend on k or ω (see Appendix A1).For $L_1 = L_2$ we have $P_i = Q_i = 0$ and one recovers the NLS equation which models the monoinductance network [10]. When $P_i \neq 0$ and $Q_i \neq 0$ equation (2.11) is a generalized Nonlinear Schrödinger (GNLS) equation, Bekki [13] has shown that a similar equation describes the modulation of ionization waves, it presents a plane wave solution. The stability of a plane wave depends on the sign of the product $\Gamma = P_r Q_r + P_i Q_i$: if $\Gamma < 0$ a plane wave is stable for the modulation, if $\Gamma > 0$ any finite amplitude plane wave becomes unstable. Here the sign of Γ depends on k and wether we consider LF or HF modes stability . For $0 \leq k < k_l$ with $k_l \approx 0{,}85$ we find that an LF plane wave mode is modulationally stable, for $k_l \leq k \leq \pi/2$ it modulationally unstable. On the other hand we have $\Gamma > 0$ for $0 \leq k \leq k_h$ (with $k_h \approx 1{,}08$) and a HF plane wave mode is unstable, and for $k_h < k \leq \pi/2$ it is stable. These results are confirmed by our numerical simulation experiments [14]. In the present physical case ($L_1 = 2L_2$) we have $P_i \approx 0$ and for small k values Q_i is very small compared to Q_r for both LF and HF modes. Under these conditions eq (2. 11) approaches the standard Non Linear Schrödinger equation with real coefficients and solitary wave modes of the envelope or hole type are respectively expected for $\Gamma > 0$ and $\Gamma < 0$. This was approximatively found in our numerical simulations and real experiments [14] only for the LF envelope modes.

When small dissipation terms $G\partial V/\partial t$ and $G\partial W/\partial t$ (where G is the conductance) are introduced in the r.h.s of equations (2.1 a,b) one gets an additionnal term $i\gamma = i\,(G/2C_0)$ in the final equation (2.11) which becomes[14] the well known Ginzburg Landau equation.

3. ANTIFERROMAGNETIC CHAIN.

3.1. Characteristic equations.

One of the most ideal one-dimensional antiferromagnetic material is perhaps, Tetramathylammonium Manganese Chloride (called "TMMC"). This system with spin $S = 5/2$ is of course quantum mechanical rather than classical in nature. However (the relative importance of quantum effects is given by the ratio of $S(S+1)$ to S^2) thus the intrinsic quantum effects scale a $1/S$. For a spin of 5/2, $1/S$ is small enough that a classical approximation appears reasonable and the spins can be represented by classical vectors. The magnetic properties of TMMC are well described by the following Hamiltonian :

$$H = \sum_n [\, 2J\, S_n S_{n+1} + A(S^z_n)^2 - g\mu_B B_x\, S^x_n \,] \tag{3.1}$$

where J is the antiferromagnetic exchange term, A the single ion anisotropy constant, $B_x = B$ the external magnetic field perpendicular to the chain axis, g and μ_B arerespectively, the Landé factor and the Bohr magneton. The dynamics of these classical spin vectors is described by the undamped Bloch equations :

$$h\left(\frac{dS_n}{dt}\right) = S_n\, [\, -J(S_{n-1} + S_{n+1}) + g\mu_B \mathbf{B} - 2AS^z_n\, \mathbf{z} \,] \tag{3.2}$$

The equation (3.2) can be transformed into the evolution equations for the angle θ_n and Φ_n defined for each spin at position z on the chain [15] . Using eqs (3a) and (3b) in ref [15] and redefining Φ as $\pi/2 + \Phi/2$ and θ as $(\pi/2 - \theta)$ (in the continuum limit) we obtain

$$\Phi_{tt} - c^2{}_0\Phi_{zz} + \frac{\omega^2{}_1}{2}\sin 2\Phi = 2\,(\, \Phi_t\theta_t - c^2{}_0\Phi_z\theta_z\,)\, \mathrm{tg}\theta - 2\omega_1\theta_t \sin\Phi \tag{3.3a}$$

$$\theta_{tt} - c^2{}_0\theta_{zz} + \frac{\omega^2{}_2}{2}\sin 2\theta = (\, c^2{}_0\Phi^2{}_z - \Phi^2{}_t\,)\frac{\sin 2\theta}{2} + \omega^2{}_1 \sin^2\Phi \frac{\sin 2\theta}{2} + 2\omega_1\Phi_t \cos^2\theta \sin\Phi\,, \tag{3.3b}$$

3.2. Coupled NLS equations.

We now consider weakly nonlinear oscillations around the ground state which is assumed to be zero [15] i.e. in the small angle limit : $\Phi \to \delta\Phi$ and $\theta \to \delta\theta$ $(\delta << 1)$.Under these conditions eqs (3a.b) reduce to

$$L\Phi = -\frac{2}{3}\omega^2{}_1\delta^2\Phi^3 + 2\varepsilon^2 (c^2{}_0\Phi_z\theta_z - \Phi_t\theta_t)\theta + 2\omega_1\varepsilon\theta_t\Phi \tag{3.4a}$$

$$L'\theta = -\frac{2}{3}\omega^2{}_2\delta^2\theta^3 + \delta^2 (\Phi^2{}_t - c^2{}_0\Phi^2{}_z)\theta - \omega^2{}_1\delta^2\Phi^2\theta - 2\omega_1\delta\Phi_t\Phi \tag{3.4b}$$

with $\omega_1 = g\mu_B H$, $\omega_2 = S4AJ$, $c_o = 2J\,S$ and where the linear operators L and L' are given by :

$$L = c^2{}_0\, \partial^2/\partial Z^2 - \partial^2/\partial t^2 - \omega_1{}^2 \tag{3.5a}$$

$$L' = c^2{}_0\, \partial^2/\partial Z^2 - \partial^2/\partial t^2 - \omega_2{}^2 \tag{3.5b}$$

We now use the multiple scale expansion technique [16]. Accordingly we use equations of the form (2.5) for the independent space and time variables and (2.6) for the corresponding operators, we write

$$\frac{\partial}{\partial x} = \frac{\partial}{\partial x} + \varepsilon\frac{\partial}{\partial X_1} + \varepsilon^2\frac{\partial}{\partial X_2} + \ldots\ldots\ , \tag{3.6a}$$

$$\frac{\partial}{\partial t} = \frac{\partial}{\partial t} + \varepsilon\frac{\partial}{\partial T_1} + \varepsilon^2\frac{\partial}{\partial T_2} + \ldots\ldots\ , \tag{3.6b}$$

and

$$\Phi = \varepsilon\Phi_1 + \varepsilon^2\Phi_2 + \varepsilon^3\Phi_3 + \ldots\ldots, \tag{3.6c}$$

with a similar expression for θ.To lowest order in ε

$$L\Phi_1 = 0\ ,\ L'\theta_1 = 0\ , \tag{3.6a,b}$$

yielding the solutions.

$$\Phi_1 = A\,(X_1, X_2, T_1,T_2)e^{i\psi_1} + C.C \tag{3.7a}$$

$$\theta_1 = B(X_1, X_2, T_1,T_2)e^{i\psi_2} + C.C \tag{3.7b}$$

with $\psi_1 = k_1x - \Omega_1t$, $\psi_2 = k_2x - \Omega_2t$, and where $\Omega^2{}_1 = \omega^2{}_1 + c^2{}_0k^2{}_1$, $\Omega^2{}_2 = \omega^2{}_2 + c^2{}_0k^2{}_2$ are the linear dispersion relations in the long wavelength limit. The solutions (7a.b.) are now introduced in the second order equations and from the removal of the secular terms we get

$$\Phi_2 = i[\, F^+ABe^{i(\psi_1+\psi_2)} - F^-AB^*e^{i(\psi_1-\psi_2)}\,] + C.C \tag{3.8a}$$

$$\theta_2 = i[\, G^{+}ABe^{i[\psi_1(k_1)+\,\psi_1(k_2))} - G^{-}AB^{*i\,[\psi_1(k_1)+\,\psi_1(k_2)]}\,] + C.C \qquad (3.8b)$$

with

$$F^{\pm} = 2\omega_1\varepsilon\Omega_2/[\, c^2{}_o(\, k_1 \pm k_2\,)^2 - (\,\Omega_1 \pm\ \Omega_2\,)^2 + \omega^2{}_1] \qquad (3.9a)$$

$$G^{\pm} = -2\omega_1\varepsilon\Omega_1/[\, c^2{}_o(\, k_1 \pm k_2\,)^2 - (\,\Omega_1(k_1) \pm\ \Omega_2(k_1))^2 + \omega^2{}_2] \qquad (3.9b)$$

To $O(\varepsilon^3)$, with the transformation $s = X_1$, $\tau = T_2$ we have

The expression (3.8a,b) are now introduced in the third order equations. After removing the secular terms two coupled Nonlinear Schrödinger (NLS) equations are obtained

$$iA_\tau + pA_{ss} + q|A|^2A + r|B|^2A = 0 \qquad (3.10a)$$

$$iB_\tau + p'B_{ss} + q'|B|^2B + r'|A|^2B = 0 \qquad (3.10b)$$

where $\partial/\partial s = \partial/\partial(\,X_1 - v_{g1}T_1) = \partial/\partial(\,X_1 - v_{g2}T_1)$ with the group velocities $v_{g1}=\partial\Omega_1/\partial k_1$, $v_{g2}=\partial\Omega_2/\partial k_2$, and $\tau = T_2$. Here p, p', q, q', r and r' are functions of k_1 , k_2 ,ω_1 , ω_2 and c_o (see Appendix A2). NLS coupled equations similar two eqs (3.10 a.b) were obtained for Langmuir and dispersive ion acoustic waves [17,18], nonlinearly coupled polarized plasma waves [19], coupled electromagnetic waves[20,21] in a dielectric and for electrical transmission lines[22], optical fibers [23]. For the system (3.10a,b) the complete integrability is established [24] for the specific parameter restrictions : p = p' ; q = q' = r = r' or p = - p' ; q = q' = - r = - r' and soliton solutions can be calculated [25].

In the present physical case these coefficients are functions of k_1,k_2, ω_1 ,ω_2 and the above constraints are not fullfilled : the equations (3.10a,b) can have solitary wave solutions [26] depending on the signs of the coefficient $a_o = (rp' - pq')/(rr' - qq')$ for A and of the coefficient $b_o = (r'p - qp')/(rr' - qq')$ for B.Experimentally a good agreement was obtained between our recent neutrons data and our description [to $O(\varepsilon^2)$] of the modes[27].

APPENDIX A-1.

$$P = -[1/2\omega C_o]\{C_oV_s{}^{2}-[\sigma_1 e^{-ik}/1]\,[2-2b_1 + b_2/2] + [\sigma_1 b_2/2L_1L_2]e^{ik}]$$

$$Q = -\,[1/2\omega Co]\,[Q_1+Q_2+Q_3+Q_1]$$

The coefficients Q_i, i = 1,2,3,4 are given by

$$Q_1 = \frac{4a^2\,C_o{}^2\,\omega^2\,V_g{}^2}{C_o\,V_g{}^2 + 2[1/L_1 + 1/L_2]/L_1{}^2 - 2/L1}\ ,$$

$$Q_2 = \frac{8a^2 C_o^2\omega^4}{D} ,$$

$$Q_3 = -\frac{4a\ C_o\ \omega^2\ C_8\ \sigma_1\ e^{-ik}}{L_1 D} , \qquad Q_4 = \frac{4a\ C_o\ \omega^2\ C_8\ \sigma_1\ e^{ik}}{L_2 D}$$

with $D = 4C_o\,\omega^2 - 1/L + \sigma_2[\,e^{-2ik}/L_1 + e^{2ik}/L_2]$

$$V_g = \frac{\pm\ \sin 2k}{C_o L_1 L_2 \omega\{[1/L_1+1/L_2]^2-[4/L_1L_2]\ \sin 2k]^{1/2}\}}$$

where the signs ± refer respectively to LF and HF modes.

$$\sigma_1 = \frac{e^{-ik}/L_2+e^{ik}/L_1}{-\omega^2 C_o+1/L_1+1/L_2}$$

$$\sigma_2 = \frac{\sigma_1^2[4\omega^2 C_o L_1 L_2-L_1-L_2]-L_1e^{-2ik}-L_2e^{2ik}}{4\omega^2 C_o L_1 L_2-L_1-L_2-\sigma_1^2[L_2e^{-2ik}+ L_1e^{2ik}]}$$

$$b_1 = \frac{\sigma_1 e^{-ik}/L_1+ e^{ik}/\sigma_1 L_1}{-C_o\omega^2+1/L_1+1/L_2} , \quad b_2 = \frac{4\sigma_1\ e^{-ik}/L_1[b_1-1] - 4i\omega C_o V_g b_1+ 4e^{ik}/\sigma_1 L_1}{-C_o\omega^2+1/L_1+ 1/L_2+\omega_1[e^{ik}/L_1+ e^{ik}/L_2]}$$

$$C_8 = \frac{2a\ C_o\omega^2[1-\sigma_2\sigma_1^*/\sigma 1]}{-\omega^2 C_o+1/L_1+1/L_2+ \sigma_1[e^{-ik}/L_1+e^{ik}/L_2]}$$

APPENDIX A2.

For $k_1 = k_2 = k$ the coefficients of the coupled nonlinear Schrödinger equations given by the following relations

$$p = \frac{c_o^2}{2\Omega_1}\,(1-\frac{c_o^2 k^2}{\omega_1^2}) , \qquad p' = \frac{c_o^2}{2\Omega_2}\,(1-\frac{c_o^2\, k^2}{\omega_2^2})$$

$$q = \frac{\varepsilon^2\omega_1^2}{\Omega_1}[(1-\frac{2\Omega_1^2}{4\omega_1^2-\omega^2})] \qquad q' = \frac{\varepsilon^2\omega_1^2}{\Omega_2}$$

$$r = \frac{\varepsilon^2}{\Omega_1}\ \frac{4\omega_1^2\omega_2^2\Omega_2^2}{\omega_2^4-4(\Omega_1\Omega_2-c_o^2k^2)^2} \qquad r' = \frac{2\varepsilon^2\omega_1^2}{\Omega_2}\ [(1-\frac{2\omega_2^2\Omega_2^2}{\omega_2^4-4(\Omega_1\Omega_2-c_o^2k^2)^2})]$$

REFERENCES.

(*) Based on a talk presented at the Workshop : " Integrable Systems and applications" . Ile d'Oléron. France. June 20-24-1988.

(**) Part 2 of this work realized in collaboration with T. KOFANE, N. FLYTZANIS and M. PEYRARD; part 3 in collaboration with J.P. BOUCHER , R. PYNN and J. L. REGNAULT.

[1] . H. C. YUEN and B. M. LAKE. Adv App Mech, 22, 67 (1982).

[2] . G.L . LAMB. Elements of Soliton Theory. [Wiley Interscience, New-York, 1980].

[3]. A. C. NEWELL. *Solitons in Mathematics and Physics.* Society for Applied and Industrial Mathematics. Philadelphia, 1985.

[4] . D. J . BENNEY and A. C. NEWELL. J Math Phys, 46,133, 1967.

[5]. V. E . ZHAKHAROV. Zh Eksp Theor Fiz, 62, 1745, (1972). [Soviet Phys JETP , 35,908, (1967)].

[6]. A. HASEGAWA and F. TAPPERT. Appl Phys Lett , 23,142, (1985).

[7]. TSURUI. Prog Theor Phys, 48,1196, (1977).

[8]. M. LAKSHMANAN. Phys Lett, 61A, 53, (1977).

[9]. J. CORONES . Phys Rev B, 16, 1763,(1977).

[10]. K. MUROYA, N. SAITOH and S. WATANABE. J. Phys Soc Japan, 51, 1024,(1982).

[11]. N. YAJIMA and J. SATSUMA. Prog Theor Phys, 62, 370 (1979).

[12] .S. PNEVMATIKOS, N. FLYTZANIS and M. REMOISSENET. Phys Rev B, 33, 2308, (1986).

[13]. BEKKI . J Phys Soc Japan, 50,659, (1981).

[14]. T. KOFANE, N. FLYTZANIS, M. REMOISSENET and M. PEYRARD. In preparation.

[15]. G. M. WYSIN, A. R. BISHOP and J. OITMAA. J Phys C : Solid State.

[16]. A. JEFFREY and T. KAWAHARA.*Asymptotic Methods in Nonlinear WaveTheory.* Pitman, Boston 1982.

[17]. K. H. SPATSCHEK. Phys Fluids, 21,1032, (1978).

[18]. B. K. SOM and M. R. GUPTA. 11,72A, (1979).

[19]. Y. INOUE. J Plasma Phys, 16,439, (1976).

[20]. Y. INOUE . J Phys Soc Japan, 43, 243, (1977).

[21]. A. L. BERKHOER and V. E . ZHAKHAROV. Soviet Phys JETP, 31, 486, (1970).

[22]. J. YOSHINAGA, N. SUGIMOTO and T. KAKUTANI. J Phys Soc Japan, 50, 82, (1981).

[23] C.R. MENYUCK. IEEE J Qant Electron, QE-23, 174, (1987).

[24]. V. E . ZHAKAROV and E. I . SCHULMAN. Physica D, 270, (1982).

[25]. R. SAHADEVAN, K. M . TAMIZHMANI and M. LAKSHMANAN. J Phys A: Math Gen, 19, 1783, (1986).

[26]. J. C. BHAKTA . Plasma Phys and Controlled Fusion. 29, 245, (1987).

[27]. J. P. BOUCHER, R. PYNN, M. REMOISSENET, L. P. REGNAULT and Y. ENDON. In Preparation.

MODULATION OF TRAPPED WAVES GIVING APPROXIMATE TWO-DIMENSIONAL SOLITONS

P.C. SABATIER
Laboratoire de Physique Mathématique
34060 MONTPELLIER Cedex - FRANCE

ABSTRACT. In a previous paper, F. Calogero and the author studied the non-linear modulation of dispersive trapped water-waves in a channel or along a beach (edge waves). They showed that the signal envelope, (which is studied with convenient time scale and length scale) obeys Nonlinear Schrödinger Equation. The result is extended here to waves trapped by the bottom geometry of a basin which is infinitely extended into all horizontal directions. One can obtain along the guided path, envelope solitons of Nonlinear Schrödinger Equation for a signal whose amplitude exponentially decreases in transverse directions. These bidimensional solitons are thus shown as asymptotic features for a large class of problems. This class is narrower than in the finite case or in the semiinfinite case because strong conditions must be set either on the linear part or on the nonlinear part of the operator in order to guarantee an exponentially decreasing behavior on both sides. In addition, the envelope solitons are in most cases unstable features, because of side-band instabilities. Nevertheless, the result suggests that the exact bidimensional solitons recently derived by Boiti et al. for special Equations are not an exceptional physical feature of two-dimensional evolutions.

1. INTRODUCTION

Like in a previous paper[1] (but with different notations), we consider the evolution equation :

$$Du = 0 \tag{1}$$

with

$$Du =: \frac{\partial^2 u}{\partial t^2} + \beta(x)\, d_y u - \frac{\partial}{\partial x}\alpha(x)\frac{\partial u}{\partial x} + \sum_{n=2} \epsilon^{n-1} F^{(n)}\left(u, \frac{\partial}{\partial x}, \frac{\partial}{\partial y}, \frac{\partial}{\partial t}\right) \tag{2}$$

$$d_y =: \sum_{m=1}^{M} (-1)\, \gamma_m \frac{\partial^{2m}}{\partial y^{2m}}, \quad \gamma_m \in \mathbb{R}$$

where $\alpha \in C_1(\mathbb{R})$, $\beta \in C(\mathbb{R})$, and the $F^{(n)}$'s are homogeneous polynomials of degree n in u and its partial derivatives, with real coefficients :

$$F^{(2)} = \sum_{j,f,\ell,m} a(0,j,f,\ell,m) \frac{\partial^{j+f} u}{\partial x^j\, \partial y^f} \frac{\partial^{\ell+m} u}{\partial x^\ell\, \partial y^m}$$

$$+ \sum_{j,f,\ell,m} a(1,j,f,\ell,m) \frac{\partial^{j+f+1}u}{\partial x^j \, \partial y^f \, \partial t} \frac{\partial^{\ell+m}u}{\partial x^\ell \, \partial y^m} \qquad (3)$$

$$F^{(3)} = \sum_{j,f,\ell,m,p,q} b(0,j,f,\ell,m,p,q) \frac{\partial^{j+f}u}{\partial x^j \, \partial y^f} \frac{\partial^{\ell+m}u}{\partial x^\ell \, \partial y^m} \frac{\partial^{p+q}u}{\partial x^p \, \partial y^q}$$

$$+ \sum_{j,f,\ell,m,p,q} b(1,j,f,\ell,m,p,q) \frac{\partial^{j+f+1}u}{\partial x^j \, \partial y^f \, \partial t} \frac{\partial^{\ell+m}u}{\partial x^\ell \, \partial y^m} \frac{\partial^{p+q}u}{\partial x^p \, \partial y^q} \qquad (4)$$

We seek an "ϵ^3-solution" of (1), ie

$$u : Du = O(\epsilon^3) \qquad (5)$$

that vanishes as $x \to \pm\infty$ (this being the present paper technical novelty), and reduces to a trapped mode $\rho(k,x) \cos[ky - \omega(k)\,t]$ in the $\epsilon = 0$ limit. Clearly, if $v = \epsilon u$, Dv is $O(\epsilon^4)$ and v is an "approximate" solution of the equation (1), or that obtained from (1) after truncating the $F^{(n)}$'s of degree > 3. Such a mathematical problem can "represent" several physical problems. Willing to be very specific in the present paper, we shall study the problem of small amplitude trapped water waves on a basin uniformly extended into the y-direction, depth $h(x)$, and we describe their propagation by means of the so-called long wave approximation, where the velocity potential obeys the equation (1) with

$$d_y = -\partial^2/\partial y^2 \quad ; \ \alpha(x) = \beta(x) = gh(x) \qquad (6)$$

$$F^{(2)} = \left(\frac{\partial^2 u}{\partial x^2} + \frac{\partial^2 u}{\partial y^2}\right) \frac{\partial u}{\partial t} + \frac{\partial}{\partial t}\left[\left(\frac{\partial u}{\partial x}\right)^2 + \left(\frac{\partial u}{\partial y}\right)^2\right] \qquad (7a)$$

$$F^{(3)} = \frac{1}{2}\left(\frac{\partial^2 u}{\partial x^2} + \frac{\partial^2 u}{\partial y^2}\right)\left[\left(\frac{\partial u}{\partial x}\right)^2 + \left(\frac{\partial u}{\partial y}\right)^2\right] + \left(\frac{\partial u}{\partial x}\right)^2 \frac{\partial^2 u}{\partial x^2} + 2\,\frac{\partial u}{\partial x}\frac{\partial u}{\partial y}\frac{\partial^2 u}{\partial x \partial y} + \left(\frac{\partial u}{\partial y}\right)^2 \frac{\partial^2 u}{\partial y^2} \qquad (7b)$$

g being the gravity acceleration.

2. THE LINEAR PROBLEM

Setting $u = \rho(k,x) \cos[ky - \omega t]$ yields for a given fixed value of k :

$$\left[\frac{d}{dx}\, gh \,\frac{d}{dx} + \omega^2 - k^2\, gh\right] \rho(k,x) = 0 \qquad (8)$$

whose solution ρ must vanish at $x \to \pm\infty$. Setting

$$gh(x) = c^2 \exp[-\,2r\,(x)] \qquad (9)$$

where $r \in C^2(\mathbb{R})$ and vanishes as $x \to \pm\infty$, and

$$s : c^{-1} \int_0^x \exp[r(t)]\, dt \quad ; \quad w(s) = P(x) \exp\left[-\frac{1}{2} r(x)\right] \tag{10}$$

we reduce (8) to the Schrödinger equation :

$$\left[\frac{d^2}{ds^2} + \omega^2 - k^2c^2 - V(s)\right] w(s) = 0 \tag{11}$$

where $V(s) = c^2\left[k^2 (e^{-2r} - 1) + \frac{1}{2}\left(r'' - \frac{3}{2} r'^2\right) e^{-2r}\right]$ is a regular potential provided r, r', r'', are $O(x^{-2-\epsilon})$ at infinity. Then a simple discussion[2] shows that the continuous spectrum is $\omega^2 > k^2c^2$ and that eigenvalues $\omega_1^2 = k^2c^2 - |E_1|$ exist if $r(x)$ is somewhere positive. Going back to (8), we can write down the spectral decomposition ("completeness relation") for an arbitrary L^2 function $g(x)$ as

$$g(x) = \int_{-\infty}^{+\infty} h(\omega^2, x)\, d\rho(\omega^2) \int_{-\infty}^{+\infty} g(y)\, h(\omega^2, y)\, \nu(y)\, dy \tag{12}$$

where

$$d\rho(\omega^2) = \begin{cases} d\,\mu(\omega^2) & \omega^2 \geqslant k^2c^2 \quad (13) \\ \sum_n C_n\, \delta\left(\omega^2 - \omega_n^2\right) & \omega^2 \leqslant k^2c^2 \quad (14) \end{cases}$$

$d\mu$ and ν are positive measures, (ν can be calculated from formulas (10)). The C_n's are normalizing constants for the eigenvalues ω_n^2. The functions $h(\omega^2, x)$ are conveniently standardized solutions of the Equation (8). For $\omega^2 = \omega_n^2$, they exponentially decrease as $x \to \pm\infty$, and they are identified to the "modes. So as to simplify the writing, one eigenvalue only, ω_o^2, is assumed in the following, and the fixed number k is generally dropped from the notations.

3. THE NONLINEAR MODULATION

We focus on the specific mode $p(x)$ ($= \rho_o(k, x)$) which correspond to ω_o^2, and we use the ansatz of Ref. (1)

$$u(x,y,t) = \psi\, e^{i\delta} + \epsilon\left(\frac{1}{2}\psi_o + \psi_2\, e^{2i\delta}\right) + \epsilon^2\, \psi_3\, e^{3i\delta} + \text{c.c.} \tag{15}$$

$$\psi(x,\xi,\tau) = p(x)\, \varphi(\xi,\tau) + i\epsilon \tilde{p}(x)\, \tilde{\varphi}(\xi,\tau) + \epsilon^2\, \Phi(x,\xi,\tau) \tag{16a}$$

$$\psi_n(x,\xi,\tau) = p_n(x)\, \varphi_n(\xi,\tau) + \epsilon\, \Phi_n(x,\xi,\tau) \qquad (n \neq 1) \tag{16b}$$

$$\delta = ky - \omega_o t \quad ; \quad \xi = \epsilon(y - vt) \quad ; \quad \tau = \epsilon^2 t \tag{17}$$

Here and below, + c.c. means that we have to add the complex conjugate of the preceding terms. Locally the complex conjugation will be denoted by a superscript *. We assume (subject to final verification) that the functions appearing in (15-16) are sufficiently differentiable. It is then easy to calculate the non linear terms which appear in $F^{(2)}$ and $F^{(3)}$, like in Ref.(1). The result is :

$$F^{(2)} = f_{1,-1} \; |\varphi|^2 + f_{1,1} \; \varphi^2 \; e^{2i\delta} +$$

$$+ \epsilon \left[g_{-1,1} \; \varphi^* \frac{\partial\varphi}{\partial\xi} + \left(f_{0,1} \; \varphi_0 \varphi + f_{2,-1} \; \varphi_2 \varphi^* \right) e^{i\delta} \right.$$

$$\left. + \left(g_{1,1} \; \varphi \frac{\partial\varphi}{\partial\xi} + h_{1,1} \; \varphi \tilde{\varphi} \right) e^{2i\delta} + f_{2,1} \; \varphi_2 \; \varphi \; e^{3i\delta} \right] + \text{c.c.} + O(\epsilon^2) \tag{18}$$

$$F^{(3)} = f_{1,1,-1} \; \varphi \; |\varphi|^2 \; e^{i\delta} + f_{1,1,1} \; \varphi^3 \; e^{3i\delta} + \text{c.c.} + O(\epsilon) \tag{19}$$

The coefficients f,g,h, that appear in these equations are readily obtained from p, $\tilde{p}$, p_n and their derivatives. They are functions of x, not of ξ and τ, with coefficients involving k, ω_o,v. They are generally complex, with the exception of $f_{1,-1}$, which is always real.

The linear terms of (1) can be calculated as well. Setting D_o for the linear part of the differential operator in D, we obtain

$$e^{i\delta} \; D_o \; [\psi \; e^{i\delta}] = \varphi \left[\beta \; d(k) - \frac{d}{dx} \alpha \frac{d}{dx} - \omega_o^2 \right] p$$

$$+ i\epsilon \left\{ \tilde{\varphi} \left[\beta \; d(k) - \frac{d}{dx} \alpha \frac{d}{dx} - \omega_o^2 \right] \tilde{p} + \frac{\partial\varphi}{\partial\xi} \; [2v \; \omega_o - \beta \; d'(k)] p \right\}$$

$$+ \epsilon^2 \left\{ \left[\beta \; d(k) - \frac{d}{dx} \alpha \frac{d}{dx} - \omega_o^2 \right] \Phi - \frac{\partial\tilde{\varphi}}{\partial\xi} \; [2v \; \omega_o - \beta \; d'(k)] \; \tilde{p} \right.$$

$$\left. + \frac{\partial^2\varphi}{\partial\xi^2} \left(v^2 - \frac{1}{2} \beta \; d''(k) \right) p - 2i \; \omega \frac{\partial\varphi}{\partial\tau} \; p \right\} + O(\epsilon^3) \tag{20}$$

and, for n = 0,2,3

$$e^{-in\delta} \; D_o \left[\psi_n \; e^{in\delta} \right] = \varphi_n \; \left[\beta \; d(nk) - \frac{d}{dx} \alpha \frac{d}{dx} - n^2 \; \omega_o^2 \right] p_n$$

$$+ \epsilon \left\{\left[\beta\, d(nk) - \frac{d}{dx}\alpha\frac{d}{dx} - n^2\,\omega_o^2\right]\Phi_n\right.$$

$$\left. + i[2nv\,\omega_o - \beta\, d'(nk)]\; p_n \frac{\partial\varphi_n}{\partial\xi}\right\} + O(\epsilon^2) \tag{21}$$

where the primes appended to d(k) denote differentiation with respect to k.

In order to satisfy the equation (2), for each term of order ϵ^n, we must successively cancel the coefficients of $e^{ip\delta}$, p = 0,1,...,(n+1), and according to (5), we may stop at n = 3. Now, the term of order ϵ^0 is already zero because of our choice of ω and p. The coefficients in the terms $O(\epsilon)$ first yield the equation

$$\frac{d}{dx}\,\alpha(x)\,\frac{d}{dx}\,p_o = 2\,f_{1,-1} \tag{22}$$

which determines p_o from $f_{1,-1,}$ ie from p and its derivatives. Actually in our specific example, $p_o = f_{1,-1} = 0$.

Cancelling the coefficient of $\epsilon\, e^{i\delta}$ yields the condition $\tilde{\varphi} = \dfrac{\partial\varphi}{\partial\xi}$ and the equation

$$\left(\beta(x)\; d(k) - \frac{d}{dx}\,\alpha\,\frac{d}{dx} - \omega_o^2\right)\tilde{p} + (2v\,\omega_o - \beta\, d'(k))p = 0 \tag{23}$$

Let us write down the spectral representation of $\tilde{p}(x)$, p(x), $\beta(x)$ p(x), along the spectrum of $\beta(x)\; d(k) - \dfrac{d}{dx}\,\alpha\,\dfrac{d}{dx}$, like in (13) (in our specific example, $\beta = \alpha$ and $d(k) = k^2$) :

$$\tilde{p}(x) = \int_{-\infty}^{+\infty} h(\omega^2,x)\; d\rho(\omega^2)\; \tilde{p}^T(\omega^2) \tag{24a}$$

$$\beta(x)\; p(x) = \int_{k^2c^2}^{\infty} h(\omega^2,x)\; d\mu(\omega^2)\; (\beta p)^T\,(\omega^2) + (\beta p)^o\; h\left(\omega_o^2,x\right) \tag{24b}$$

$$p(x) = p^o\; h\left(\omega_o^2,x\right) \tag{24c}$$

where the generalized Fourier coefficients, denoted by the subscript T, (continuous spectrum) or O (eigenvalue ω_o^2) can be identified by comparing with (13). Notice that $h\left(\omega_o^2,x\right)$ can be identified to p(x). Inserting (24) into (23), we see that v and the "continuous spectrum" components of $\tilde{p}(x)$ are derived :

$$2v\,\omega_o\; p^o = d'(k)\; (\beta p)^o \tag{25}$$

$$\tilde{p}^T(\omega^2)\left(\omega^2 - \omega_o^2\right) + d'(k)\; (\beta p)^T\,(\omega^2) = 0 \tag{26}$$

In other words, $\tilde{p}(x)$ is determined if we set to zero its component along $h\left(\omega_o^2, x\right)$ - this being possible according to the argument already used in Ref (1). Notice also that it is not obvious from (26) that $\tilde{p}(x)$ goes rapidly to zero as x goes to infinity. That it is true for our specific example will be shown below.

The coefficient of $\epsilon\, e^{2i\delta}$ yields the condition $\varphi_2 = \varphi^2$ and the equation

$$\left[\beta(x)\, d(2k) - \frac{d}{dx}\alpha\frac{d}{dx} - 4\,\omega_o^2\right] p_2 + f_{1,1} = 0 \qquad (27)$$

which determine $p_2(x)$ from p(x) (the only function involved in $f_{1,1}$) provided that $4\,\omega_o^2$ is not an eigenvalue of the operator $\beta(x)\, d(2k) - \frac{d}{dx}\alpha\frac{d}{dx}$. This condition, and the possibility of deriving a function $p_2(x)$ that rapidly goes to zero as $x \to \pm\infty$, are studied below for our specific example.

The next step is cancelling the coefficient of ϵ^2. It yields the condition $\Phi_o = \varphi^* \frac{\partial\varphi}{\partial\xi}\tilde{p}_o(x)$, where $\tilde{p}_o(x)$ is again determined from p(x) by a differential equation :

$$-\frac{1}{2}\frac{d}{dx}\alpha\frac{d}{dx}\tilde{p}_o(x) + g_{-1,1} = 0 \qquad (28)$$

whose discussion again is postponed.

The step of cancelling the coefficient of $\epsilon^2 e^{i\delta}$ is the most important one. Taking into account the results obtained above, we see that only three functions of ξ and τ are involved, namely $\frac{\partial\varphi}{\partial\tau}$, $\frac{\partial^2\varphi}{\partial\xi^2}$, $|\varphi|^2\,\varphi$. Setting

$$\Phi(x,\xi,\tau) = \int_{k^2c^2}^{\infty} h(\omega^2,x)\, d\mu(\omega^2)\left[\alpha(\omega^2)\frac{\partial^2\varphi}{\partial\xi^2} + \beta(\omega^2)\,|\varphi|^2\,\varphi\right] \qquad (29)$$

we readily show that the cancellation conditions are

$$\begin{cases}\alpha(\omega^2) = \left(\omega^2 - \omega_o^2\right)^{-1} A^T(\omega^2)\\ \beta(\omega^2) = \left(\omega^2 - \omega_o^2\right)^{-1} B^T(\omega^2)\end{cases} \qquad (30)$$

$$2i\,\omega_o\frac{\partial\varphi}{\partial\tau} + A^o\frac{\partial^2\varphi}{\partial\xi^2} + B^o\,\varphi|\varphi|^2 = 0 \qquad (31)$$

where $A^T(\omega^2)$, A^o are the expansion coefficients of

$$[2v\,\omega_o - \beta(x)\, d'(k)]\,\tilde{p}(x) - \left[v^2 - \frac{1}{2}\beta(x)\, d''(k)\right] p(x) \qquad (32)$$

and $B^T(w^2)$, B^o those of $-(f_{0,1} + f_{2,-1} + f_{1,1,-1})$.
The equation (31) is the Nonlinear Schrödinger Equation. It rules the evolution of φ as a function of the "slow variables" ξ and τ that describes the signal envelope. As we have seen, all the other functions of ξ and τ already derived are derived from φ. This remains true in the two last steps. Cancelling the coefficient of $\epsilon^2\ e^{2i\delta}$ yields $\tilde{\Phi}_2 = p_2(x)\ \varphi \frac{\partial\varphi}{\partial\xi}$ and the equation

$$\left[\beta\ d(2k) - \frac{d}{dx}\ \alpha\ \frac{d}{dx} - 4\ \omega_o^2\right] \tilde{p}_2 + 2i\left[4v\ \omega_o^2 - \beta\ d'(2k)\right]p_2 + g_{1,1} + h_{1,1} = 0 \tag{33}$$

whereas cancelling the coefficient of $\epsilon^2\ e^{3i\delta}$ yields $\varphi_3 = \varphi^3$ and

$$\left[\beta\ d(3k) - \frac{d}{dx}\ \alpha\ \frac{d}{dx} - 9\ \omega_o^2\right]p_3 + f_{2,1} + f_{1,1,1} = 0 \tag{34}$$

Postponing the discussion below, we see that all the functions can be determined in such a way that our ansatz for u yields $Du = 0\ (\epsilon^3)$. In fact, it would be possible to continue the expansion in such a way that N being a fixed integer, the constructed expansion for u would give

$$Du = O(\epsilon^N)\ . \tag{35}$$

This is due to the fact that all further steps involve quantities which can be constructed recurrently, and all the equations can be solved provided $n^2\omega_o^2$ is not an eigenvalue of the operator $\beta\ d(nk) - \frac{d}{dx}\ \alpha\ \frac{d}{dx}$, and provided φ obeys the nonlinear equation (31). It remains to prove that the asymptotic motion is really confined in the transverse direction, ie that the functions of x go to zero as $x \to \pm\infty$. We now discuss this problem.

4. DISCUSSION

We begin the discussion with our example (§4.1), where it is almost possible but not possible to obtain a true asymptotic solution. This becomes possible (§ 4.2) if the nonlinear terms are modified to belong to a certain class or (§ 4.3) if $\alpha(x)$ is allowed to increase rapidly at ∞. Connected problems are finally discussed (§ 4.4).

4.1. In our example, $d(k) = k^2$, $\alpha(x) = \beta(x) = gh(x)$ is given by (9). We must find solutions that decrease exponentially at $x \to \pm\infty$ for three kinds of equations. The first kind is,

$$\left[\frac{d}{dx}\ \alpha(x)\ \frac{d}{dx} + n^2\left(\omega_o^2 - k^2\ \alpha(x)\right)\right]\ u_n(x) = h(x) \tag{36}$$

where $n = 2,3,\ldots$, and the right-hand side $h(x)$ is a continuous function exponentially decreasing as $|x| \to \infty$. The equations (27), (33) and (34) are of this kind. We have assumed that $n^2 \omega_o^2$ is not an eigenvalue of $\frac{d}{dx}\alpha(x)\frac{d}{dx} - n^2k^2\alpha(x)$ (this assumption can be easily checked for instance if α is a square well). Because of it and the inequality $\omega_o^2 < k^2c^2$, we can define two linearly independent solutions of the homogeneous equation by the conditions :

$$\begin{cases} f_+\left(n,\omega_o^2,x\right) \to \exp\left[-n\sqrt{k^2 - \omega_o^2/c^2}\; x\right] & (x \to +\infty) \\ f_-\left(n,\omega_o^2,x\right) \to \exp\left[n\sqrt{k^2 - \omega_o^2/c^2}\; x\right] & (x \to -\infty) \end{cases} \tag{37}$$

and build out of them the Green's function :

$$G(x,y) = W^{-1}\begin{cases} f_+(x)\; f_-(y) & (x > y) \\ f_+(y)\; f_-(x) & (y > x) \end{cases} \tag{38}$$

where $W = \alpha(y)\left[f_-(y)\frac{\partial}{\partial y}f_+(y) - f_+(y)\frac{\partial}{\partial y}f_-(y)\right]$ does not depend on y, and the labels n, ω_o^2 have been omitted. It is clear that $\int_{-\infty}^{+\infty} G(x,y)\,h(y)\,dy$ is the required solution of (36).

The second kind of equation is

$$\left[\frac{d}{dx}\alpha(x)\frac{d}{dx} + \omega_o^2 - k^2\alpha(x)\right]u_1(x) = h(x) \tag{39}$$

where h(x) is orthogonal to the regular solution f(x) of the homogeneous solution. Since ω_o^2 is an eigenvalue, $f_+\left(\omega_o^2, x\right)$ and $f_-\left(\omega_o^2, x\right)$ reduce to $C^+ f(x)$ and $C^- f(x)$ and are no longer independent. The "mode" p(x), and f(x) are proportional, and we achieve their identification by assuming they are normalized to 1 in $L^2(\mathbb{R})$. We define an independent solution g(x) of the homogeneous equation :

$$g(x) = f(x)\int_b^x \frac{dt}{\alpha(t)\,f^2(t)} \tag{40}$$

where b is arbitrary. f and g yield the "generalized Green's function" G'(x,y), which for x > y is

$$G'(x,y) = f(y)\,g(x)\int_x^{\infty} f^2(t)\,dt - f(x)\,g(y)\int_{-\infty}^{y} f^2(t)\,dt + f(x)\,f(y)\left[\int_a^x f(t)\,g(t)\,dt + \int_a^y f(t)\,g(t)\,dt - \gamma\right] \tag{41}$$

where $\gamma = 2\int_{-\infty}^{+\infty} f^2(y)\,dy\int_a^y f(t)\,g(t)\,dt$, and is symmetric : G'(x,y) = G'(y,x). G'(x,y) enables us to construct the unique solution of (39) which is itself orthogonal to

$f(x)$, as $\int_{-\infty}^{+\infty} G'(x,y)\, h(y)\, dy$. It obviously has the desired properties and is nothing but the solution already defined for this kind of equation in (26) and (30).

It remains to study the seemingly simplest kind of equation

$$\frac{d}{dx}\, \alpha(x)\, \frac{d}{dx}\, u_o(x) = h(x) \tag{42}$$

and we readily see a condition which is necessary to derive a twice differentiable exponentially decreasing solution $u_o(x)$: the condition

$$\int_{-\infty}^{+\infty} h(x)\, dx = 0 \tag{43}$$

that must be satisfied by the right-hand side of (22) and (28). Unfortunately, this condition cannot be satisfied in our example by $g_{-1,1}$:

$$g_{-1,1} = -\, p^2 (2k\omega + k^2 v) - v\, pp'' - 2v\, p'^2 \tag{44}$$

Indeed, $pp'' + p'^2$ can be suppressed, and the remainder is everywhere negative ! It follows that Φ_o goes to ∞ (like x or x^2) as $x \to \infty$. One may argue that Φ_o contributes a term of relative order $O(\epsilon^2)$ (at finite x) and also that it does not really carry a signal, since it depends on the slow variables only. It remains that even if the N.L.S. Equation carries over a soliton, so that a confined solution is a true asymptotic behavior in the finite or the half-infinite case, and remains a limit solution (up to relative order $O(\epsilon)$) in the fully infinite case, it cannot remain a true asymptotic solution, (ie up to all orders), and, in addition, the order $O(\epsilon^2)$ of the "slow variables" background diverges ! It is easy to see that these negative results generally hold even if $\beta(x) \neq \alpha(x)$, or for other nonlinear terms, unless they satisfy sufficient conditions like those stated below.

4.2. We keep the previous assumptions on the linear problem, so that we know that equations of the first kind and of the second kind have solutions with the required properties. In addition, we assume that $\alpha(x)$ is even and all nonlinear terms $F^{(n)}$, are exact derivatives :

$$F^{(n)} \left(u, \frac{\partial}{\partial x}, \frac{\partial}{\partial y}, \frac{\partial}{\partial t} \right) = \frac{\partial}{\partial x}\, f^{(n)} \left(u, \frac{\partial}{\partial x}, \frac{\partial}{\partial y}, \frac{\partial}{\partial t} \right) \tag{45}$$

where $f^{(n)}$ is of the form described in (3) and (4) (or similar ones in the case $n > 3$). With this assumption, $f_{1,-1}$, $g_{-1,1}$, and more generally the coefficient of ϵ^n in the terms that do not contain any positive power of $e^{i\delta}$, and hence $h(x)$ in (42), are themselves exact x-derivatives of terms made of $p(x)$ or previously calculated from other coefficients which exponentially decrease on both sides. Hence $h(x)$ has an integral which is an odd function and exponentially vanishes as $x \to \pm\infty$, and so is its product by $[\alpha(x)]^{-1}$. Hence we can construct from (43) a function $u_o(x)$ that goes to zero exponentially on both sides _ Furthermore, this can be done not only up to $O(\epsilon^4)$ but also $O(\epsilon^N)$ for any fixed N. If the N.L.S. Equation carries over an enveloppe soliton, we obtain in this way a truly confined two-dimensional envelope solution.

The N.L.S. Equation carries a soliton if A°, B°, are real and their product is positive. A° is real. B° is real for instance if $F^{(2)}$ and $F^{(3)}$ are the derivatives (or their opposite) of those given by (7a) and (7b). Hence there is certainly a very large class of equations whose solution has a true asymptotic behavior showing a confined two-dimensional envelope soliton.

4.3. In order to keep the solutions confined, we can also try to modify $\alpha(x)$. Assume that $[\alpha(x)]^{-1}$ is even and in $L^1(\mathbb{R})$. Even if $\int_{-\infty}^{+\infty} h(x)\, dx \neq 0$, the function

$$u_o(x) = - \int_x^\infty \frac{ds}{\alpha(s)} \int_o^s h(x)\, dt = - \int_x^\infty \frac{ds}{\alpha(s)} \left[\int_o^\infty - \int_s^\infty h(t)\, dt \right] \tag{46}$$

is an even solution of (42) which goes to zero as $x \to \pm\infty$. The modes corresponding to linear problems that involve $\alpha(x)$ with these properties can be studied - For instance, the case $\alpha(x) = c_o^2$ for $x \leqslant a$ and $c_o^2 x^2/a^2$ for $x \geqslant a$ can be managed by matching $\cos\left[\sqrt{\frac{\omega^2}{c_o^2} - k^2}\; x\right]$ for $x \leqslant a$ and $K_{i\sqrt{\frac{1}{4} + \frac{\omega^2 a^2}{c_o^2}}}(kx)$ for $x \geqslant a$, the matching being possible only for an infinite set of separate modes ω_n^2. For this problem, what we called the equations of first and second kind have indeed exponentially decreasing solutions. But the confinement of the ϵ^2 and higher $O(\epsilon^n)$ terms which depend only on slow variables is not an exponential confinement : the corresponding function $u_o(x)$ goes to zero only as x^{-1}.

4.4. Gathering the results, we see that for very large classes of equations - at least those with nonlinear terms of the form (45), there are true asymptotic solutions where the signal envelope is governed by Non Linear Schrödinger Equation in the y-direction, and exponentially confined in the x direction. When the N.L.S. Equation carries over a soliton (not an isolated case !), the two-dimensional asymptotic signal is exponentially confined in $\mathbb{R}^2$. This result yields four remarks.

First, physical Non Linear Schrödinger solitons will be difficult to observe, because of side band instabilities (3).Next, the result suggests that the two-dimensional exact exponentially confined solitons recently derived (4) by Boiti et al. are not an isolated feature in non linear partial differential equations. Next, the fact that our result is not such an universal limit as in the finite case, essentially because the non-modulated part of the signal may be not confined, may be somewhat related to the fact that Boiti results need flow conditions at ∞. Finally, it would be of much interest to do the same asymptotic analysis of two dimensional signals that are transversally trapped along a guiding path (here the y direction) and governed along this path by the Korteweg-de Vries Equation. The success - or failure - of such an analysis would give evidence in favor of - or against the existence of exponentially confined two-dimensional KdV-like solitons.

REFERENCES

(1) F. Calogero and P.C. Sabatier "Non linear modulation of a transversally trapped mode". p.307-318 in "Topics in SOliton Theory and Exactly solvable Nonlinear Equations" M. Ablowitz, B. Fuchssteiner, M. Kruskal Ed. - World Scientific (Singapore) 1987.

(2) P.C. Sabatier "Nonlinearity in dispersive trapped waves" in "Nonlinear Topics in Ocean Physics", Course at the International School of Physics Enrico Fermi of Varenna, 1989.

(3) A.C. Newell "Solitons in Mathematics and Physics" S.I.A.M. Philadelphia (1985).

(4) M. Boiti, J. JP Leon, L. Martina, F. Pempinelli "Two-dimensional Solitons" (Same Proceedings).

SOLVABLE NONLINEAR EQUATIONS AS CONCRETE REALIZATIONS OF THE SAME ABSTRACT ALGEBRA

P.M. Santini
Dipartimento di Fisica, Università "La Sapienza", Roma
I.N.F.N., Sezione di Roma, Italy

ABSTRACT

We introduce an elementary operator structure and we show that well-known examples of integrable systems like the Korteweg-de Vries, the Nonlinear Schrodinger, the Benjamin-Ono, the Chiral fields, the Kadomtsev-Petviashvili, the Davey-Stewartson and the self-dual Yang Mills equations are generated by different concrete realizations of it. We also show that the simplest realization of this structure gives rise to nonlinear algebraic equations which share with their differential analogues the basic features of integrability and therefore are examples of solvable nonlinear algebraic systems.

1. INTRODUCTION

The algebraic properties of integrable systems are conveniently described introducing the so-called *recursion operator* Φ which satisfies the following properties:

i) Φ generates hierarchies of evolution equations associated with a given spectral problem (observation made first by Lenard for the Korteweg-de Vries (KdV) equation

$$u_t = u_{xxx} + 6uu_x, \tag{1}$$

and by Ablowitz, Kaup, Newell and Segur for the nonlinear Schrodinger (NLS) equation [1])

$$iu_t + \frac{1}{2}u_{xx} - u|u|^2 = 0; \tag{2}$$

ii) a suitable extension of Φ generates a class of Backlund Transformations (BT) of the given equations [2]);

iii) its adjoint is the squared eigenfunction operator [1]);

iv) Φ generates non-Lie point symmetries of the given equations [3]) and these symmetries commute [4]);

vi) Φ is factorizable in terms of two hamiltonian operators and then the associated equations are bi-hamiltonian [5]);

vii) Φ generates a hierarchy of constants of the motion in involution with respect to each of the hamiltonian operators [5]).

In a series of recent papers [6-10]) Fokas and the author have shown that the above algebraic properties, suitably extended, are also satisfied by integrable equations in 2+1 dimensions like the Kadomtsev-Petviashvili (KP) [6])

$$u_t = u_{xxx} + 6uu_x + \alpha^2\partial_x^{-1}u_{yy}, \tag{3}$$

and by integro-differential equations like the intermediate long wave (ILW) [9],

$$u_t = \hat{T}u_{xx} + 2uu_x, (\hat{T}f)(x) := (\eta)^{-1}P\int_{-\infty}^{+\infty} coth((\pi/\eta)(\xi - x))f(\xi)d\xi, \qquad (4a)$$

the Benjamin-Ono (BO) [10]

$$u_t = \hat{H}u_{xx} + 2uu_x, \quad (\hat{H}f)(x) := \pi^{-1}P\int_{-\infty}^{+\infty} (\xi - x)^{-1}f(\xi)d\xi, \qquad (4b)$$

equations and their 2+1 dimensional extensions [11]

$$u_t = \hat{T}D^2u + 2uDu, \quad D :== \alpha\partial_y, \qquad (5a)$$

$$u_t = \hat{H}D^2u + 2uDu. \qquad (5b)$$

This theory
i) describes integrable systems regardless of the number of dimensions;
ii) incorporates the notion of BT's into the scheme;
iii) requires the essential use of a formalism for generalized functions (distributions) with bilocal argument and, correspondly, the introduction of
a) a new operation, the *d-derivative*, which generalizes the usual Frechet derivative to the space of bilocal operators,
b) a new notion, *the extended symmetry*, which characterizes the *regular* coefficients of distributions and unifies the concepts of symmetries and BT's of a given system.

It turns out that nonlinear integrable systems are elements of hierarchies of integrable equations that can be represented in the following *bilocal* form [6–12]

$$\delta(x - x')u'_t = \beta_n\delta(x - x')K^{(n)}(u, u'), \quad u = u(x,t), u' := u(x',t), \qquad (6a)$$

$$K^{(n)}(u, u') := \Phi^n\hat{K}^0 \cdot 1, \qquad (6b)$$

which obviously implies

$$u_t = \beta_n\int_{\Re^\nu} dx'\delta(x - x')K^{(n)}(u, u') = \beta_nK^{(n)}(u, u) =: \beta_nk^{(n)}(u), \qquad (6c)$$

where x, x' are ν - dimensional vectors of components x_i, x'_i, i=1,..,ν,

$$\delta(x - x') = \prod_{i=1}^{\nu}\delta(x_i - x'_i)$$

(δ is the Dirac function), $K^{(n)}$ belongs to a suitable space S of polynomials in $u = u(x,t)$, $u' := u(x',t)$ and in their integrals and derivatives, and the recursion operator Φ and the "starting operator" $\hat{K}^0$ are operator valued functions on S. Through this paper m and n are nonnegative integers.

As a natural application of this theory, the author has recently introduced an approach [11], the *dimensional deformations*, in which the richness of the algebraic structure of a given integrable system is explored to produce its integrable multidimensional generalizations through an elementary deformation of the associated recursion operator. Acting at the level of simple operator structures, this approach has allowed to derive, with essentially *no* calculations involved, all the known examples of integrable multidimensional systems and some novel ones. In particular it was shown that the KP, ILW in 2+1 dimensions and the self-dual Yang-Mills (SDYM) equation

$$(g^{-1}g_t)_x = \alpha(g^{-1}g_y)_z \tag{7}$$

are dimensional deformations of the KdV, ILW and Chiral fields (CF) equation

$$(g^{-1}g_t)_x = \alpha(g^{-1}g_x)_t \tag{8}$$

respectively.

This approach has given a further contribution toward a unification of the theory describing the algebraic properties of integrable systems; indeed the possibility of deforming the operator structure of, say, KdV to obtain the operator structure of, say, KP can be interpreted, from a different and complementary point of view, in the following way: KdV and KP are *different* concrete realizations of the *same* operator structure (see §6 of [11])! With these motivations, it was possible to introduce in [13,14] a unified approach to integrable systems, identifying an abstract and elementary operator structure and showing that different bilocal realizations of this structure give rise to a universe of celebrated integrable systems in different dimensions like the KdV, NLS, BO, ILW, CF, KP, Davey-Stewartson (DS)

$$iu_t + \frac{1}{2}(u_{x_1x_1} + \alpha^2 u_{x_2x_2}) = uv \tag{9a}$$

$$v_{x_1x_1} - \alpha^2 v_{x_2x_2} = (|u|^2)_{x_1x_1} + \alpha^2(|u|^2)_{x_2x_2}, \tag{9b}$$

N-wave interaction (NWI)

$$u_{ij_t} = (a_i - a_j)^{-1}\sum_{l=1}^{\nu}((e_l + b_l a_i)c_j - (e_l + b_l a_j)c_i)u_{ij_{x_l}} - \sum_{k=1}^{N}\left(\frac{c_i - c_k}{a_i - a_k} - \frac{c_k - c_j}{a_k - a_j}\right)u_{ik}u_{kj}. \tag{10}$$

and the SDYM equations. It was finally possible to show that the simplest bilocal realization of this elementary algebra generates examples of nonlinear algebraic systems [15]. These systems share with their differential analogues (1-5,7-10) the basic features of integrability and therefore are examples of "solvable nonlinear algebraic equations".

Here we present a short summary of these recent developments; detailes and proofs can be found in [13,14,15].

2. THE OPERATOR STRUCTURES UNDERLYING INTEGRABILITY

The operator structures underlying integrability consist of *basic*, *linear* operators q_R and q_L (right and left multiplication operators) acting on a suitable functional space. Here we use their canonical representation in terms of *nonlocal* operators, i.e.

$$(q_L f)(x) := \int dx'' q(x,x'') f(x''), \tag{11a}$$

$$(q_R f)(x) := \int dx'' f(x'') q(x'',x), \tag{11b}$$

and q_R, q_L are then defined in terms of their kernel q which takes values (for instance) in $\Re^\nu$ (ν is a natural number) or in Z^ν (if $x \in Z^\nu$, then $\int dx$ is replaced by $\sum_\nu$). Instead of right and left multiplications, it is perhaps convenient to choose as basic operators the commutation and anticommutation operators $q^\pm$, defined by

$$q^\pm := q_L \pm q_R. \tag{12}$$

In order to appreciate the generality captured by (11,12), let's consider few concrete realizations of the abstract operator q^-:

a) if $q = \delta(x - x')a$, a is an NxN matrix independent of x, x', then q^- reduces to the usual commutation operator between matrices

$$q^- f = \hat{a} f := af - fa, \tag{13a}$$

b) If $\nu = 1$ and $q = \delta'(x - x')$, then q^- reduces to the bilocal differential operator

$$q^- = \partial_{x_1} + \partial_{x_1'}. \tag{13b}$$

c) If $\nu = 1$ and $q = -(1/2)\delta(x + i\eta - x'), \eta > 0$, then q^- reduces to a rational combination of shift operators

$$q^- = (1 - ee')(2e')^{-1}, \tag{13c}$$

$$(ef)(x,x') := f(x + i\eta, x'), \quad (e'f)(x,x') := f(x, x' + i\eta)$$

d) If $\nu = 2$ and $q = \delta'(x_1 - x_1')\delta(x_2 - x_2') + \delta(x_1 - x_1')\delta''(x_2 - x_2')$, then q^- reduces to a bilocal differential operator in multidimensions

$$q^- = \partial_{x_1} + \partial_{x_1'} + \partial_{x_2}{}^2 - \partial_{x_2'}{}^2. \tag{13d}$$

a)-d) provide *different* bilocal realizations of the operator structure q^-!

The commutator q^- is the simplest (nonconstant) hamiltonian operator, namely

a) q^- is skew-symmetric:

$$(q^-)^* = -q^-, \tag{14}$$

where L^* denotes the adjoint of L with respect to the symmetric bilinear form

$$< g, f > := tr \int_{\Re^{2\nu}} dx dx' g(x',x) f(x,x') \tag{15}$$

(the trace operation is dropped if g and f are scalars);

b) q^- satisfies the Jacobi identity w.r.t. the bracket

$$\{a,b,c\} :=< a,\ (q^-)'[b]c >; \tag{16}$$

where $(q^-)'[g]$ denotes the Frechet derivative of the operator q^- in the direction of the integral operators g^-:

$$q^{-\prime}[g(x,x')]f(x,x') = \int_{\Re^\nu} dx''(g(x,x'')f(x'',x') -$$

$$-f(x,x'')g(x'',x')) =: (g^- f)(x,x'); \tag{17a}$$

consequently:

$$(L(q^-))'[g^-] = \partial_\epsilon L(q^- + \epsilon g^-)|_{\epsilon=0}. \tag{17b}$$

If we deform the hamiltonian operator q^- in the following way:

$$q^- \to q^- + \lambda d^-,\quad (d^-)' = 0, \tag{}$$

we obviously produce a new hamiltonian operator $\forall\ \lambda$, then q^- and d^- are two *compatible* hamiltonian operators and the following results can be proven in a straightforward manner.

i) The factorization

$$\Phi = q^-(d^-)^{-1}, \tag{19}$$

of q^- and d^- is a hereditary or Nijenhuis operator, namely

$$\Phi'[\Phi f]g - \Phi\Phi'[f]g \textit{ is symmetric w.r.t. } f \textit{ and } g; \tag{20}$$

ii) The starting point

$$\hat{K}^0 H = q^- H,\quad d^- H = 0, \tag{21}$$

is a strong symmetry for Φ or, equivalently, the Lie derivative of Φ in the direction $\hat{K}^0 H$ is zero, namely

$$\Phi'[\hat{K}^0 H] + \Phi(\hat{K}^0 H)' - (\hat{K}^0 H)'\Phi = 0. \tag{22}$$

iii) If $\tilde{\text{H}}$ is the Lie algebra of functions $H(x,x')$ endowed with the bracket

$$[H^{(1)}, H^{(2)}]_I := \int_{\Re^\nu} dx''(H^{(1)}(x,x'')H^{(2)}(x'',x') -$$

$$-H^{(2)}(x,x'')H^{(1)}(x'',x')), \tag{23}$$

and containing the *abelian* subalgebra $\tilde{\text{h}}$ of functions $h = h(x-x')$, then the vector space

$$\kappa_{\tilde{H}} := \{X:\ X = \sum_{n,j} \alpha_{n,j}\Phi^{n_j} q^- H_j,\, H_j \in Ker\ d^-\} \tag{24}$$

is an abstract infinite dimensional Lie algebra (hereafter indicated as canonical) endowed with the bracket

$$[X^{(1)},X^{(2)}]_d := X^{(1)\prime}[X^{(2)}] - X^{(2)\prime}[X^{(1)}]. \tag{25}$$

Equations (23,24) imply that the subspace $\kappa_{\tilde{h}}$ is an abelian subalgebra of $\kappa_{\tilde{H}}$.

iv) The eigenfunctions of $\Phi^* = (d^-)^{-1}q^-$ are the "bilocal squared eigenfunctions" of the associated linear problem, namely if

$$\Phi^* w = \lambda w, \quad w = w(x,x'), \tag{26a}$$

then

$$w = v(x)\tilde{v}(x'), \tag{26b}$$

where v and $\tilde{v}$ are the right and left eigenfunctions of the associated spectral problem

$$\lambda d_L v = q_L v, \quad \lambda d_R \tilde{v} = q_R \tilde{v}. \tag{27}$$

The spectral problem (27) plays a central role in any method of solution of the associated nonlinear equations.

If $d = \delta(x-x')a$, where a is a constant NxN diagonal matrix, then q^- reduces to the usual commutation operator between matrices $d^- f = \hat{a} f := af - fa$ and we can derive the following interesting "off-diagonal" reduction of the operator structure (19,21)

$$\Phi = (q_D{}^- + (1-\Pi)q_O{}^- - q_O{}^-(q_D{}^-)^{-1}\Pi q_O{}^-)\hat{a}^{-1}, \tag{28a}$$

$$\hat{K}^O = q_O{}^- C_D, \tag{28b}$$

where q_O and q_D are the off-diagonal and diagonal parts of q, Πf is the diagonal part of f and C_D is a diagonal matrix. The operator structures (28) enjoy properties i)-iv) of this section, being reductions of the elementary structures (19,21)[16].

Through a sequence of elementary steps we have built an abstract canonical algebra $\kappa_{\tilde{H}}$ whose "constraction bricks" are the basic operators q^-, d^-; now we show that different concrete realizations of these basic operators give rise, through formulae (6), to a universe of well-known integrable systems.

Concrete realizations.

1) If $\nu = 1, d = \delta'(x_1 - x_1'), q = \delta(x_1 - x_1')u' + \delta''(x_1 - x_{1'})$ and u is a scalar, then $d^- = \partial_{+1}$, $q^- = u - u' + \partial_{1+}\partial_{1-} =: u^-$, where

$$\partial_{i\pm}{}^j := (\partial_{x_i} \pm \partial_{x_{i'}})^j,$$

and (19,21) reduces to the bilocal representation of the KdV class; the KdV equation (1) is the fourth member (n=3,$\beta_3 = 1/2$) of the class (6) [8,11].

2) If $\nu = 2, d = \delta_1^1, q = \delta u' + \delta_1^2 + \alpha\delta_2^1$, where

$$\delta_i^j := \delta^j(x_i - x_i') \prod_{s \neq i} \delta(x_s - x_s'),$$

$$\delta^j(x_i - x_i') := \partial^j \delta(x_i - x_i')/\partial x_i{}^j$$

and u is a scalar, then $d^- = \partial_{1+}, q^- = u - u' + \partial_{1+}\partial_{1-} + \alpha\partial_{2+} =: u^-$ and (19,21) reduces to the bilocal representation of the KP class; the KP equation (3) is its fourth member (n=3, $\beta_3 = 1/2$)[8,10,11].

3) If $\nu = 1, d = -(1/2)\delta(x_1 + i\eta - x_1')$, $q = \delta(x_1 - x_1')u' + i\delta'(x_1 - x_1')$ and u is a scalar, then

$$d^- = (1 - ee')(2e')^{-1}, \quad q^- = u - u' + i\partial_{1+} := u^-,$$

and (19,21) reduces to the bilocal representation of the ILW class; the ILW equation (4a) is its third member (n=2, $\beta_2 = 1/4i$) [9,11].

4) If $\nu = 2, d = (-1/2)\delta(x_1 + i\eta - x_1')$, $q = \delta u' + \alpha\delta_2^1$, then $d^- = (1 - ee')(2e')^{-1}$, $q^- = u - u' + \alpha\partial_{2+} := u^-$ and (19,21) becomes the bilocal representation of the 2+1 dimensional ILW equation (5), corresponding to n=2, $\beta_2 = 1/4i$ in (6) [11].

5) If $\nu = 1, d = \delta'(x_1 - x_1')$, $q = \delta(x_1 - x_1')u'$, and u is a matrix, then $d^- = \partial_{1+}$, $q^- f = uf - fu' := u^-$ and we obtain the bilocal representation of the CF algebra; if $\nu = 2, d = \delta^1{}_1$, $q = \delta u' + \delta^1{}_2$, then $d^- = \partial_{1+}$, $q^- f = uf - fu' + \alpha\partial_{2+} f =: u^-$ and we obtain the bilocal representation of the SDYM algebra. The CF and SDYM equations (8,7) are generated by these algebras via elementary deformations [11].

6) If N=2, $\nu = 1, d = \delta a$, $a = diag(1, -1)$, $q_0 = \delta u_0'$, $q_D = I\delta'(x_1 - x_1')$, then $d^- f = \hat{a} f, q_D{}^- = \partial_{1+}$, $q_0{}^- f = uf - fu' =: u_0{}^-$ and (28) reduces to the bilocal representation of the NLS class; the NLS equation (2) comes from (6) for n=2,$\beta_2 = 1/4i, \bar{u}_{12} = u_{21} = u$ [8,12].

7) If N=2, $\nu = 2, d = \delta a$, $a = diag(1, -1)$, $q_0 = \delta u_0'$, $q_D = \delta_1{}^1 I + \alpha\delta_2{}^1 a$, then $q_D{}^- f = \partial_{1+} f + \alpha(af_{x_2} + f_{x_2'}a) =: u_D{}^-$ and (28) reduces to the representation of the DS class; the DS equation (9) comes from (6) for n=2, $\beta_2 = 1/4i$ [6,7,11].

8) If N and ν are arbitrary, $d = \delta a$, a is diagonal,$q_0 = \delta u_0', q_D = \delta^1{}_1 I + \sum_{i=2}^{\nu} \delta^1{}_i J_i$ ($J_i := e_i I + b_i a, e_i$ and b_i are scalars), then $q_d{}^- = \sum_{i=1}^{\nu}(J_i f_{x_i} + f_{x'_i} J_i), J_1 = I$ and (28) reduces to the representation of the NWI class; the NWI equation (10) comes from (6) for n=1, $\beta_1 = 1$.

9) If N is arbitrary and $d = \delta(x - x')a$, $q = \delta(x - x')u$, then [15] $q^- f = \hat{u} f$, $d^- f = \hat{a} f$, $\hat{g} f := gf - fg$ and the algebraic structure (19,21) becomes

$$\Phi = \hat{u}(\hat{a})^{-1}, \; K^{(0)} = \hat{u}H, \; H \in Ker\; \hat{a} = \{H, \; H = \sum_{j=0}^{N-1} c_j a^j, \; c_j \in C\}. \tag{29}$$

The associate class of matrix evolution equations reads [15]

$$u_t = K_N(u) := \sum_{i=1}^{N-1}\sum_{j=1}^{i}(-1)^j c_{ij}\Phi^{j-1}\hat{u}a^i =$$

$$= \hat{a}\sum_{i=1}^{N-1}\sum_{j-1}^{i} c_{ij}\Gamma^{(i-j,j)} = -\hat{u}\sum_{i-1}^{N-1}\sum_{j=1}^{i} c_{ij}\Gamma^{(i-j+1,j-1)}, \tag{30}$$

where c_{ij}'s are arbitrary scalars and matrices $\Gamma^{(i,j)}$'s are defined by

$$\Gamma^{(i,j)} := \sum_{l} \prod_{k=1}^{i+j} \chi_{l_k},$$

where $\chi_0 = a$, $\chi_1 = u$ and $\sum_l$ indicates the sum over all possible i+j-dimensional vectors l whose components l_k consist of i zeroes and j ones.

$\Gamma^{(i,j)}$'s are the *symmetric nonabelian generalizations* of scalar monomials and therefore we refer to them as *symmetric nonabelian monomials* (SNM)[15]; the first few instances of SNM's are

$$\Gamma^{(0,j)} = u^j, \quad \Gamma^{(i,0)} = a^i,$$

$$\Gamma^{(1,1)} = au + ua, \quad \Gamma^{(1,2)} = au^2 + uau + u^2a,$$

$$\Gamma^{(1,3)} = au^3 + uau^2 + u^2au + u^3a,$$

$$\Gamma^{(2,2)} = a^2u^2 + au^2a + auau + uaua + ua^2u + u^2a^2.$$

3. ALGEBRAIC PROPERTIES OF INTEGRABLE SYSTEMS

As we have seen in §2, the evolution equations (1-5,7-10,30) are obtained choosing q and d to be generalized functions; correspondingly the basic operator q^- becomes a bilocal operator u^-, and Φ and $\hat{K}^0$ assume a bilocal form involving function u evaluated at two different points x and x' (see the concrete realizations 1)-9) of §2).

Since $\kappa_{\tilde{H}}$ is canonical with respect to the Frechet derivative of the nonlocal operator q^-, *any* particular choice (like the one given in examples 1)-9) of §2) of $q(x,x')$ preserves the canonicity of the obtained bilocal algebra with respect to the so called *directional d-derivative*

$$u^-{}_d[g(x,x')] := q^{-\prime}[g(x,x')] = g^- \tag{31}$$

of the bilocal operator u^-.For this reason any bilocal realization of a canonical Lie algebra $\kappa_{\tilde{H}}$ is hereafter called canonical too.

The d-derivative of the bilocal operator u^- satisfies the following *projective* formula

$$u^-{}_d[\delta(x-x')g(x,x')]f(x,x') = g(x,x)f(x,x') - f(x,x')g(x',x') =: u^-{}_f[g]f(x,x'), \tag{32}$$

where the subscript f denotes the usual Frechet derivative with respect to u, namely

$$L_f(u,u')[g] = \partial_\epsilon L(u + \epsilon g(x,x), u' + \epsilon g(x',x'))|_{\epsilon=0}. \tag{33}$$

The d-derivative and the usual Frechet derivative are the main tools of the theory.

Since equations (1-5,7-10,30) are different bilocal realizations of the same operator structure (19,21), then they inherit from it the same algebraic properties illustrated in the following fundamental Theorem (whose proof can be found, for instance, in [6]):

Theorem.

Let $\Theta^{(1)}, \Theta^{(2)}$ and $\hat{K}^0$ be the bilocal realizations of the operators d^-, q^- and q^- respectively; define $\Phi := \Theta^{(2)}(\Theta^{(1)})^{-1}$ and $\hat{\Gamma}^0 := (\Theta^{(1)})^{-1}\hat{K}^0$;

then

1) Φ is a hereditary or Nijenhuis operator;

2) $\Phi^m \Theta^{(1)}$ are hamiltonian operators;

3) $\kappa_{\tilde{H}}$ is a bilocal canonical Lie algebra endowed with the Lie bracket $[,]_d$.

Moreover if, in addition,

i) the distribution $\delta(x-x')K^{(n)}(u,u')$ belongs to the abelian sub-algebra $\kappa_{\tilde{h}}$, then

4) $\Sigma^{(m)} := \Phi^m \hat{K}^0 \cdot 1$ and $\Gamma^{(m)} := (\Theta^{(1)})^{-1}\Sigma^{(m)} = (\Phi^*)^m \hat{\Gamma}^0 \cdot 1$ are extended symmetries and extended gradients of the conserved quantities I_n in involution respectively, for equation (6), namely

$$\Sigma^{(m)}{}_f[K^{(n)}(u,u)] = (\delta K^{(n)})_d[\Sigma^{(m)}], \tag{34a}$$

$$\Gamma^{(m)}{}_f[K^{(n)}(u,u)] = -(\delta K^{(n)})_d{}^*[\Gamma^{(m)}], \tag{34b}$$

$$(\Gamma^{(m)})_d = (\Gamma^{(m)})_d{}^* \Leftrightarrow I_{md}[f] =< \Gamma^{(m)}, f >, \tag{34c}$$

$$\{I_m, I_n\}_i :=< \delta\Gamma^{(m)}, \Theta^{(i)}\Gamma^{(n)} >, \; i = 1,2; \tag{34d}$$

5) equations (6) are *extended bi-hamiltonian systems*, since they can be written in the following two extended hamiltonian forms

$$\delta(x-x')u'_t = \delta(x-x')\Theta^{(1)}\Gamma^{(n)} = \delta(x-x')\Theta^{(2)}\Gamma^{(n-1)}; \tag{35}$$

6) $\sigma^{(m)} = \sigma^{(m)}(u) := \Sigma^{(m)}(u,u)$ and $\gamma^{(m)} = \gamma^{(m)}(u) := \Gamma^{(m)}(u,u)$ are commuting symmetries and gradients of the conserved quantities I_m in involution respectively for equations (6), namely

$$\sigma^{(m)}{}_f[k^{(n)}] = k^{(n)}[\sigma^{(m)}], \tag{36a}$$

$$\gamma^{(m)}{}_f[k^{(n)}] = -k^{(n)}{}_f{}^+[\gamma^{(m)}], \tag{36b}$$

$$\gamma^{(m)}{}_f = \gamma^{(m)}{}_f{}^+ \Leftrightarrow I_{m\,f}[f] = (\gamma^{(m)}, f) := \int_{\Re^\nu} dx \gamma^{(m)} f; \tag{36c}$$

7) the equations $\Sigma^{(m)} = \Sigma^{(m)}(u,u') = 0$ are auto-BT's of equations (6).

Now some remarks:

i) this theorem shows that symmetries and BT's of an integrable system originate from the same entity: the *extended symmetry*, defined in equation (34a). In this equation two *different* operations on bilocal structures appear: the usual Frechet derivative and the novel d-derivative, which takes account of the nonlocal origin of bilocal operators. Moreover when we interpret the equation $\Sigma(u,u') = 0$ as BT,, the functions u and u' must be viewed as two different solutions $u = u(x)$, $u' = u'(x)$ of $u_t = k(u)$ and, correspondly, in equations (34) the even derivative ∂_{i+} of the operator $(\delta K^{(n)})_d$ is replaced by ∂x_i; the odd derivative ∂_{i-} is absent (see theorem 4.1 of [6]).

ii) If the hypothesis i) of the theorem is satisfied, namely if

$$\delta(x-x')K^{(n)}(u,u') = \sum_{l=1}^{n} b_{n,l}\Phi^{n-l}\hat{K}^0 \cdot h^{(l)}, \;\; h^{(l)} \in \tilde{h}, \tag{37}$$

for some constants $b_{n,l}$, then the operators $(\delta K^{(n)})_d$, appearing in equation (34), are well defined and read

$$(\delta K^{(n)})_d = \sum_{l=1}^{n} b_{n,l}(\Phi^{n-l}\hat{K}^0 \cdot h^{(l)})_d. \tag{38}$$

It is straightforward to show that the concrete bilocal realizations which generate the evolution equations (1-5,7-10) satisfy equation (37) (see §4 for details); therefore these evolution equations enjoy the integrability properties described in the above theorem.

Equations (30) play a distinct role in the universe of the integrable systems generated by the abstract structure (19,21). Precisely we have the following properties.

The recursion operator $\Phi = \hat{u}(\hat{a})^{-1}$ is *nihilpotent* on the subspace of its starting symmetries, namely $\Phi^j \hat{u} a^i = 0$, if $i \geq j$; therefore it generates a finite number (which depends on the rank N of the matrix) of commuting flows and constants of the motion in involution, consistently with the fact that the evolution equations (30) are finite dimensional.

The spectral problem associated with equations (32) is given by the following matrix eigenvalue problem

$$\lambda \hat{a} v = u v; \tag{39}$$

the associated inverse scattering (or spectral) (IST) formalism, which is the tool for solving the initial value problem for equations (30), is presently under investigation.

The stationary versions

$$\sum_{i=0}^{N-1} \sum_{j=0}^{i} c_{ij} \Gamma^{(i,j)} = 0 \tag{40}$$

of equations (30) are examples of solvable nonlinear algebraic systems (see for example Reference [17] for a discussion on the integrability properties of the stationary version of the KdV class). Since the $\Gamma^{(i,j)}$'s are gradients (see the theorem of the previous section), then equations (40) are Lagrangian: $\sum_{i=0}^{N-1} \sum_{j=0}^{i} c_{ij} \Gamma^{(i,j)} = grad\ L.$

If u and a are scalars, the SNM $\Gamma^{(i,j)}$ reduces to a scalar monomial of degree j and the nonlinear algebraic equations (40) reduce to scalar polynomial equations. Thus equations (40) provide a *solvable* nonabelian generalization of scalar polynomial equations; it is interesting to remark that, among all possible nonabelian extensions of scalar polynomial equations, the solvable one (equation (40)) is also the symmetric one, being constructed by the SNM's $\Gamma^{(i,j)}$! The linearizing transform for these algebraic systems, based on (39), is presently under investigation.

4. DOES EVERY CONCRETE REALIZATION OF (19,21) GIVE RISE TO INTEGRABLE SYSTEMS?

We have shown that the different bilocal realizations 1)-9) of §2 give rise to the integrable systems (1-5,7-10,30). An obvious question arises at this point: does *every* choice of the kernels q and d of the nonlocal operators q^-, d^- give rise to integrable systems and, if not, what are the necessary constraints? This question can be answered exaustively using the ideas and results of the deformation approach [11].

A nonlinear evolutionary system $u_t = k(u)$ is integrable iff it is associated with a canonical Lie algebra $\kappa_{\tilde{H}}$ in the way prescribed by the fundamental Theorem of the previous section, namely the following *two* conditions must be simultaneously satisfied:

A) there exists an underlying canonical Lie algebra $\kappa_{\tilde{H}}$ generated by the bilocal operators Φ and $\hat{K}^0$;

B) the evolution equation takes the distributional form (6), and $\delta K^{(n)}(u,u')$ belongs to the canonical algebra $\kappa_{\tilde{h}}$, namely equation (37) holds.

On the other hand the previous analysis shows that the bilocal operators obtained by *any* choice of q and d give rise to bilocal canonical algebras, namely condition A is always satisfied. Condition B is satisfied if the obtained equations belong to the associated canonical algebra and this property can be checked very easily just computing the commutators $[\Phi,\delta], [\hat{K}^0,\delta]$. Let's illustrate this point in two simple examples:

1) the realization of example 2) of §2 gives rise to the following commutations

$$[\Phi,\delta] = 2\delta_1^1,\ [\hat{K}^0,\delta] = 2\delta_1^1\partial_{1+},$$

which can be used to show that the class of equations (6) belongs to the associated canonical algebra, since

$$\delta(x-x')K^{(n)}(u,u') = \delta\Phi^n\hat{K}^0\cdot 1 =$$

$$= \sum_{l=1}^{n} b_{n,l}\Phi^{n-l}\hat{K}^0\cdot\delta^{(l)} \in \kappa_{\tilde{h}}, \qquad b_{n,l} = (-2)^l\binom{n+1}{l};$$

we have obtained the KP class.

2) For the choice $q = \delta u' + {\delta_1}^2 + \alpha{\delta_2}^2, d = {\delta_1}^1$, we have

$$[\Phi,\delta] = 2({\delta_1}^1 + \alpha{\delta_2}^1\partial_{2+}{\partial_{1+}}^{-1}), [\hat{K}^0,\delta] = 2({\delta_1}^1\partial_{1+} + \alpha{\delta_2}^1\partial_{2+}),$$

and the corresponding class (6) does not belong to the associated canonical algebra; for example

$$\delta K^{(1)}(u,u') = \delta\Phi q^-\cdot 1 = \Phi q^-\cdot\delta - 4q^-\cdot\delta_1^1 - 4\partial_{2+}\partial_{1+}^{-1}q^-\cdot\delta_2^1 \notin \kappa_{\tilde{h}},$$

and *none* of the beautiful properties of the canonical algebra can be inherited by the evolution equation!

We remark that satisfying condition B is *equivalent* to the existence of a Lax pair for the evolution equations under scrutiny. The often cumbersome and tedius calculations involved in the check of the Lax compatibility are replaced, in this framework, by the trivial calculation of a commutator [18)]! Moreover the simplicity of the operation involved in this check allows to isolate and focus the technical and conceptual reasons for the almost total absence of integrable equations in more than 2+1 dimensions [11)].

We conclude pointing out that

i) the existence of the abstract nonlocal structure (19,21) underlying the integrability of so (apparently) different systems like equations (1-5,7-10, 30), emphasizes their "algebraic" equivalence!

ii) The evolution equations (30) and the associated nonlinear algebraic equations (40) play a fundamental role in the theory of integrable systems, being the most elementary (but non trivial) instances of nonlinear equations which enjoy the basic properties underlying solvability! It is worthwhile to remark that the algebraic versions of integrable differential systems like the Burgers [19] equation $u_t = u_{xx} + 2uu_x$ can also be constructed , and their general solution can be obtained via an algebraic analogue of the Cole-Hopf transformation [15].

REFERENCES

1. M.J.Ablowitz,D.J.Kaup,A.C.Newell and H.Segur, Stud.Appl.Math.**53**, 249(1984).
2. F.Calogero, Lett. Nuovo Cimento **14**, 443, (1975). F.Calogero, Lett.Nuovo Cimento **14**, 537 (1975);
3. P.J.Olver,J.Math.Phys. **18** 1212 (1977).
4. B.Fuchssteiner, Nonlinear Anal. **3**,849 (1979). B.Fuchssteiner and A.S.Fokas,Physica **4D**,47 (1981).
5. F.Magri,J.Math.Phys.**19**,1156(1978).
6. P.M.Santini and A.S.Fokas,Comm.Math.Phys.**115**,375 (1988).
7. A.S.Fokas and P.M.Santini,Comm.Math.Phys.**116**,449 (1988).
8. P.M.Santini and A.S.Fokas,The bi-hamiltonian formulations of integrable evolution equations in multidimensions, in Nonlinear Evolutions, Proceedings of the IV Workshop on Nonlinear Evolution Equations and Dynamical Systems; edited by J.Leon, World Scientific Publishing Company, Singapore (1988).
9. P. M. Santini, Bi-hamiltonian formulations of the Intermediate Long Wave equation, Preprint INS 80,Clarkson University,1987;Inverse Problems (in press).
10. A.S.Fokas and P.M.Santini, J.Math.Phys.**29**,604 (1988).
11. P.M.Santini, Dimensional deformations of integrable systems, an approach to integrability in multidimensions.I, Preprint 586 Dipartimento di Fisica, Universita' di Roma I,1988; Inverse Problems (in press).
12. B.G.Konopelchenko, Inverse Problems **4**,785 (1988).
    M.Boiti,J.Leon and F.Pempinelli, Stud.Appl.Math., **78**, 1 (1988).
13. P.M.Santini, The algebraic structures underlying integrability, PreprintPM/88-55, Montpellier, 1988. Phys.Lett. (submitted to).
14. P. M. Santini, Algebraic properties and symmetries of integrable evolution equations, Proceedings of the Symposium on Symmetries in Sciences III, Landes-Bildungszentrum, Austria, edited by B. Gruber. Preprint n. 636, Dipartimento di Fisica, Roma, 1988.
    A.S.Fokas and P.M.Santini, A unified approach to recursion operators, Preprint INS 101, Clarkson University, 1988.
15. P.M.Santini, Solvable nonlinear algebraic equations, Preprint PM/88- 56, Montpellier 1988. Phys. Lett. (submitted to).
16. F.Magri and C.Morosi, A geometrical characterization of integrable hamiltonian systems through the theory of Poisson-Nijenhuis manifolds, Preprint Universita' di Milano, 1984;
    F.Magri, C.Morosi and O.Ragnisco, Comm.Math.Phys. **99**, 115 (1985).

17. S.Novikov,S.V.Manakov, L.P.Pitaevskii and V.E.Zakharov, Theory of Solitons, The Inverse Scattering Method, Contemporary Soviet Mathematics, Consultant Bureau. New York and London, 1984.
18. We thank M.Boiti for pointing out in a private conversation the role of the commutators $[\Phi, \delta], [\hat{K}^0, \delta]$ in connection with the Lax compatibility.
19. J.M.Burgers, The nonlinear diffusion equation, Reidl, Dordrecht, 1974.
E. Hopf, Comm.Pure Appl. Math. **3**, 201 (1950).
J.D.Cole, Q. Appl. Math. **9**, 225 (1950).

Properties of Solutions of Dispersive Equations

Jean-Claude Saut

Mathématiques, Université Paris XII and
Laboratoire d'Analyse Numérique, CNRS and Université Paris XI,
Bâtiment 425, 91405 Orsay (France)

Abstract. We discuss several recent issues concerning the solutions of nonlinear dispersive equations, based on the strong effects of dispersion on the short-wave components : local smoothing properties, dispersive blow-up.

Introduction. This paper is concerned with qualitative properties of solutions of rather general dispersive equations, which are based on the strong effects of dispersion on the short-wave components. Included will be Korteweg-de Vries, Schrödinger equations and various of their generalizations, local or non local.

The first property will be explored in the relatively narrow context of generalized Korteweg-de Vries equations of the form

$$u_t + u^p u_x + u_{xxx} = 0, \tag{1.1}$$

where p is a nonnegative integer and $u = u(t,x)$ is a real valued function of the two real variables x and t. Though the equations are nonlinear for positive p, it is noteworthy that the loss of smoothness suffered by some solutions is associated only with the linearized dispersion relation possessed in common by all the equations (1.1), and hence the term "dispersive blow up" will be used as a descriptive label.

As far as we know, the gist of the idea that comes to the fore herein first appeared in an extended remak in the paper of Benjamin, Bona and Mahony [5] concerning the linear equation (1.1) $p = 0$. Demanding that a simple harmonic wave train of the form $cos(kx - \omega t)$ be a solution leads to the dispersion relation for the frequency ω as a function of wavenumber k, namely

$$w = \omega(k) = k(1 - k^2) \tag{1.2}$$

Especially important is the fact that the group velocity $c_g(k) = \omega'(k)$ and the phase velocity $c(k) = \omega(k)/k$ are both unbounded, assigning arbitrary large values to short-wave components. Because of this property, it is possible using Fourier's principle to specify initial data arranged in such a way that infinitely many, widely spaced, short-wave components wille coalesce at a single point of some given time and thereby create some loss of spatial smoothness in the solution at that time. Thus it was shown in the last cited paper that for (1.1) with $p = 0$, and infinitely differentiable, $L^2(\mathcal{R})$ solution can become unbounded at a single point in space-time.

We will review in Section 2 recent work with J. Bona [6][7] which shows that a similar dispersive blow up occurs for the nonlinear equations (1.1). Our work relates in a general way to papers of Cohen [10], Kato [21] (see also below), wherein it is shown that in certain function classes, the KdV equation is smoothing. From a smoothing result, it is adduced at once by running time backward that certain solutions form singularities in finite time. The techniques employed here are quite different from those of Cohen, who uses the inverse scattering transform in a careful way to deduce her results, and those of Kato, which are more generally applicable to the generalized KdV equations, but do not lead to the sort of specific conclusions were are able to draw here.

The method whereby our results are obtained is quite simple. The evolution equation is viewed, via Duhamel's principle, as a linear equation forced by its nonlinearity. Looked at this way, the equation may be solved by linear techniques, so resulting in an equivalent integral equation in which the solution of the linearized initial value problem appears explicitly. Initial data is then specified along the lines suggested by Benjamin *et al* [5] that features loss of regularity for the solution of the linearized equation. It is then shown that the nonlinear term in the integral equation remains smooth, and so the full solution of the equation is inferred to form the same singularity that was present for the linear equation. The main ingredient that is used to establish control of the nonlinear term in the integral equation is an existence theory for the evolution equation in certain weighted Sobolev spaces which has its own interest.

The second problem we want to adress in this lecture is concerned with local smoothing properties of evolution partial differential equations which are reversible and conservative. Such a property is excluded for the wave equation. On the other hand, Kato [21] has shown a local smoothing property of the Korteweg-de Vries equation : the solution of the initial value problem is, locally, one derivative smoother than the initial datum. Kato's proof uses in a crucial way, the algebraic properties of the symbol for the Korteweg-de Vries equation and the fact that the underlying spatial dimension is one. Actually, judging from the way several integrations by parts and cancellations conspire to reveal a smoothing effect, one would be inclined to believe this was a special property of the Korteweg-de Vries equation. This is not, however, the case. Part 3 of this paper will attempt to describe a recent work with P.Constantin [11][12] which proves a general local smoothing effect for dispersive equations and systems, in arbitrary dimension. All the physically significant strongly dispersive equations and systems known to us have linear parts displaying this local smoothing property. To mention a few, the Korteweg-de Vries, Benjamin-Ono, intermediate long wave, various Boussinesq, and Schrödinger equations are included. We will study thus, equations and systems of the form

$$\frac{\partial u}{\partial t} + iP(D)u = F \tag{1.3}$$

$$u(0,x) = u_0(x) \tag{1.4}$$

where $u(t,x)$, $t \in \mathcal{R}$, $x \in \mathcal{R}^n$, $D = \frac{1}{i}\left(\frac{\partial}{\partial x_1}, ..., \frac{\partial}{\partial x_n}\right)$, and $P(D)u$ is defined via a real symbol $p(\xi)$ in the scalar case (or a matrix with real entries in the case of systems),

$$P(D)u = \int_{\mathcal{R}^n} e^{2i\pi<x,\xi>} p(\xi)(\mathcal{F}_2 u)(\xi)d\xi \tag{1.5}$$

where $\mathcal{F}_2$ is the Fourier transform with respect to the x variables.

The assumptions on $p(\xi)$, reflecting the strong dispersive nature of (1.3) are that, roughly speaking, $p(\xi)$ behaves like $|\xi|^m$ for $|\xi| \to \infty$, with $m > 1$:

$$\begin{cases} p \in L^\infty_{loc}(\mathcal{R}^n, \mathcal{R}) \ \textit{and is continuously} \\ \textit{differentiable for} \ |\xi| > R, \ \textit{for some} \ R \geq 0 \end{cases} \tag{1.6}$$

There exists $m > 1$, $c_1 > 0$, $c_2 > 0$ such that

$$|p(\xi)| \leq c_1(1 + |\xi|)^m \ \textit{for all} \ \xi \in \mathcal{R}^n \tag{1.7}$$

$$\left|\frac{\partial p}{\partial \xi_j}(\xi_j)\right| \geq c_2(1 + |\xi|)^{m-1} |\xi_j| / |\xi|, \quad \textit{for all} \ \xi \in \mathcal{R}^n, \ |\xi| > R, \ j = 1, ..., n \tag{1.8}$$

In applications, two cases are more frequently encountered. In the first case, (1.3) is of Korteweg-de Vries type, where u is real valued, $n = 1$ and $p(\xi) = \xi q(\xi^2)$. This included the (linearized) Benjamin-Ono ($p(\xi) = \xi|\xi|$), intermediate long wave ($p(\xi) = \xi^2 coth\delta\xi - \xi/\delta$, $\delta > 0$), Smith ($p(\xi) = \xi(\sqrt{1+\xi^2} - 1)$) equations. In the second case, (1.3) is of Schrödinger type ; u is complex valued, n is arbitrary, and $p(\xi) = q(|\xi|^2)$. Note that when

$m = 2$, (1.8) does not imply the ellipticity of the symbol $p(\xi)$; it holds for instance when $p(\xi) = \xi_1^2 - \xi_2^2$. $(n = 2)$.

The exact assumptions for systems are a little more complicated since we allow non selfadjoint matrices $P(D)$. Our hypothesis for system ensure well-posedness of the Cauchy problem in Sobolev spaces ; the dispersive character is provided by properties (1.6), (1.7) and (1.8) required to hold for certain combinations of the matrix entries which play the role of propagators.

The equation (1.3) is conservative and time reversible : the solution defines a continuous unitary group on every Sobolev space $H^s(\mathcal{R}^n)$. (In the case of non selfadjoint systems, our assumptions imply that the Sobolev norms of the solutions are controlled for all times by those of the initial data). Thus a global smoothing effect is excluded in Sobolev spaces.

A typical result that we obtain is : if u_0 belongs to $H^s(\mathcal{R}^n)$, then, for almost every $t \neq 0$, the solution $u(t, \cdot)$ belongs to $H_{loc}^{s+d}(\mathcal{R}^n)$ where $d = (m-1)/2$; that is d depends on the order of the operator $P(D)$ but not on the spatial dimension n. The higher the order m, the more dispersive the equation is and stronger the local smoothing effect becomes.

The proofs rely on a new restriction lemma for the Fourier transform. Unlike the classical one of Stritchartz [27], its local nature allows smoothing effects.

Results which overlap with some of [12] were obtained independently by Sjölin [26] and Vega [29]. Indeed one of the inequalities of [26] and [29] amounts to a local smoothing property for the free Schrödinger equations, essentially the same as the one we obtain for that equation.

While global smoothing cannot occur in L^2-Sobolev spaces, there exists the possibility of global smoothing in different spaces. For instance global smoothing holds provide the initial data decays sufficiently fast at infinity (see e.g. Cohen [10] for the Korteveg-de Vries equations by inverse scattering techniques, Kato [21] also for the Korteveg-de Vries equation and Hayashi, Nakamizu and Tsutsumi [19] [20] for a class of nonlinear Schrödinger equation, using the conformal invariance). On the other hand, under the assumptions of initial data in $L^1(\mathcal{R}^n)$, Balabane and Emami Rad [3] [4] obtain a global smoothing in Sobolev spaces $W^{p,k}(\mathcal{R}^n)$ for linear equations of Schrödinger type of high order enough provided the spatial dimension n is corresponding large enough.

2. Dispersive blow-up of solutions of generalized Korteweg-de Vries equations.

The generalized KdV equation (GKdV henceforth) introduced earlier

$$u_t + u^p u_x + u_{xxx} = 0 \tag{2.1}$$

will be posed for $x \in \mathcal{R}$, $t > 0$ subject to the initial condition

$$u(0,x) = \psi(x) \tag{2.2}$$

To start with we re going to review some known and extensions of known results concerning the Cauchy problem (2.1), (2.2). We refer to [7] for precise attributions and some proofs (See also the forthcoming papers of Ginibre, Tsutsumi and Velo [17] [18] for other very nice results concerning solutions of GKdV with non polynomial nonlinearities).

Throughout this paper $\| \cdot \|_s$ will stand for the norms in the Sobolev space H^s.

THEOREM 2.1. Let $\psi \in H^k(\mathcal{R})$. Then the following conclusions obtain

(i) *If $k = 0$ and $p < 4$, then there exists a solution u of (2.1)(2.2) which for any $T > 0$ and $R > 0$, lies in $L^\infty(\mathcal{R}_+; L^2(\mathcal{R})) \cap L^2(0,T; H^1(-R,R))$. The norm of u in this space depends only on T, R and $\| \psi \|_0$.*

(ii) *If $k = 1$ and $p < 4$, then there exists a solution u of (2.1) (2.2) which, for any $T > 0$ and $R > 0$ lies in $L^\infty(\mathcal{R}_+; H^1(\mathcal{R})) \cap L^2(0,T; H^2(-R,R))$. The norm of u in this space depends only on T, R and $\| \psi \|_1$.* If $p \geq 4$ and $\| \psi \|_1$ is sufficiently small, then the same result holds.

THEOREM 2.1. Let $k \geq 2$ be an integer and let $\psi \in H^k(\mathcal{R})$. Then the following holds true.

(i) *If $p < 4$, there exists a unique solution u of GKdV corresponding to the initial value ψ which, for any $T > 0$ and $R > 0$, belongs to $C(0,T; H^k(\mathcal{R})) \cap L^2(0,T; H^{k+1}(-R,R))$. Moreover, the correspondance $\psi \to u$ is continuous from $H^k(\mathcal{R})$ into $C(0,T; H^k(\mathcal{R})) \cap L^2(0,T; H^{k+1}(-R,R))$ for any $T > 0$.*

(ii) *If $p = 4$, the same conclusions as those in* (i) *hold provided that $\| \psi \|_0$ is not too large.*

(iii) *If $p \geq 4$ and if $\| \psi \|_1$ is not too large, then the same conclusions enunciated in* (i) *continue to be valid.*

(iv) *If* $p \geq 4$, *but* ψ *is unrestricted in size, then there exists a positive* $T^* = T^*(\| \psi \|_1)$ *such that the conclusions in* (i) *hold for all* T *in the interval* $(0, T^*)$.

It is worth noticing that recent numerical simulations of the initial value problem (2.1) with $p > 4$ indicate that solutions need not remain in the class $H^k(\mathcal{R})$ for all time (see Bona et al [8] [9]). Nevertheless, this sort of singularity formation subsists essentially on the nonlinearity, and consequently we term it "nonlinear blow up" to distinguish it from the dispersive blow-up that is the forms for the present.

Some technical theorems that are central to our main line of argument will be stated, namely an existence theory for the initial-value problem (2.1) (2.2) set in weighted, L^2-based Sobolev spaces. The results established here are similar to those established by Kato [21] and Krushkov and Faminski [23] although the precise results obtained appear to be new in case $p > 1$.

First we introduce $w_\sigma = w_\sigma(x)$, a non-decreasing C^∞ weight function depending on a positive parameter σ for which

$$w_\sigma(x) = \begin{pmatrix} 1 & for\ x < 0\ and \\ (1+x^2)^\sigma & for\ x > 1 \end{pmatrix} \tag{2.3}$$

The class $L^2(\mathcal{R}, w)$ is the class of measurable functions which are square integrable with respect to the measure $w^2(x)dx$. The class $H^k(\mathcal{R}, w)$ is the subspace of $L^2(\mathcal{R}, w)$ consisting of all these elements whose first k distributional derivatives also lie in $L^2(\mathcal{R}, w)$.

THEOREM 2.3. Let p *and* k *be non-negative integers and let the parameter* σ *associated with the weight* w *be non-negative, but otherwise arbitrary. Suppose the initial data* ψ *in (2.2) to lie in* $H^k(\mathcal{R}, w)$.

(i) *If* $k = 0$ *or 1 and* $p < 4$, *then there exists a solution* u of *(2.1) corresponding to* ψ *such that, for any* $T > 0$, u *belongs to* $L^\infty(0, T; H^k(\mathcal{R}, w)) \cap L^2(0, T; H^{k+1}_{loc}(\mathcal{R}))$.

(ii) *If* $k \geq 2$ *and* $p < 4$, *then there exists a unique solution* u of *(2.1) corresponding to* ψ *such that, for any* $T > 0$, u *belongs to* $C(0, T; H^k(\mathcal{R}, w)) \cap L^2(0, T; H^{k+1}_{loc}(\mathcal{R}))$. *Moreover, the mapping that associates* u *to* ψ *is continuous from* $H^k(\mathcal{R}, w)$ *into* $C(0, T; H^k(\mathcal{R}, w)) \cap L^2(0, T; H^{k+1}_{loc}(\mathcal{R}))$.

(iii) *If* $k \geq 1$ *and* $p \geq 4$, *then corresponding to each* ψ *there is a* $T^* = T^*(\| \psi \|_1)$ *such that for any* $T \in (0, T^*)$, *there is a solution* u *of (2.1) associated to* ψ *which lies in*

$L^\infty(0,T;H^k(\mathcal{R},w)) \cap L^2(0,T;H^{k+1}_{loc}(\mathcal{R}))$. If $k \geq 2$, u lies in $C(0,T;H^k(\mathcal{R},w)) \cap L^2(0,T;H^{k+1}_{loc}(\mathcal{R}))$ and is unique within its function class. Moreover, the mapping that associates u to ψ is continuous. If $\| \psi \|_1$ is small enough, then $T^* = +\infty$.

The proof of Theorem 2.3 proceeds by deriving suitable a priori estimates for the function $v = wu$ which satisfies the equation

$$v_t + v_{xxx} + v\left(6\frac{w_x w_{xx}}{w^2} - 6\frac{w_x^3}{w^3} - \frac{w_{xxx}}{w}\right) + v_x\left(6\frac{w_x^2}{w^2} - 3\frac{w_{xx}}{w}\right)$$
$$-3\frac{w_x}{w}v_{xx} + \frac{1}{w^p}v^p v_x - \frac{w_x}{w^{p+1}}v^{p+1} = 0.$$

Details can be found in [7].

The results concerning dispersive blow-up will now be stated. The proofs involve relatively careful calculations of integral of products of the Airy function

$$Ai(\xi) = \frac{1}{\pi}\int_0^\infty \cos\left(\frac{1}{3}\theta^3 + \theta\xi\right) d\theta \tag{2.4}$$

We recall (see [15] for instance) that the Airy function of a real argument is a bounded, real-analytic function that tends to zero at $\pm\infty$. On the positive real axis it decreases monotonically and exponentially to zero. In fact, the Airy function and its first derivative satisfy the inequalities

$$\begin{aligned} Ai(x) &\leq \frac{1}{2\pi^{1/2}x^{1/4}}e^{-\xi}, \\ |\, Ai'(x) \,| &\leq \frac{x^{1/4}}{2\pi^{1/2}}e^{-\xi}\left(1 + \frac{7}{72\xi}\right) \end{aligned} \tag{2.5}$$

for $x \geq 0$, when $\xi = \frac{2}{3}x^{3/2}$. On the other hand, as x tends to $-\infty$, Ai only decreases algebraically, but it oscillates fiercely, having the form

$$\begin{cases} Ai(-x) = \dfrac{1}{2\pi^{1/2}x^{1/4}}\cos\left(\xi - \dfrac{1}{4}\pi\right)\left(1 + O\left(\dfrac{1}{\xi}\right)\right) \\ Ai'(-x) = \dfrac{x^{1/4}}{2\pi^{1/2}}\sin\left(\xi - \dfrac{1}{4}\pi\right)\left(1 + O\left(\dfrac{1}{\xi}\right)\right) \end{cases} \tag{2.6}$$

Now, it is easily checked that every function $u \in L^\infty(0,T;L^2(\mathcal{R}))$, such that $u^{p+1} \in L^\infty(0,T;L^1(\mathcal{R}))$, which satisfies the GKdV (1.1) and $u(0) = \psi \in L^2(\mathcal{R})$ can be represented as :

$$u(t,x) = \frac{1}{t^{1/3}} \int_{-\infty}^{\infty} Ai\left(\frac{x-y}{t^{1/3}}\right)\psi(y)dy + \\ + \frac{1}{p+1}\int_0^t \int_{-\infty}^{\infty} \frac{1}{(t-s)^{1/3}} Ai'\left(\frac{x-y}{(t-s)^{1/3}}\right) u^{p+1}(s,y)dyds \tag{2.7}$$

The next two lemmas report on estimates for the integrals on the right-hand side of (2.7).

LEMMA 2.1. Let $k \geq 0$ and $\psi \in H^k(\mathcal{R}; w_\sigma)$, where $\sigma \geq \frac{1}{16}$ and the weight w_σ is as defined in (2.3). Let $u \in L^\infty(0,T;H^k(\mathcal{R};w_\sigma) \cap L^2(0,T;H^{k+1}_{loc}(\mathcal{R}))$ be the solution of GKdV corresponding to the initial data ψ constructed in Theorem 2.3. If $k = 0$ and $p = 1$ or if $k \geq 1$ and p is arbitrary, then the integral

$$\Lambda(t,x) = \Lambda_p(t,x) = \\ \int_0^t \int_{-\infty}^{\infty} \frac{1}{(t-s)^{1/3}} Ai'\left(\frac{x-y}{(t-s)^{1/3}}\right) u^{p+1}(s,y)dsdy \tag{2.8}$$

is k times differentiable with respect to x for (t,x) in the strip $(0,T) \times \mathcal{R}$ and $\partial_x^k \Lambda(t,x)$ is bounded on compact subsets of this strip.

Next, attention is fixed upon the first integral on the right hand side of (2.7). We propose to give explicit initial data ψ so that its convolution with the Airy function kernel develops particular singularities at a given point in space-time.

LEMMA 2.2. let k be a non-negative integer and let

$$\psi(y) = \frac{Ai^{(k)}(-\beta y)}{(1+y^2)^m} \tag{2.9}$$

where β, $m > 0$. If m lies in the interval $(\frac{1}{8} + \frac{k}{2}, \frac{1}{4} + \frac{k}{2}]$ then $\psi \in H^k(\mathcal{R}) \cap C^\infty(\mathcal{R})$ and the function

$$\psi(t,x) = \frac{1}{t^{1/3}} \int_{-\infty}^{\infty} Ai\left(\frac{x-y}{t^{1/3}}\right)\psi(y)dy \tag{2.10}$$

lies in $C(\mathcal{R}_+; H^k(\mathcal{R}))$, and its kth derivative with respect to x is continuous everywhere except at the point $(\beta^{-3}, 0)$.

The dispersive blow-up results will follow by combining Lemma 2.1 and 2.2. Here is a typical result ([7]).

THEOREM 2.4. Let $T > 0$ and (t_*, x_*), $0 < t_* < T$ be given. For $k = 0$ and $p = 1$ or $k \geq 1$ and p arbitrary, there exists $\psi \in H^k(\mathcal{R})$ and a solution u of GKdV corresponding to ψ which satisfies $u \in L^\infty(0,T; H^k(\mathcal{R})) \cap L^2(0,T; H^{k+1}_{loc}(\mathcal{R}))$.

$$\partial_x^k u \text{ is continuous on } (0,T) \times \mathcal{R} \backslash \{t_*, x_*\}$$

$$\lim_{x \to x_*, t \to t_*} | \partial_x^k u(t,x) | = +\infty.$$

The proof of Theorem 2.4 follows immediately from Lemma 2.1 and 2.2. We take $(t_*, x_*) = (\beta^{-3}, 0)$ for simplicity.

Let ψ be given by (2.9) with $\frac{3}{16} + \frac{k}{2} < m \leq \frac{1}{4} + \frac{k}{2}$. It is readily seen that $\psi \in H^k(\mathcal{R}; w_\sigma)$ with $\sigma = \frac{1}{16}$ so that $\Lambda(t,\cdot)$ is C^k for every $t > 0$. On the other hand the first integral in the right hand side of (2.7) blows up in C^k precisely at $(\beta^{-3}, 0)$.

Extensions and applications of Theorem 2.4 can be found in [7].

3. Local Smoothing Effects for Dispersive Equations.

As was mentionned in the Introduction, we report here on recent work with P. Constantin [11], [12].

Here is a typical result concerning local smoothing properties of the group associated to equation (1.3). We introduce $\chi : \mathcal{R}^{n+1} \to \mathcal{R}$ of the type

$$\chi(t,x) = \chi_0(t)\chi_1(x_1)\ldots\chi_n(x_n), \text{ where } x = (x_1, \ldots, x_n)$$

and

$$\chi_j \in C_0^\infty(\mathcal{R}) \quad j = 0, 1, \ldots, n.$$

Theorem 3.1. *We assume that the symbol p of P satisfies (1.6) and (1.8). Let $S \geq -(m-1)/2$ and $u_0 \in H^S(\mathcal{R}^n)$. Then the solution u of (1.3), (1.4) satisfies*

$$\int_{\mathcal{R}^{n+1}} \chi^2(t,x) \, | (I - \Delta)^{(m-1+2S)/4} u(t,x) |^2 \, dx dt \leq C_\chi^2 \, \| u_0 \|^2_{H^S(\mathcal{R}^n)} \tag{3.1}$$

where C_χ is a constant which can be estimated as follows :

$$C_\chi \leq C \left[\prod_{j=0}^{n} \| \chi_j \|_{L^2} + \| \chi_0 \|_{L^2} \sum_{j=1}^{n} \| \chi_j \|_{L^2} \| \chi_1 \|_{L^\infty} \ldots \| \tilde{\chi}_j \|_{L^\infty} \ldots \| \chi_n \|_{L^\infty} \right]$$

(~ means that the corresponding term is omitted, and C is an absolute constant).

In particular, $u \in L^2\left(-T,T; H_{loc}^{S+(m-1)/2}(\mathcal{R}^n)\right)$, *for every* $T > 0$.

Proof. We proceed by duality. Let $f \in S(\mathcal{R}^{n+1})$. We set

$$I = \left| \int_{\mathcal{R}^{n+1}} \chi(t,x)(I-\Delta)^{(m-1+2S)/4} u(t,x)\, \overline{f(t,x)}\, dtdx \right|$$

Since $u(t,\cdot) = e^{-itP(D)}u_0$, we can write by transposition

$$I = \left| < u_0, (I-\Delta)^{(m-1+2S)/4} \int_{\mathcal{R}} e^{itP(D)}(\chi f) dt > \right|$$

where $< F, G >= \int_{\mathcal{R}^n} F\bar{G}dx$. Because of the Parseval identity we obtain also (denoting or $\mathcal{F}_2$ the Fourier transform with respect to the x variable

$$\begin{aligned} I &= \left| < \hat{u}_0, (1+|\xi|^2)^{(m-1+2S)/4} \int_{\mathcal{R}} e^{itp(\xi)} \mathcal{F}_2(\chi f) dt > \right| \\ &= \left| < \hat{u}_0, (1+|\xi|^2)^{(m-1+2S)/4} \mathcal{F}(\chi f)(-p(\xi)/2\pi, \xi) > \right| \\ &\leq \left(\int_{\mathcal{R}^n} (1+|\xi|^2)^S \,|\hat{u}_0(\xi)|^2\, d\xi \right)^{1/2} \\ &\times \int_{\mathcal{R}^n} (1+|\xi|^2)^{(m-1)/2} \,|\mathcal{F}(\chi f)(-p(\xi)/2\pi, \xi)|^2\, d\xi)^{1/2}, \end{aligned}$$

where we have denote $\mathcal{F}$ the Fourier transform with respect to (t,x).

In view of this last inequality, Theorem 3.1 will follow from the following result on the restriction of the Fourier transform. We state it in a form more general than needed in Theorem 3.1 in view of applications on smoothing effects in L^p-based Sobolev spaces (see [12]).

To start with, let a real q, $1 \leq q \leq 2$ and a real α be such that

$$\begin{cases} 2\alpha < m - 1 - ((2-q)/q)n & if\ 1 \leq q < 2 \\ 2\alpha \leq m-1 & if\ q = 2 \end{cases} \tag{3.2}$$

Theorem 3.2. Assume that p satisfies (1.6), (1.8) if $q = 2$ and moreover (1.7) if $1 \leq q < 2$. Let a be given by (3.2), and $\chi \in C_0^\infty(\mathcal{R}^{n+1})$ as in Theorem 3.1. Then there exists a constant C_χ (see Theorem 3.1) such that for every $f \in S(\mathcal{R}^n)$

$$\left(\int_{\mathcal{R}^n} (1+|\xi|^2)^{\alpha q/2} \,|\mathcal{F}(\chi f)(p(\xi), \xi)|^q\, d\xi \right)^{1/q} \leq C_\chi \| f \|_{L^2(\mathcal{R}^{n+1})} \tag{3.3}$$

For a proof of Theorem 3.2, see [12]. A version of Theorem 3.2 was proven independently by P. Sjölin [21].

We consider now the inhomogeneous equation

$$\frac{\partial u}{\partial t} + iP(D)u = F \ in \ \mathcal{R} \times \mathcal{R}^n \tag{3.4}$$

$$u(0,x) = u_0(x) \ in \ \mathcal{R}^n \tag{3.5}$$

We introduce the potential space $H^{s,p}(\mathcal{R}^n)$, $s \in \mathcal{R}$, $1 < p < \infty$,

$$H^{s,p}(\mathcal{R}^n) = \{f \in S'(\mathcal{R}^n),\ (I-\Delta)^{s/2} f \in L^p(\mathcal{R}^n)\}.$$

Using the Duhamel representation formula

$$u(t,\cdot) = e^{-itP(D)}u_0 + \int_0^t e^{-i(t-\tau)P(D)}F(\tau,\cdot)d\tau \tag{3.6}$$

one can prove the

Theorem 3.3. *We assume that p satisfies* (1.6), (1.7), (1.8). *Let $s \geq -(m-1)/2$, $u_0 \in H^s(\mathcal{R}^n)$, $F \in L^1_{loc}(\mathcal{R}; H^{s,q}(\mathcal{R}^n))$, $1 \leq q \leq 2$. Let α be given by* (3.2). *Then the solution u of* (3.4), (3.5) *belongs to $L^2(-T,T; H^{s+\alpha}_{loc}(\mathcal{R}^n))$ for every $T > 0$.*

Combining Theorem 3.3 and recent estimates of K. Yajima [28] one can obtain smoothing effects for Schrödinger equations with a real time dependent potential $V(t,x)$, namely

$$i\frac{\partial u}{\partial t} + \Delta u + V(t,x)u = 0 \ \ in \ \mathcal{R} \times \mathcal{R}^n \tag{3.7}$$

$$u(0,x) = u_0(x) \tag{3.8}$$

The next Proposition corrects Corollary 3.1 and Remark 3.1 in [12] (where we misunderstood Yajima's notations).

Proposition 3.1. *Let $p > n$, $\beta > \frac{p}{p-\frac{n}{2}}$. Assume that $V \in L^\beta_{loc}(\mathcal{R}; L^p(\mathcal{R}^n))$. Then the solution u of* (3.7)(3.8) *with $u_0 \in L^2(\mathcal{R}^n)$ satisfies $u \in L^2(-T,T; H^\delta_{loc}(\mathcal{R}^n))$, for $\delta < \frac{1}{2} - \frac{n}{2p}$ and every $T > 0$.*

If $V \in L^1_{loc}(\mathcal{R}; L^\infty(\mathcal{R}^n))$, *then* $u \in L^2(-T,T; H^{1/2}_{loc}(\mathcal{R}^n))$ *for every* $T > 0$.

In a work in preparation [13] we prove local smoothing effects for (3.7) with a potential $V(x)$ which is short range [2].

We indicate now how Theorem 3.1 can be extended to a class of systems. We will therefore study linear dispersive systems of the following type

$$\begin{aligned} &\frac{\partial u}{\partial t} + i\, P(D)u = 0 \ \ in\ \mathcal{R} \times \mathcal{R}^n \\ &u(0) = u_0 \end{aligned} \tag{3.9}$$

where $u = u(t,x)\ :\ \mathcal{R} \times \mathcal{R}^n \to \mathcal{R}^2$ or $\mathcal{C}^2$ and $P(\xi)$ is a *real* matrix symbol

$$P(\xi) = \begin{pmatrix} p_1(\xi) & p_3(\xi) \\ p_4(\xi) & p_2(\xi) \end{pmatrix}, \quad \xi \in \mathcal{R}^n.$$

The coefficients $p_i(\xi)$ will satisfy

$$p_i \in L^\infty_{loc}(\mathcal{R}^n) \quad 1 \le i \le 4 \tag{3.10}$$

$$\begin{aligned} &\frac{(p_1(\xi) - p_2(\xi))^2}{4} + p_3(\xi)p_4(\xi) = b^2(\xi) > 0 \\ &for\ all\ \xi \in \mathcal{R}^n,\ |\,\xi\,| \ge M \ge 0 \end{aligned} \tag{3.11}$$

There exists $A^2 > 0$ such that

$$\begin{aligned} &(p_3(\xi) - p_4(\xi))^2 \le A^2 \left(\frac{(p_1(\xi) - p_2(\xi))^2}{4} + p_3(\xi)p_4(\xi) \right) \\ &for\ all\ \xi \in \mathcal{R}^n,\ |\,\xi\,| \ge M \ge 0 \end{aligned} \tag{3.12}$$

There exists $R \ge M$ such that the functions

$$\tau_\pm(\xi) = \frac{p_1(\xi) + p_2(\xi)}{2} \pm b(\xi) \tag{3.13}$$

are differentiable for $|\,\xi\,| > R$ and satisfy (1.7) (1.8) (with p replaced by $\tau_\pm$).

The assumption (3.11) is to insure that the Cauchy problem for (3.9) is well posed (no instabilities to short waves).

Many linearized systems arising in the theory of dispersive long waves of small amplitude satisfy hypothesis (3.10)-(3.13). Another example is the (linearized) Boussinesq equation

$$u_{tt} + \Delta^2 u - \Delta u = 0$$

which is equivalent to the system

$$\begin{aligned} u_t + i(\Delta v - v) &= 0 \\ v_t + i\Delta u &= 0 \end{aligned}$$

We are now ready to state a smoothing property similar to Theorem 3.1.

Theorem 3.4. We assume that $P(\xi)$ satisfies (3.10)-(3.13). Let $s \geq -\frac{m-1}{2}$. Then, the solution of (3.9) associated to $u_0 \in H^s(\mathcal{R}^n)^2$ satisfies $u \in L^2(-T,T; H_{loc}^{s+(m-1)/2}(\mathcal{R}^n))^2$ for every $T > 0$.

We refer to [12] for other smoothing properties of equations (1.3) or systems (3.9), for instance in a L^p, $p \neq 2$, setting.

We would like to conclude this paper by some results concerning nonlinear equations, and related works.

In view of Theorem 3.3, we get immediately a local smoothing for nonlinear Schrödinger equations, as soon as $u_0 \in H^s(\mathcal{R}^n)$, $s > \frac{n}{2}$. For instance we consider the Cauchy problem for the nonlinear Schrödinger equations

$$\begin{cases} i\dfrac{\partial u}{\partial t} + \Delta u + F(u) = 0 \quad x \in \mathcal{R}^n, \ n \leq 3 \\ u(0,x) = u_0(x) \end{cases} \tag{3.14}$$

The assumptions ont he function F will be similar to those of Kato [22].

$$F \in C^2(\mathcal{C},\mathcal{C}) \ ; \ F(0) = 0, \ | F'(z) | \leq M \,| z |^{p-1} \ for \ | z | \geq 1 \tag{3.15}$$

where

$$1 < p < 5 \ if \ n = 3, \ 1 < p < \infty \ if \ n \leq 2 \tag{3.16}$$

Under these assumptions, Kato [22] has proved the existence of a unique local solution $u \in C(-T,T; H^2(\mathcal{R}^n))$ provided $u_0 \in H^2(\mathcal{R}^n)$. Since $F(u) \in L^\infty(-T,T; H^2(\mathcal{R}^n))$, Theorem 3.3 implies that $u \in L^2(-T,T; H_{loc}^{5/2}(\mathcal{R}^n))$.

One can also obtain local smoothing for nonlinear Schrödinger equations by combining Theorem 3.3 and the $L^p - L^q$ estimates of Strichartz type ([27], [30]). Let us consider for instance the nonlinear Schrödinger equation

$$\begin{cases} i\dfrac{\partial u}{\partial t} + \Delta u + \lambda \mid u \mid^p u = 0 \quad x \in \mathcal{R}^n \\ u(0,x) = u_0(x) \end{cases} \tag{3.17}$$

where $\lambda \in \mathcal{R}$, $p \in \mathcal{R}_+$.

PROPOSITION 3.2. Assume $u_0 \in L^2(\mathcal{R}^n)$, and that $p < \frac{4}{n}$ *if* $n \leq 4$ *or* $p < \frac{2}{n-2}$ *if* $n \geq 5$. *then the unique solution* u *of* (3.17) *satisfies* $u \in C(\mathcal{R}_+; L^2(\mathcal{R}^n)) \cap L^2_{loc}(\mathcal{R}_+; H^{1/2}_{loc}(\mathcal{R}^n))$. *If* $p = \frac{4}{n}$ *and* $n \leq 3$, *there exists a unique maximal solution* u *of* (3.17) *on* $[0, T^*)$ *such that*

$$u \in C([0,T]; L^2(\mathcal{R}^n)) \cap L^2(0,T; H^{1/2}_{loc}(\mathcal{R}^n)), \quad T < T^*.$$

Proof. If $p < \frac{4}{n}$, a result of Tsutsumi [31] shows that (3.17) possesses a unique global solution u in $C(\mathcal{R}_+; L^2(\mathcal{R}^n)) \cap L^q(\mathcal{R}_+; L^r(\mathcal{R}^n))$ where $r \in [2, 2n/(n-2))$ ($r \in [2,\infty)$ if $n = 1$, and $r \in [2, \infty)$ if $n = 2$) and q satisfies $\frac{2}{q} = n\left(\frac{1}{2} - \frac{1}{r}\right)$. This result is true locally in time if $p = \frac{4}{n}$ (Cazenave and Weissler [32]).The proof is now reduced to checking that under the hypothesis of Proposition 3.2, one has $\mid u \mid^p u \in L^1_{loc}(\mathcal{R}_+; L^2(\mathcal{R}^n))$ in order to apply Theorem 3.3.

Concerning smoothing properties of nonlinear equations of KdV type, apart from a general example in [12], recent results have been obtained for the Benjamin-Ono equation

$$\frac{\partial u}{\partial t} + u\frac{\partial u}{\partial x} - H\frac{\partial^2 u}{\partial x^2} = 0 \tag{3.18}$$

where H is the Hilbert transform.

Ponce [24] proved a local smoothing effect for (3.17) in $H^s(\mathcal{R})$, $s > 3/2$ (see also Mudi [25] for the case $s = 3/2$). More recently Ginibre and Velo derived the general smoothing effect in $H^s(\mathcal{R})$, $s \geq 0$ [16]. Their results apply as well to perturbations of (3.17) such as the Intermediate Long Wave and Smith equations (see [1] for a review on nonlocal dispersive wave equations). These papers follow the approach of Kato, multiplying the equation by a suitable weight function. Ponce uses a rather delicate commutator estimate in the spirit of Coifman and Meyer, while Ginibre and Velo use a more direct estimate in Fourier space.

Finally we would like to mention the very interesting work of Craig and Goodman [14] who proves in particular global smoothing properties for equations of the form

$$\frac{\partial u}{\partial t} + a(x,t)\frac{\partial^3 u}{\partial x^3} = 0 \tag{3.19}$$

References

[1] L. Abdelouhab, J.L. Bona, M. Felland, J.C. Saut, Nonlocal models for nonlinear, dispersive waves, to appear.

[2] S. Agmon, Spectral properties of Schrödinger operators and scattering theory, *Ann. Sc. Norm. Sup. di Pisa (4), 2,* (1975) 151-218.

[3] M. Balabane, On a regularizing effect of Schrödinger type groups, Pré-publications Mathématiques, Université Paris-Nord, n° 68, 1986.

[4] M. Balabane, M.A. Emani-Rad, L^p estimates for Schrödinger evolution equations, *Trans. Amer. Math. Soc. 292* (1985), 357-373.

[5] T.B. Benjamin, J.L. Bona, J.J. Mahony, Model equations for long waves in nonlinear dispersive media, *Phil. Trans. Roy. Soc. London A, 272* (1972), 47-78.

[6] J.L. Bona, J.C. Saut, Singularités dispersives de solutions d'équations du type Korteweg-de Vries, *C. R. Acad. Sci. Paris, Série I, 303* (1986), 101-103.

[7] J.L. Bona, J.C. Saut, Dispersive blow-up of solutions of generalized Korteweg-de Vries equations, to appear.

[8] J.L. Bona, V.A. Dougalis, O.A. Karakashian, Fully discrete Galerkin methods for the Korteweg-de Vries equations, *Comp. Math. with Appl., 12A,* (1986), 859-884.

[9] J.L. Bona, V.A. Dougalis, O.A. Karakashian, Conservative high order numerical schemes for the generalized Korteweg-de Vries equations, to appear.

[10] A. Cohen, Solutions of the Korteweg-de Vries equations for irregular data, *Duke Math. J., 45,* (1978), 149-181.

[11] P. Constantin, J.C. Saut, Effets régularisants locaux pour des équations dispersives générales, *C. R. Acad. Sci. Paris, Série I, 304*, (1987), 407-410.

[12] P. Constantin, J.C. Saut, Local smoothing properties of dispersive equations, *J. A.M.S., 1,2* (1988), 413-439.

[13] P. Constantin, J.C. Saut, Local smoothing properties of Schrödinger equations with short range potential, to appear.

[14] W. Craig, J. Goodman, Dispersive equations, Preprint 1988.

[15] M. Fedoriouk, Méthodes asymptotiques pour les équations différentielles ordinaires linéaires, Editions Mir, Moscou (1987).

[16] J. Ginibre, G. Velo, Propriétés de lissage et existence de solutions pour l'équation de Benjamin-Ono généralisée, Prépublication, (1988).

[17] J. Ginibre, Y. Tsutsumi, Uniqueness of solutions for the generalized Korteweg-de Vries equations, to appear.

[18] J. Ginibre, Y. Tsutsumi, G. Velo, Existence and uniqueness of solutions for the generalized Korteweg-de Vries equations, to appear.

[19] N. Hayasmi, K. Nakamitsu, M. Tsutsumi, On solutions of the initial value problem for the nonlinear Schrödinger equation in one space dimension, *Math. Z.*, *192*, (1986), 637-650.

[20] N. Hayashi, K. Nakamitsu, M. Tsutsumi, On solutions of the initial value problem for the nonlinear Schrödinger equation, *J. Funct. Anal.*, *71*, (1987), 218-245.

[21] T. Kato, On the Cauchy problem for the (generalized) Korteweg-de Vries equations, *Stud. Appl. Math. Adv. in Math.*, Supplementary Study 18, (1983), 93-128.

[22] T. Kato, On nonlinear Schrödinger equations, *Ann. Inst. H. Poincaré Phys. Théor.*, *46*, (1987), 113-129.

[23] S.N. Krushkov, A.V. Faminskii, Generalized solutions to the Cauchy problem for the Korteweg-de Vries equation, *Math. U.S.S.R. Sbornik*, *48*, (1984), 93-138.

[24] G. Ponce, Smoothing properties of solutions to the Benjamin-Ono equations, Preprint, 1988.

[25] M. Tom Mudi, Ph.D. Thesis, in preparation, Penn. State University.

[26] P. Sjölin, Regularity of solutions to the Schrödinger equations, *Duke Math. J., 55*, (1987), 699-715.

[27] R.S. Strichartz, Restrictions of the Fourier transforms to quadratic surfaces and decay of solutions of wave equations, *Duke Math. J.*, *44*, (1977), 705-714.

[28] K. Yajima, Existence of solutions of Schrödinger evolution equations, *Comm. Math. Phys., 110* (1987), 415-426.

[29] L. Vega, Schrödinger equations : pointwise convergence to the initial data, *Proc. A.M.S. 102, 4*, (1988), 874-878.

[30] J. Ginibre, G. Velo, The global Cauchy problem for the nonlinear Schrödinger equation revisited, *Ann. Inst. Henri Poincaré, Analyse Non Linéaire 2*, (1985), 309-327.

[31] Y. Tsutsumi, L^2–solutions for nonlinear Schrödinger equations and nonlinear groups, *Funkcialj Ekvacioj, 30, 1*, (1987), 115-125.

[32] T. Cazenave, F.B. Weissler, Some remarks on the nonlinear Schrödinger equation in the critical case, Preprint, 1988.

MULTICHANNEL NONLINEAR SCATTERING THEORY FOR NONINTEGRABLE EQUATIONS

A. Soffer*
Department of Mathematics
Princeton University
Princeton, NJ 08544

M. I. Weinstein**
Department of Mathematics
University of Michigan
Ann Arbor, MI 48109

Abstract

We consider a class of nonlinear equations with localized and dispersive solutions; we show that for a ball in some Banach space of initial conditions, the asymptotic behavior (as $t \longrightarrow \pm\infty$) of such states is given by a linear combination of a periodic (in time), localized (in space) solution (nonlinear bound state) of the equation and a purely dispersive part (with free dispersion). We also show that given data near a nonlinear bound state of the system, there is a nonlinear bound state of nearby energy and phase, such that the difference between the solution (adjusted by a phase) and the latter disperses to zero. It turns out that in general the time-period (and energy) of the localized part is different for $t \longrightarrow +\infty$ from that for $t \longrightarrow -\infty$. Moreover, the solution acquires an extra constant phase $e^{i\gamma\pm}$.

Section 1. Introduction

This paper deals with the scattering theory of a class of nonlinear dispersive equations admitting more than one channel. By this we mean that the asymptotic behavior is given by a linear combination of localized (in space), periodic (in time) wave (solitary or standing wave) and a dispersive piece. For nonlinear flows

*Alfred P. Sloan Fellow
**Supported in part by NSF Grant #DMS 88-0185

which are completely integrable (e.g. 1 dimensional cubic nonlinear Schrödinger and Korteweg-de Vries equations), some analysis of the asymptotic system of e.g. localized part (solitons) + dispersion can be carried out by inverse scattering theory [G-G-K-M, Z-S, Lax, C-K]. The inverse scattering transform decouples the localized from the dispersive part. The cases we consider are not completely integrable (we will mainly consider the nonlinear Schrödinger equation (NLS) in 2 and 3 dimensions) and the main new feature here is that the localized and dispersive parts are interacting at all times.

The spatially localized part that emerges as $t \longrightarrow \pm\infty$ is identified with an exact solitary wave solution or nonlinear bound state of the full nonlinear equation. For the above-mentioned integrable systems the analogue of the solitary wave is the one-soliton.

Our main results are:

(i) (Stability) Given initial conditions which lie in a neighborhood of solitary wave of energy E_0 and phase γ_0, the asymptotic state of the system $(t \longrightarrow \pm\infty)$ is given by a solitary wave of nearby energy $E^{\pm}$ and phase $\gamma^{\pm}$ plus a remainder which disperses to zero, i.e. the solution converges asymptotically to a solitary wave. (e.g. in some L^p norm, $p > 2$).

(ii) (Scattering) There is a ball about the origin in a Banach space of initial conditions for which the asymptotic behavior $(t \longrightarrow \pm\infty)$ of all solutions is given by a linear combination of a solitary wave of energy $E^{\pm}$ and phase $\gamma^{\pm}$, plus a remainder which is dispersive. The remainder is <u>purely dispersive</u> in the sense that it satisfies the usual local decay and L^p decay estimates of the linear theory.

Previous results on the stability of solitary waves involves the use of energy norms, e.g. H^1 (see for example Ca-Li, Sh-Str, We 1, We 2, Ro-We, G-S-S). A typical result of this type is that if the solution begins in some neighborhood of the solitary wave orbit, then it remains in a neighborhood. Since energy norms are insensitive to dispersive behavior one cannot conclude, as above, that solutions asymptotically converge to a solitary wave.

Previous work on nonlinear scattering has focused on the situation where there are no bound states. In the above terminology, these are problems with a single (dispersive) channel (see for example Str 1, Str 3, G-V).

Cast into precise mathematical form, we prove that for a class of initial conditions for NLS with a potential term, the solution $\Phi(t)$ at large times is given by

$$\Phi(t) = e^{-i\int_0^t E(s)ds + i\gamma(t)} \psi(E(t)) + \phi_d(t)$$

where $\psi(E)$ is a spatially localized solution of the nonlinear bound state equation (with energy E) and $\phi_d(t)$ is a purely dispersive wave. $\gamma(t)$ is an extra phase that appears during the scattering process. Like Berry's geometrical phase [Ber], it cannot be fully accounted for by dynamical considerations (e.g. semiclassical approximation). As $t \longrightarrow \pm\infty$ we have that $E(t) \longrightarrow E^{\pm}$ and $\gamma(t) \longrightarrow \gamma^{\pm}$. In the completely integrable case, one has $E(t) = E^+ = E^-$ for all times. Here in general, $E^+ \neq E^-$. The mapping $(\gamma^-,E^-) \longrightarrow (\gamma^+,E^+)$ is a part of the S-matrix of the problem.

While there has been a lot of progress in understanding the linear multichannel scattering theory (see [En, Sig-Sof] and those cited therein) in the past ten years, little is known about the corresponding nonlinear situations. Questions like when a bound state (temporally periodic, spatially localized solution) breaks down due to nonlinear (repulsive) interaction, the scattering theory of localized waves in the presence of impurities and inhomogeneous media are not understood beyond heuristic considerations or finite time approximations.

Our approach to the problem begins with the simple physical observation that if one starts with the linear Schrödinger equation which describes a bound state and a dispersive wave (corresponding to the continuous spectral part of the Hamiltonian), then adding a small nonlinear term should not change the qualitative behavior that much, i.e. we should still see a localized part which asymptotically decouples from the dispersive part. We then make an Ansatz which incorporates this observation, from which we derive equations governing the interaction of the two channels. One set describes the variation of the energy and phase of the nonlinear bound state (respectively, $E(t)$ and $\gamma(t)$) of some linear time-dependent Hamiltonian; the second, is a nonlinear equation which describes a purely dispersive wave moving under the effect of the nonlinearity, as well as the effective potential coming from the presence of the localized part. We then show that, $\frac{dE}{dt}, \frac{d\gamma}{dt} \in L^1(dt)$ if the remainder wave is dispersive (with sufficient decay rate) and that the remainder is dispersive if $\frac{dE}{dt}, \frac{d\gamma}{dt} \in L^1(dt)$. Therefore, solving the coupled equations gives the required results.

The modulating energy and phase of the nonlinear bound state, $E(t)$ and $\gamma(t)$, which govern the localized part of the nonlinear evolution are sometimes referred to by physicists as collective coordinates. Equations for collective coordinates have been derived using various formalisms (e.g. averaging of conservation laws, direct perturbation theory [K-A, K-M, Ne]). These equations are sometimes referred to as modulation equations. In [We 1] their validity was studied in the linear approximation for certain systems which are conservative or small perturbations of conservative systems (e.g. weakly dissipative). We believe that our present results are the first rigorous justification of the collective coordinate description on an infinite time interval for nonintegrable systems.

A final remark is that the problem we consider can be viewed as a kind of restricted three body scattering, where the localized part corresponds to a bound pair and the dispersive part is the "third particle" moving away as $|t| \longrightarrow \infty$. It is hoped that such an analogy can be developed further and may allow the application of some powerful methods of phase space analysis developed for the linear N-body case.

Notation: All integrals are assumed to be taken over $\mathbb{R}^n$ unless otherwise specified.

Δ = Laplacian on $L^2(\mathbb{R}^n)$

$\langle x\rangle = (1+|x|^2)^{\frac{1}{2}}$ where $x \in \mathbb{R}^n$

$\langle f,g\rangle = \int f^* g$

$L^p = \{f: \|f\|_p < \infty\}$, $\|f\|_p = (\int |f(x)|^p)^{\frac{1}{p}}$

$H^s = \{f: (I-\Delta)^{s/2} f \in L^2\}$

$B = \{f: f \in H^1$, $\langle x\rangle^{1+a} f \in L^2\}$, $\|f\|_B = \|f\|_{H^1} + \|\langle x\rangle^{1+a} f\|_2$

Section 2. Definitions, Notations, Preliminaries

I. We will mainly consider the nonlinear Schrödinger equation (NLS) with a potential term, in spatial dimensions 2 and 3:

$$i\,\frac{\partial\Phi(t)}{\partial t} = [-\Delta + f(x,|\Phi(t)|)]\,\Phi(t) \tag{2.1}$$

Here $\Phi(t)$ is considered as an element of $H^1(\mathbb{R}^n)$, where n is the spatial dimension. Consequently (2.1) is understood in the sense of an equivalent integral equation. The well-posedness for the initial value problem in H^1 and in spaces with specified spatial decay rates has been considered for general nonlinearities in [G-V, K, H-N-T, C-W].

In the sequel $f(x,\xi)$ will be chosen so that the global existence of solutions of (2.1), perhaps under some restrictions on Φ_0, is known. We specialize here to the case where

$$f(x,\xi) = V(x) + \lambda|\xi|^{m-1} \qquad 2 < m < \frac{n+2}{n-2} \tag{2.2}$$

although the analysis holds for more general choices.

For the choice (2.2), the existence theory implies:

(i) $\lambda > 0$ (repulsive nonlinearity) global solutions for all $\Phi_0 \in H^1$ i.e. $\Phi \in C((-\infty,\infty);\, H^1)$

(ii) $\lambda < 0$ (attractive nonlinearity)

(a) $m < 1+4/n$, global solutions for all $\Phi_0 \in H^1$

(b) $m \geq 1+4/n$, global solutions for all Φ_0 such that $\|\Phi_0\|_{H^1}$ is sufficiently small.

We shall require the following of the linear potential $V(x)$:

Hypothesis (V)

Let $V(x) : \mathbb{R}^n \longrightarrow \mathbb{R}$ be a smooth function satisfying

i) $\langle x\rangle^{3+k+\varepsilon} \left|\frac{\partial^k V(x)}{\partial x^k}\right| \leq C_k < \infty$ for all $k \geq 0$.

ii) $-\Delta+V$ has only one bound state (isolated eigenvalue) on $L^2(\mathbb{R}^n)$ with strictly negative value, E_* .

Our approach reduces the study of (2.1) to essentially two independent problems. The first is the study of existence and certain decay properties of the nonlinear bound states of (2.1) (solitary waves) where many results are known. Then, one has to study the evolution equation for the dispersive part of the solution which one gets by linearizing around a certain time-dependent bound state. At least in the small data case, this reduces to linear spectral analysis of a Schrödinger Hamiltonian.

II. The solitary wave and its properties

We seek a time periodic, localized solution of (2.1) of the form $e^{-iEt}\psi_E(x)$. Then ψ_E satisfies the equation:

$$-\Delta\psi_E(x)+f(x,|\psi_E(x)|)\,\psi_E(x) = E\psi_E(x) \qquad \psi_E \in H^2(\mathbb{R}^n) . \tag{2.3}$$

We call an H^2 solution of (2.3) a nonlinear bound state or solitary wave profile. The solutions of (2.3) are much studied in the literature (see for example [Str 2, Be-Li] and those cited therein). We will concentrate on the case (2.2) with $2 < m < 3$ and $m > 2$ for $n = 2$. We take $V(x)$ to be spherically symmetric. The result we now state follows from variational and bifurcation methods.

Theorem 2.1

Let for $\lambda > 0$, $E \in (E_*,0)$, and for $\lambda < 0$, $E < E_*$. Then there exists a solution $0 < \psi_E$ of (2.3), $\psi_E \in H^2(\mathbb{R}^n)$ and, moreover

a) $|\psi_E(x)| < c_\varepsilon(E)e^{-(|E|-\varepsilon)|x|}$ for all $\varepsilon > 0$.

b) $\langle x\rangle^k \frac{\partial\psi_E}{\partial E} \in L^p \cap H^2$ $p \geq 1$ and all $k \geq 0$.

c) $\langle x\rangle^k \frac{\partial^2\psi_E}{\partial E^2} \in L^p \cap H^2$ $p \geq 1$ and all $k \geq 0$.

The solution curve $E \longmapsto \psi_E$ bifurcates from the zero solution at $E = E_*$.

Theorem 2.1 summarizes our requirements on the solutions of the time independent problem, (2.3). These conditions are not at all optimal; they are dictated by the known local decay estimates for the Schrödinger propagator associated with $-\Delta+V$ (on its continuous spectral part) which are, at present, far from being optimal. We believe this situation will be rectified soon, enabling us to relax our conditions on $f(x,\xi)$ considerably (e.g. remove the assumption of spherical symmetry, certain limitations on m etc.) The proof of existence of solutions of eq. (2.3) follows from the theory of bifurcation from a simple eigenvalue (see for example C-R, Nir). Part a) follows from arguments used in [Str 2]. To derive b) and c) we use the following relations: let

$$L \equiv -\Delta+V+\lambda m|\psi_E(x)|^{m-1}-E \quad \text{on} \quad L^2(\mathbb{R}^n) \tag{2.4}$$

then we get by differentiation

$$\frac{\partial\psi_E}{\partial E} = L^{-1}\,\psi_E \tag{2.5a}$$

$$\frac{\partial^2\psi_E}{\partial E^2} = L^{-1}\left[2\,\frac{\partial\psi_E}{\partial E} - \lambda m(m-1)|\psi_E|^{m-2}\left(\frac{\partial\psi_E}{\partial E}\right)^2\right]. \tag{2.5b}$$

One can check that

$$\langle x\rangle^k L^{-1}\langle x\rangle^{-k}: H^2(\mathbb{R}^n) \longrightarrow H^2(\mathbb{R}^n)$$

by successive commutation of $\langle x\rangle$ through L^{-1}. Also, for E near E_*

$$\|L^{-1}\|_2 < \frac{\text{const.}}{|E-E_*|}.$$

Remark. In the repulsive case $(\lambda > 0)$ the desired estimates on ψ_E follows

$$\sup_{E\in(E_*,E^*)} |\psi_E(|x|)| = |\psi_{E^*}(|x|)|$$

which is a consequence of the fact that L^{-1} is positivity preserving in this case.

Corollary 2.2. (a) If $\lambda > 0$ then $\|\psi_E\|_{H^2} \leq C_\Omega \|\psi_E\|_2$, where $E \in \Omega$, any compact subinterval of $(E_*,0)$.

(b) If $\lambda < 0$ then $\|\psi_E\|_{H^2} \leq C_\Omega \|\psi_E\|_2$ if $E_C \leq E < E_*$ for some critical E_C.

III. Some linear estimates

Let $L \equiv -\Delta + V$ on $L^2(\mathbb{R}^n)$ and assume V satisfies Hypothesis (V). We denote by $P_c(L)$ the projection on the continuous spectral part of L $(\chi_{(0,\infty)}(L))$.

We assume that V satisfies the following Non-Resonance (NR) condition:

(NR) $L\phi = 0$ and $\Delta\phi \in L^p$ $p > n/2$ implies $\phi \in L^2(\mathbb{R}^n)$.

Then we have the following local decay estimate [K-J, Mu]

$$\|\langle x\rangle^{-\sigma} e^{-iLt} P_c(L)g\|_2 \leq \frac{C(V)}{\langle t\rangle^{1+\delta}} \|\langle x\rangle^{1+a} P_c(L)g\|_2 \qquad n > 2 \tag{2.6}$$

where $C(V)$ is a constant which depends continuously on the $\|\langle x\rangle^{2+a} V\|_2$, $\sigma > 2$, $a > 0$, and $\delta > 0$.

For $n = 2$, $\langle t\rangle^{1+\delta}$ should be replaced by $\langle t\ell n^2 t\rangle$ in (2.6). Furthermore, the following L^p estimates hold with $2 < p < \frac{2n}{n-2}$ $(n \geq 3)$, and $p > 2$ for $n = 2$.

$$\|e^{-iLt} P_c(L)g\|_p \leq \frac{C(V)}{|t|^{\frac{n}{2}-\frac{n}{p}}} (\|P_c(L)g\|_q + \|\langle x\rangle^{1+a} P_c(L)g\|_2),$$

and,

$$\|e^{-iLt} P_c(L)g\|_p \leq \frac{C(V)}{\langle t\rangle^{\frac{n}{2}-\frac{n}{p}}} (\|P_c(L)g\|_q + \|g\|_{H^1} + \|\langle x\rangle^{1+a} P_c(L)g\|_2)$$

for some $1 \gg a > 0$.

These estimates can be derived from the local decay estimate and by writing equation (2.1) as an integral equation. We then use

Lemma

$$\int_0^t \frac{ds}{|t-s|^{\alpha}\langle s\rangle^{\beta}} < \frac{C(\alpha,\beta)}{\langle t\rangle^{\min(\alpha,\alpha+\beta-1)}}$$

if $\alpha < 1$.

Section 3. The equations for the localized and dispersive parts

Equation (2.1) together with our special choice of nonlinearity f can be written as

$$i\,\frac{\partial\Phi(x,t)}{\partial t} = (-\Delta+V(x)+\lambda|\Phi(x,t)|^{m-1})\,\Phi(x,t) \tag{3.1}$$

$$\Phi(x,0) = \Phi_0(x) \in H^1(\mathbb{R}^n) \quad n > 2 .$$

To distinguish between the localized the dispersive parts of Φ we need an

Ansatz:

(α) Decomposition:

$$\Phi(x,t) = e^{-i\theta}\psi + e^{-iE_0t}\,\phi(x,t) \tag{3.2}$$

$$\Phi(x,0) = e^{-i\gamma_0}\,\psi_{E_0} + \phi(x,0)$$

$$\theta \equiv \int_0^t E(s)ds-\gamma(t) \qquad E(0) = E_0,\ \gamma(0) = \gamma_0 .$$

$$\psi \equiv \psi(x,E(t)) \equiv \psi(E(t)) = \psi_{E(t)} .$$

Here, $\psi(E)$ is the ground state of (2.3):

$$H(E)\psi(E) \equiv (-\Delta+V+\lambda|\psi(E)|^{m-1})\psi(E) = E\psi(E) \tag{3.3}$$

$$\psi(E) \in H^2 \ , \quad \psi > 0$$

$$\text{for } \ E \in (E_*,0) \ \text{ if } \ \lambda > 0$$

$$E \in (-\infty,E_*) \ \text{if} \ \lambda < 0 \ ,$$

where $$0 > E_* \equiv \inf \operatorname{spec}(-\Delta+V) \ .$$

(β) Orthogonality condition

$$\begin{aligned} &\langle\psi(E_0),\phi(\cdot,0)\rangle = 0 \\ \text{and } \ &\frac{d}{dt}\langle\psi(E_0),\phi(\cdot,t)\rangle = 0 \ . \end{aligned} \tag{3.4}$$

The orthogonality condition ensures that $\phi(\cdot,t)$ lies in Range $P_c(H(E_0))$, where H is defined in (3.3).

Using the above Ansatz, we derive the following equation for ϕ :

$$i\frac{\partial\phi}{\partial t} = (H(E_0)-E_0)\phi + [\psi\dot{\gamma}-i\frac{\partial\psi}{\partial E}\dot{E}]e^{-i\theta+iE_0t} + F \tag{3.5}$$

Here $E \equiv E(t)$, $\dot{E} = \frac{dE(t)}{dt}$, $\dot{\gamma} = \frac{d\gamma(t)}{dt}$, and $F = F(\phi,\psi(E))$.

To impose (β), we multiply (3.5) by ψ_{E_0} and integrate over all space, equate the real and imaginary parts to zero (condition (β)) to get a coupled system for E and γ :

$$\frac{d}{dt}\begin{pmatrix} E(t) \\ \gamma(t) \end{pmatrix} = M\begin{pmatrix} G_1 \\ G_2 \end{pmatrix} \tag{3.6}$$

$$G_{1,2} = G_{1,2}(E,\gamma,\phi)$$

$$M(t) = \begin{pmatrix} \dfrac{\cos\Omega(t)}{\langle\frac{\partial\psi}{\partial E},\psi_{E_0}\rangle} & \dfrac{-\sin\Omega(t)}{\langle\frac{\partial\psi}{\partial E},\psi_{E_0}\rangle} \\ \\ \dfrac{\sin\Omega(t)}{\langle\psi_E,\psi_{E_0}\rangle} & \dfrac{\cos\Omega(t)}{\langle\psi_E,\psi_{E_0}\rangle} \end{pmatrix}$$

$$\Omega(t) \equiv \int_0^t (E(s)-E_0)ds-\gamma(t) = \theta(t)-E_0t$$

Section 4. The Main Results

We assume, as before, that $n = 2$ or 3, V satisfies our condition (V) and we let $f(|x|,\Phi) = V(|x|) + \lambda|\Phi|^{m-1}$. We define the B-norm of a function g by $\|g\|_B \equiv \|g\|_{H^1} + \|\langle x\rangle^{1+a} g\|_2$.

Theorem 4.1 (Scattering)

Let $m > 1 + \frac{1}{n} + \sqrt{\frac{2}{n} + \frac{1}{n^2}}$

and if $n = 3$ we assume that $m < 3$.
Then, if there exists a small number δ_0 such that if $\Phi(0) \equiv \Phi_0(|x|)$ and

i) $\|\Phi_0\|_B < \delta_0$

ii) $\min_{E,\theta} \|\Phi_0 - e^{i\theta}\psi_E\|_B$ subject to the constraint

$$\langle e^{i\theta}\psi_E, \Phi_0 - e^{i\theta}\psi_E\rangle = 0$$

has a non-zero solution $\psi_{E_0} e^{i\theta_0}$, i.e. $E_0 \neq E_*$.

iii) $V+\lambda|\psi_{E_0}|^2$ satisfies the (NR) condition of section 2,

then

$$\Phi(t) = e^{-i\int_0^t E(s)ds + i\gamma(t)}\psi(E(t)) + e^{-iE_0 t}\phi(t)$$

with $\frac{dE(t)}{dt} \in L^1(dt)$ (so that $\lim_{t\to\pm\infty} E(t) \equiv E^{\pm}$ exist)

$\frac{d\gamma(t)}{dt} \in L^1(dt)$ (so that $\lim_{t\to\pm\infty} \gamma(t) \equiv \gamma^{\pm}$ exist),

and $\phi(t)$ is purely dispersive in the sense that

$$\|\langle x\rangle^{-\sigma}\phi(t)\|_2 = O(\langle t\rangle^{-1-\delta}) \text{ for } \sigma > 2 \text{ and some } \delta > 0 \text{ if } n = 3$$

and $\|\langle x\rangle^{-\sigma}\phi(t)\|_2 = O(\langle t\,\ell n^2 t\rangle^{-1})$ for $n = 2$.

Moreover, $\|\phi(t)\|_{2m} = O(t^{-\frac{n}{2}+\frac{n}{2m}})$.

Remark. The use of the 2m-norm is dictated by the dependence of the linear local decay estimates on the weighted norm $\|\langle x\rangle^{1+a}f\|_2$ (cf. section 2). This is the source of the restriction to the spherically symmetric case and to $m < 3$ (if $n = 3$). It is believed that the estimates for the linear theory hold with the weighted norms replaced by L^p norms. Such estimates would extend our results to the non-spherically symmetric case and $m < \frac{n+2}{n-2}$ for $n > 3$.

The following is a related stability result which says that if the initial data for (2.1) lies near a particular nonlinear bound state of energy E_0 and phase γ_0, then the solution $\Phi(t)$ converges, as $t \longrightarrow \pm\infty$, to a nearby nonlinear bound state of energy $E^{\pm}$ and phase $\gamma^{\pm}$.

Theorem 4.2 (Stability)

Let m and n be as in Theorem 4.1. Let $\Omega_\eta = (E_*, E_*+\eta\,\mathrm{sgn}(\lambda))$ where η is positive and sufficiently small. Then for all $E_0 \in \Omega_\eta$ and $\gamma_0 \in [0,2\pi]$, there is a positive number $\varepsilon(\eta,E_0)$ such that if $\Phi(0) = \psi(E_0)e^{i\gamma_0}+\phi(0)$, where $\|\phi(0)\|_B < \varepsilon$, then $\Phi(t)$ decomposes into localized and dispersive parts as in (4.1) where $\frac{dE}{dt}$, $\frac{d\gamma}{dt}$ are in $L^1(dt)$ and $\phi(t)$ obeys the linear dispersive and local decay estimates.

Section 5. A priori estimates of the Coupled Equations

I. We let $L \equiv H(E_0)-E_0$ and write the equation for $\phi(t)$ as an integral equation:

$$\phi(t) = e^{-iLt}\phi(0) + \lambda \int_0^t e^{-iL(t-s)} P_c(L)F(s)ds \tag{5.1}$$

where $F(s)$ is a complicated expression which depends on $\phi(s)$, $\dot{E}(s)$ $\dot{\gamma}(s)$ and $\psi(E(s))$.

For $p > 2$ we can derive from (5.1) the following inequality:

$$||\phi(t)||_p \le ||e^{-iLt}\phi(0)||_p + \int_0^t \frac{ds}{|t-s|^{1-\varepsilon_p}} \Big[\sum_{j=2}^{m} c_j(\psi,\phi)\,||\phi(s)||_p^j \tag{5.2}$$

$$+ c_1(\psi,\phi)||\langle x\rangle^{-\sigma}\phi(s)||_2 + c_0(\psi)[|\dot{E}(s)|+|\dot{\gamma}(s)|]\Big]$$

Similarly we get:

$$||\langle x\rangle^{-\sigma}\phi(t)||_2 \le ||\langle x\rangle^{-\sigma}e^{-iLt}\phi(0)||_2 + \int_0^t \frac{ds}{\langle t-s\rangle^{1+\delta}} \Big[\sum_{j=2}^{m} \bar{c}_j(\psi,\phi)||\phi(s)||_p^j \tag{5.3}$$

$$+ \bar{c}_1(\psi,\phi)||\langle x\rangle^{-\sigma}\phi(s)||_2 + \bar{c}_0(\psi)[|\dot{E}(s)|+|\dot{\gamma}(s)|]\Big] .$$

Here $c_i(\psi,\phi)$ and $\bar{c}_i(\psi,\phi)$ are constants which depend on the weighted norms of $\psi(E(s))$ and the H^1 norm of $\phi(s)$. The above inequalities are derived by the help of the local decay and L^p estimates on $e^{-iHt}\,P_c(L)$ described in section 2. The weighted norms are controlled by the following estimate for radial functions, $g(r)$, [Str 2]

$$|g(r)| \le c\, r^{\frac{1-n}{2}}\, ||g||_{H^1} \tag{5.4}$$

the uniform bound on the H^1 norm (in time) of the full solution, combined with local existence for the coupled equations.

Remark. One does not know, á priori, that the decomposition of $\phi(t)$ into localized and dispersive waves is well defined. Therefore, we must first show that there exists a local solution of the coupled system governing $E(t)$, $\gamma(t)$ and $\phi(t)$. This is achieved by a straightforward contraction mapping argument. The continuation of the local solution to a global one with appropriate asymptotic behavior follows from the decay estimates which we derive in this section.

Next we estimate the E , γ equations to get:

$$|\dot{E}(t)| \le \frac{C_E(\psi_E)\,|\lambda|}{|\langle \psi_{E_0}, \frac{\partial\psi}{\partial E}\rangle|} \Big[||\langle x\rangle^{-\sigma}\phi(t)||_2 + ||\phi(t)||_p^{2+\varepsilon_m} \Big] \tag{5.5a}$$

$$|\dot{\gamma}(t)| \le \frac{C_\gamma(\psi_E)\,|\lambda|}{|\langle \psi_{E_0}, \psi_E\rangle|} \Big[||\langle x\rangle^{-\sigma}\phi(t)||_2 + ||\phi(t)||_p^{2+\varepsilon_m} \Big] . \tag{5.5b}$$

We choose p so that $(2+\varepsilon_m)(1-\varepsilon_p) > 1+\delta$ (where $1-\varepsilon_p = \frac{n}{2} - \frac{n}{p}$) . (5.5c)

It turns out that with the local decay estimates we use, it is natural to choose $p = 2m$. Better local decay estimates would permit $p = m+1$ and would improve the range of validity of the above results considerably.

II. Large Time Estimates

Next let

$$\begin{aligned} M_1(T) &\equiv \sup_{|t| \leq T} \langle t\rangle^{1-\varepsilon_p} \|\phi(t)\|_p \\ M_2(T) &\equiv \sup_{|t| \leq T} \langle t\rangle^{1+\delta} \|\langle x\rangle^{-\sigma}\phi(t)\|_2 \end{aligned} \tag{5.6}$$

(when $n = 2$ replace $\langle t\rangle^{1+\delta}$ by $\langle t\,\ell n^2 t\rangle$).
Then (5.2)-(5.6) imply

$$\begin{aligned} &\text{(a)} \qquad M_2(T) \leq d_1(\phi(0),\psi_{E_0})+d_2(\psi_{E_0})M_1^{m-1}(T) \\ &\text{(b)} \qquad M_1(T) \leq \bar{d}_1(\phi(0),\psi_{E_0})+\bar{d}_2(\psi_{E_0})M_1^{m-1}(T) \end{aligned} \tag{5.7}$$

provided $E(t) \subset \Omega$. Recall Ω is a compact sub-interval contained in $(E_*,0)$ for $\lambda > 0$ and in (E_c,E_*) for $\lambda < 0$. The main ingredient in the proof is to choose the constants in the equation for $\dot{E}(t)$ so small so that

$$\int_0^\infty |\dot{E}(s)|ds < \bar{\delta} \text{ , with some small } \bar{\delta} .$$

This is possible since $(2+\varepsilon_m)(1-\varepsilon_p) > 1$.

The smallness of $\bar{\delta}$ implies that $E(t) \in [E_0-\bar{\delta},E_0+\bar{\delta}] \cap \Omega$ hence all the constants are uniformly bounded in time by our estimates on $\psi(E(t))$. From inequalities (5.7) and assumptions on smallness of $M_1(0)$ and $M_2(0)$, we obtain bounds on $M_1(T)$ and $M_2(T)$ which are independent of T . Passing to the limit, $T \longrightarrow \infty$, we conclude the desired asymptotic behavior of $E(t)$, $\gamma(t)$ and $\phi(t)$. Theorem 4.2 then follows. Theorem 4.1 follows from the above and the following discussion.

III. Choosing $E(0)$, $\gamma(0)$

Next, we want to explain how, given an initial condition $\Phi(0)$, we choose the decomposition at $t = 0$.

We take $\min_{E,\theta} \|e^{i\theta}\psi(E)-\Phi(0)\|_B$ subject to the constraint

$$<\Phi(0)-\psi(E)e^{i\theta},\psi(E)e^{i\theta}> = 0 .$$

The constraint implies $\|\Phi(0)\|_2 > \|\psi(E)\|_2$, hence $(E,\theta) \in$ closed rectangle in $\mathbb{R}^2$. Now, we observe that when $E \in \Omega$, for any norm on ψ , introduced in section 2 we have

$$\|\psi(E)\|_\beta < C_\beta\|\psi(E)\|_2 < C_\beta\|\Phi(0)\|_2$$

$$\|\phi\|_\beta < \|\Phi(0)\|_\beta + \|\psi(E)\|_\beta < C_\beta(\|\Phi(0)\|_2+\|\Phi(0)\|_\beta) .$$

In this way the smallness of all constants is reduced to smallness condition on the $\|\Phi(0)\|_\beta$ norm of the initial data.

An interesting question arises here, whether we can derive the $\dot{E}$, $\dot{\gamma}$ equations from a variational problem we solve for the initial data. That may help in understanding in a geometric way the dynamical evolution for all times (of $\Phi(t)$) and may further elucidate the role of $\gamma(t)$ in the analysis.

IV. Scattering Theory

In scattering theory one would like to construct the operator that maps the state of system at $t \longrightarrow -\infty$ onto that at $t \longrightarrow \infty$. This operator is called the Scattering Matrix (see [R-S]). In our case the S-matrix maps (γ^-,E^-) on (γ^-,E^+) , and since

$$\phi(t) \sim e^{-i\Delta t}\phi_\pm \qquad \text{for } t \longrightarrow \pm\infty$$

we have $$S\,\phi_- \longrightarrow \phi_+ .$$

That $\phi(t)$ is asymptotically free follows from the existence and completeness theory for the Hamiltonian $H(E_0)$.

References

[Ben] Benjamin, T.B., The stability of solitary waves, Proc. R. Soc. London A328, 1972, p. 153.

[Ber] Berry, M.V., Quantal phase factors accompanying adiabatic changes, Proc. Roy. Soc. London Ser. A 392, 1984, p. 45.

[Be-Li] Berestycki, H. and Lions, P.L., Nonlinear scalar field equations I - Existence of a ground state, Arch. Rat. Mech. Anal. 82 , 1983, pp. 313-345.

[Ca-Li] Cazenave, T., Lions, P.L., Orbital stability of standing waves for some nonlinear Schrödinger equations", Comm. Math. Phys. 85, 1982, pp. 549-561.

[C-K] Cohen, A., Kappeler, T., preprint.

[C-R] Crandall, M., Rabinowitz, P., Bifurcation of simple eigenvalues and linearized stability, Arch. Rat. Mech. Anal. 52, 1973, pp. 161-181.

[C-W] Cazenave, T., Weissler, F.B., The Cauchy problem for the nonlinear Schrödinger equation in H^1, preprint.

[En] Enss, V., Quantum Scattering Theory of Two and Three Body Systems with Potentials of Short and Longe Range in Schrödinger Operators, ed. by S. Graffi, Lecture Notes in Mathematics, 1159, Springer, 1985.

[G-G-K-M] Gardner, C.S., Greene, J.M., Kruskal, M.D., Miura, R.M., Method for solving the Korteweg-de Vries equation, Phys. Rev. Lett. 19, 1967, pp. 1095-1097,

[G-S-S] Grillakis, M., Shatah, J., and Strauss, W., Stability theory of solitary waves in the presence of symmetry, I, to appear in Jour. Func. Anal.

[G-V] Ginibre, J. and Velo, G., On a class of nonlinear Schrödinger equations I, II, J. Func. Anal. 32, 1979, pp. 1-71.

[H-N-T] Hayashi, N., Nakamitsu, K., Tsutsumi, M., On solutions of the initial value problem for the nonlinear Schrödinger equations, J. Func. Anal., 71, 1987, pp. 218-245.

[J-K] Jensen, A., Kato, T., "Spectral properties of Schrödinger operators and time decay of the wave functions", Duke Math. J. 46, 1979, pp. 583-611.

[K] Kato, T., On nonlinear Schrödinger equations, Ann. Inst. Henri Poincaré, Physique Théorique, 46, 1987, pp. 113-129.

[K-A] Kodama, Y., Ablowitz, M.J., Perturbations of solitons and solitary waves, Stud. in Appl. Math. 64, 1981, pp. 225-245.

[K-M] Keener, J.P., McLaughlin, D.W., Solitons under perturbations, Phys. Rev. A 16, 1977, pp. 777-790.

[Lax] Lax, P.D., Integrals of nonlinear equations of evolution and solitary waves, Commun. Pure Appl. Math. 21, 1968, pp. 467-490.

[Li] Lions, P.L., in Nonlinear Problems: Present and Future, North Holland Math. Studies 61, A. Bishop, D. Campbell, B. Nicolaenko, eds.

[Mu] Murata, M., Rate of decay of local energy and spectral properties of elliptic operators, Japan J. Math. 6, 1980, pp. 77-127.

[Ne] Newell, A.C., Near-integrable systems, nonlinear tunneling and solitons in slowly changing media, in Nonlinear Evolution Equations Solvable by the Inverse Spectral Transform, F. Calogero ed., London: Pitman, 1978, pp. 127-179.

[Nir] Nirenberg, L., Topics in Nonlinear Functional Analysis, Courant Institute Lecture Notes, 1974.

[R-S] Reed, M., Simon, B., Methods of Mathematical Physics III: Scattering Theory, Academic Press.

[Ro-We] Rose, H.A., Weinstein, M.I., On the bound states of the nonlinear Schrödinger equation with a linear potential, Physica D 30, 1988, pp. 207-218.

[Str 1] Strauss, W.A., Dispersion of low energy waves for two conservative equations, Arch. Rat. Mech. Anal. 55, 1974, pp. 86-92.

[Str 2] Strauss, W.A., Existence of solitary waves in higher dimensions, Commun. Math. Phys. 55, 1977, pp. 149-162.

[Str 3] Strauss, W.A., Nonlinear scattering theory at low energy, J. Funct. Anal. 41, 1981, pp. 110-133.

[Sh-Str] Shatah, J. and Strauss, W., Instability of nonlinear bound states, Comm. Math. Phys. 100, 1985, pp. 173-190.

[Sig-Sof] Sigal, I.M., Soffer, A., The N-particle scattering problem: asymptotic completeness for short range systems, Ann. Math. 126, 1987, pp. 35-108.

[We 1] Weinstein, M.I., Modulational stability of ground states of nonlinear Schrödinger equations, Siam. J. Math. Anal. 16, #3, 1985, pp. 472-491.

[We 2] Weinstein, M.I., Lyapunov stability of ground states of nonlinear dispersive evolution equations, Commun. Pure Appl. Math. 39, 1986, pp. 51-68.

[Z-S] Zakharov, V.E., Shabat, A.B., Exact theory of two-dimensional self-focusing and one dimensional self modulation of waves in nonlinear media, Sov. Phys. J.E.T.P. 34, 1972, pp. 62-69.

ON THE LINEAR STABILITY OF SOLITARY WAVES IN HAMILTONIAN SYSTEMS WITH SYMMETRY

Joachim Stubbe[(+)]
Département de Physique Théorique
Université de Genève
24, quai Ernest-Anserment
CH-1211 Genève 4 (Suisse)

1. Introduction

I have studied abstract Hamiltonian systems of the form

$$\frac{du}{dt} = J\,E'(u) \tag{H}$$

which are locally well-posed in a real Hilbert space X, where E denotes the energy functional on X, E' its "derivative with respect to u" and J is a skew-symmetric linear operator. I assume that the system (H) is invariant under a representation $T(\cdot)$ of the group $G = (\mathbb{R},t)$. By a solitary wave I mean a solution of the form

$$u(t) = T(\omega t)\phi_\omega, \qquad \phi_\omega \in X \tag{1.1}$$

and I assume that such solutions of (H) exist for an interval of "frequencies" ω. For the stability analysis of solitary waves there are many different notions of stability which have been used in the literature, e.g.

Nonlinear (orbital) stability: In view of the invariance one considers the stability of the ϕ_ω-orbit

$$O_\omega \equiv \{T(s)\phi_\omega,\ s \in \mathbb{R}\}. \tag{1.2}$$

The ϕ_ω-orbit is called stable if a solution u(t) of (H) exists for all $t \geq 0$ and remains near O_ω measured by the norm of X provided its initial datum u(0) is sufficiently close to O_ω.

Energetic stability: Because of the symmetry there is another conserved functional Q. The vector ϕ_ω is a critical point of

$$L_\omega \equiv E - \omega Q.$$

(+) On the leave from Fakultät für Physik, Universität Bielefeld, Postfach 8640, D-4800 Bielefeld 1, FRG

Now one says that ϕ_ω is energetically stable, if it (locally) minimizes the energy E subject to constant Q.

<u>Linear stability:</u> If I linearize the system (H) around a solitary wave of the form (1.1), I shall obtain the linear Hamiltonian system

$$\frac{dw}{dt} = J H_\omega w \qquad (H_{lin}, \omega)$$

where H_ω is the "linearized Hamiltonian" given by

$$H_\omega \equiv L_\omega''(\phi_\omega). \qquad (1.4)$$

Now roughly spoken, ϕ_ω is called linearly stable if any solution w of (H_{lin}, ω) remains bounded.

A basic problem is to find the relations between these notions of stability. In general one cannot expect an answer to this question (see e.g. [1,2]. However, under certain assumptions there are useful implications like the Lagrange-Dirichlet theorem for finitedimensional systems [3]. So one can ask: Are there conditions which guarantee the equivalence of these notions of stability? Indeed, this problem was solved even within this abstract framework in two recent papers by Grillakis, Shatah and Strauss [4] and by myself [5]. The basic condition is an assumption on the critical point ϕ_ω of L_ω which can be expressed as a property of the linearized operator H_ω: Let H_ω have at most one negative eigenvalue. Now, the invariance implies that zero is also in the spectrum of H_ω. The rest of the spectrum has to be positive and bounded away from zero.

In the following I shall only discuss the case of one negative eigenvalue since this is the more interesting one. Using this assumption Grillakis, Shatah and Strauss show that the ϕ_ω-orbit is stable if and only if the scalar function

$$d(\omega) \equiv L_\omega(\phi_\omega) \qquad (1.5)$$

is convex at ω. The main intermediate step is to show that ϕ_ω is energetically stable if and only if $d(\omega)$ is convex. Hence, under the above assumption on the linearized operator the nonlinear stability and the energetic stability are equivalent.

Within the same abstract framework I could solve the linear stability problem provided $d''(\omega)$ is nonzero [5]. The main difficulty is the following: Obviously the definition of linear stability cannot hold for (H_{lim},ω) since the critical point around which we linearize is degenerate. To explain this fact we observe that the functions

$$\psi(t,\theta,\omega) \equiv T(\omega t+\theta)\ \phi_\omega \tag{1.6}$$

form a two-parameter family of solitary wave solutions of the Hamiltonian system (H). Formally, the derivatives of ψ with respect to the parameters θ and ω are solutions of the linearized system which (may) grow linearly in time. These "secular modes" exist for all solitary waves (independently of all other stability properties). In [5] I constructed a subspace W of X (with codimension two) where the secular modes are removed. This space W is invariant for the linear evolution equation and the linear stability of ϕ_ω under perturbations in W is equivalent to the orbital (resp. energetic) stability of ϕ_ω. This linear analysis is also useful in studying the effects of structural perturbations and is known in the literature as "parameter modulation approach" [2,6,7,8].

In this contribution I want to propose an alternative approach to linear stability of solitary waves in systems with symmetries. The basic idea is to factor out the symmetry (at least locally around the ϕ_ω-orbit) and then to linearize system. More precisely, I try to construct a functional $s = s(u)$ such that

$$d(T(s)u,\phi_\omega) = d(u,O_\omega) \tag{1.7}$$

where d represents a distance (usually the norm distance) in X. Then I want to study the evolution of

$$v(t) \equiv T(s(u(t)))\ u(t) \tag{1.8}$$

instead of u(t), i.e. I investigate the evolution of the element of the u-orbit $\{T(\theta)u(t)\}$ which is closest to the ϕ_ω-orbit with respect to d. In this sense I factor out the symmetry group. I shall then linearize the evolution equation for v(t) around its stationary solution ϕ_ω to study its linear stability. This approach has, at least, two advantages: First of all, no secular mode will occur. In fact, in some sense the function v(t) given by (1.8) represent the collective coordinate ansatz used in field theory. Furthermore, I can now treat the open case $d''(\omega) = 0$. While for $d''(\omega)$ nonzero I shall find again a one to one correspondence between linear and energetic (resp. orbital) stability, this does not hold in the critical case $d''(\omega) = 0$.

In order to avoid the technical difficulties of this new approach in the abstract framework, I shall only study a simple example arising from many applications, namely the one-dimensional nonlinear Schrödinger equation

$$iu_t + u_{xx} + |u|^{p-1}u = 0 \qquad \text{on } \mathbb{R}, \quad p > 1 . \tag{NLS}$$

The abstract results will follow along the same lines and will be published elsewhere.

2. Stability analysis for a nonlinear Schrödinger equation

I consider the one-dimensional nonlinear Schrödinger equation

$$iu_t + u_{xx} + |u|^{p-1}u = 0 \qquad \text{on } \mathbb{R} \tag{NLS}$$

where $u = u(x,t)$ is a complex valued function and $p > 1$. (NLS) is a Hamiltonian system with energy

$$E(u) = \frac{1}{2}\int_{\mathbb{R}} |u_x|^2 dx - \frac{1}{p+1}\int_{\mathbb{R}} |u|^{p+1} dx . \tag{2.1}$$

(NLS) is invariant under the translation group and under the phase transformation $T(s) = \exp(is)$. In order to simplify my analysis I consider only functions u which are symmetric with respect to the origin so that there is only the global gauge symmetry. The space X where (NLS) is well-defined is then $H^1_s(\mathbb{R})$, the (complex) subspace of symmetric functions in $H^1(\mathbb{R})$, with real inner product. I define an identification map $I : X \to X^*$ by

$$\begin{aligned} \langle I(u),v\rangle &= \mathrm{Re}(u,\bar{v}) \\ &= \mathrm{Re}\int_{\mathbb{R}} u\bar{v}dx \end{aligned}$$

where $\bar{v}$ denotes the complex conjugate of v.

Then I can write (NLS) as

$$I\frac{du}{dt} = -iE'(u) \tag{H}$$

i.e., at least formally, $J = -iI^{-1}$. The charge Q associated to the invariance is given by

$$Q(u) = -\frac{1}{2}\langle I(u),u\rangle = -\frac{1}{2}\int_{\mathbb{R}} |u|^2 dx . \tag{2.3}$$

It is well-known that for each $\omega > 0$, (NLS) possesses a unique solitary wave (iuX) of the form (1.1), i.e.

$$u(x,t) = \phi_\omega(x)\exp(i\omega t) \tag{2.4}$$

with $\phi_\omega > 0$ [9]. Furthermore, the linearized operator H_ω satisfies the spectral assumption made in the introduction (one negative eigenvalue, nullspace spanned by $i\phi_\omega$ and rest of the spectrum positive, bounded away from zero) [4,7] and therefore the orbital (resp. energetic) stability is determined by the behaviour of $d(\omega)$ which can be calculated explicitely:

$$d(\omega) = \omega^{\frac{1}{2} + \frac{2}{p-1}} d_o \, , \quad d_o > 0 \, .$$

Hence we have the following result [4,10]:

Proposition 2.1

If $1 < p < 5$, then $d''(\omega) > 0$ and all solitary waves are orbitally stable.
If $p > 5$, then $d''(\omega) < 0$ and all solitary waves are orbitally unstable.
If $p = 5$ (the critical case) then $d''(\omega) = 0$.
All solitary waves are orbitally unstable.

A conditional stability result for the critical case including dilations of ϕ_ω is given in [11].

The linear stability result for the linearized Schrödinger equation by removing the secular modes can also be found in [7] where a detailed analysis of the spectrum of H_ω is presented.

Now I start with my new approach sketched in the introduction. I construct the functional $s(u)$ such that $v = u \exp(is(u))$ is orthogonal to the orbit $O_\omega = \{\phi_\omega \exp(is), s \in \mathbb{R}\}$ at ϕ_ω with respect to $<,>$, i.e.

$$<I(u \exp(is)), i\phi_\omega> = 0 \, . \tag{2.6}$$

I obtain

$$s(u) = - \arctan \frac{<I(u), i\phi_\omega>}{<I(u), \phi_\omega>} \quad . \tag{2.7}$$

Then $v = u \exp(is(u))$ is orthogonal to the nullspace of H_ω (which is spanned by $i\phi_\omega$) with respect to $<,>$. It is not orthogonal with respect to the inner product of X so that in minimizes the X-distance. But the construction via (2.6) will be enough for my purpose.

Let $u(t)$ be a solution of (NLS) (resp. (H)). After some simple computations I obtain the following evolution equation for $v(t) = u(t) \exp(is(u(t)))$

$$I \frac{dv}{dt} = i \frac{<E'(v), \phi_\omega>}{<I(v), \phi_\omega>} Iv - iE'(v) \, . \tag{2.9}$$

Obviously E and Q are also conserved for (2.9). In addition, $\langle I(v), i\phi_\omega\rangle = 0$ by construction. The latter conservation law shows that (2.9) is a (nonlinear!) projection of (H) onto a subspace of codimension one in X.

The linearization $v(t) = \phi_\omega + z(t)$ will give

$$I\,\frac{dz}{dt} = i\,\frac{\langle H_\omega z, \phi_\omega\rangle}{(\phi_\omega, \phi_\omega)}\, I\phi_\omega - iH_\omega z\ . \qquad (2.10)$$

An explicit representation of this system can be given by splitting z in its real and imaginary part $z = a + ib$ (confer also [7]). In this representation we have

$$H_\omega = \begin{pmatrix} L_+ & 0 \\ 0 & L_- \end{pmatrix} = \begin{pmatrix} -\partial_x^2 + \omega - p\phi_\omega^{p-1} & 0 \\ 0 & -\partial_x^2 + \omega - \phi_\omega^{p-1} \end{pmatrix} \qquad (2.11)$$

and therefore

$$\begin{aligned} a_t &= L_- b \\ b_t &= \frac{\langle L_+ a, \phi_\omega\rangle}{(\phi_\omega, \phi_\omega)}\,\phi_\omega - L_+ a\ . \end{aligned} \qquad (2.12)$$

In the following I shall always work with equation (2.9). Formally, it has the following conserved quantities:

$$\frac{1}{2}\langle H_\omega z, z\rangle = \text{ete} \qquad (2.13a)$$

$$\langle I(z), \phi_\omega\rangle = \text{ete} \qquad (2.13b)$$

$$\langle I(z), i\phi_\omega\rangle = \text{ete}\ . \qquad (2.13c)$$

As pointed out in the introduction I use the following definition of linear stability.

<u>Definition:</u> ϕ_ω is called linearly stable if any solution of the linearized system (2.10) remains bounded in X.

<u>A. The case $d''(\omega) > 0$</u>

The basic fact I use is the energetic stability of ϕ_ω. More precisely, we have the following result [4,7]:

Lemma 2.2: Let $d''(\omega) > 0$. If $\langle Ig,\phi_\omega\rangle = \langle Ig,i\phi_\omega\rangle = 0$ then $\langle H_\omega g,g\rangle \geq c\,\|y\|^2_{H^1}$.

In view of the conservation laws (2.13) it is easy to see that any solution $z(t)$ of (2.10) with initial values z_o satisfying $\langle Iz_o,\phi_\omega\rangle = \langle Iz_o,i\phi_\omega\rangle = 0$ remains bounded. In the following theorem I show that all solutions of (2.10) remain bounded.

Theorem 2.3: Let $d''(\omega) > 0$. Then ϕ_ω is linearly stable.

Proof: $i\phi_\omega$ and $\phi'_\omega \equiv \frac{\partial\phi_\omega}{\partial\omega}$ are stationary solutions of (2.10) since $H_\omega(i\phi_\omega) = 0$ and $H_\omega\phi'_\omega = I\phi_\omega$. $i\phi_\omega$ and ϕ'_ω are linearly independent since $\langle H_\omega\phi'_\omega,\phi'_\omega\rangle = -d''(\omega) \neq 0$.

If z is a solution of (2.10) so is $y = z + a_1 i\phi_\omega + a_2\phi'_\omega$. Choosing a_1, a_2 such that $\langle Iy,\phi_\omega\rangle = \langle Iy,i\phi_\omega\rangle = 0$ (again I use the fact that $d''(\omega) \neq 0$) the Theorem follows immediately.

B. The case $d''(\omega) < 0$

In this case the energetic instability is crucial.

Lemma 2.4 [4]: Let $d''(\omega) < 0$. There exists $y \in X$ such that $\langle Iy,\phi_\omega\rangle = \langle Iy,i\phi_\omega\rangle = 0$ and $\langle H_\omega y,y\rangle < 0$.

Taking $A(z) = \langle Iz,iy\rangle$ with y from Lemma 2.4 as a Liapunov functional I can prove as in [5] the following result:

Theorem 2.5: Let $d''(\omega) < 0$. Then ϕ_ω is linearly unstable.

C. The case $d''(\omega) = 0$

I show that in this case the useful relations between energetic, linear and nonlinear stability break down.

Although $\langle H_\omega y,y\rangle \geq 0$ for all y satisfying $\langle Iy,\phi_\omega\rangle = \langle Iy,i\phi_\omega\rangle = 0$, ϕ_ω is orbitally unstable as shown by M. Weinstein [10,11]. The "weak" energetic stability is not sufficient to guarantee orbital stability. However, if $d''(\omega) = 0$ and $d(\omega)$ is strictly convex then we have orbital stability [4,12].

Here I prove that $d''(\omega) = 0$ implies linear instability. The advantage of my method is that it does not make use of certain particular properties of the "critical" nonlinear Schrödinger equation (NLS).

Theorem 2.6: Let $d''(\omega) = 0$. Then there exist solutions of the linearized system (2.10) which are unbounded.

Proof: I choose $z_o \in X$ such that $<I\phi_\omega, z_o> > 0$ and $<iI\phi_\omega', z_o> = 0$.

Since $<I\phi_\omega, \phi_\omega'> = - d''(\omega) = 0$ the solution $z(t)$ of (2.10) with initial value z_o satisfies

$$\frac{d}{dt} <iI\phi_\omega', z(t)> = <I\phi_\omega, z_o> > 0 \tag{2.14}$$

by (2.13b) which implies the statement of the Theorem.

3. Summary

For a class of nonlinear Schrödinger equations we saw that there is a one to one correspondence between orbital, energetic and linear stability of solitary waves provided $d''(\omega)$ is nonzero. The case $d''(\omega) = 0$ is critical in the sense that this equivalence breaks down. This result will also extend to the abstract theory provided the linearized operator satisfies the assumptions on its spectrum listed in the introduction. In addition, a more detailed analysis of the critical case will show that for linear stability a conditional stability. result can be obtained which corresponds to the (nonlinear) stability results by M. Weinstein [11].

Acknowledgement:

I thank the organizers M. Balabane, P. Lochak and C. Sulem of the Conference "Integrable Systems and Applications" for inviting me to this very stimulating meeting.

References:

[1] D.D. Holm, J.E. Marsden, T. Ratin and A. Weinstein, Phys. Rep. 123, 1 (1985)
[2] J. Stubbe and L. Vazquez, to appear in "Mathematics + Physics. Lectures on recent results" Vol. 3, L. Streit, editor, World Scientific, Singapore
[3] C.L. Siegel and J.K. Moser, "Lectures on Celestial Mechanics", Springer 1971
[4] M. Grillakis, J. Shatah and W. Strauss, J. Funct. Anal. 74, 160 (1987)
[5] J. Stubbe, Portugagaliae Mathematica, to appear
[6] D.W. McLaughlin and A.C. Scott, Phys. Rev. A18, 1652 (1978)
[7] M. Weinstein, Siam J. Math. Anal. Vol. 16, 472 (1985)
[8] J. Stubbe, to appear in the Proceedings of the IVth German-French meeting on Mathematical Physics, Marseille 1988
[9] H. Berestycki and P.L. Lions, Arch. Rat. Mech. Anal. 82, 313(1983)
[10] M. Weinstein, Comm. Math. Phys. 87, 567 (1983)
[11] M. Weinstein, Comm. Part. Diff. Eq. 11, 545 (1986)
[12] J. Shatah, Comm. Math. Phys. 91, 313 (1983)

NONLINEAR SCHRÖDINGER EQUATIONS WITH MAGNETIC FIELD EFFECT: EXISTENCE AND STABILITY OF SOLITARY WAVES

Joachim Stubbe (+)(*)
Département de Physique Théorique
Université de Genève
24, quai Ernest-Anserment
CH-1211 Genève 4 (Suisse)

Luis Vázquez (**)
Departamento de Física Teórica
Facultad de Ciencias Fisicas
Universidad Complutense
28040-Madrid (Spain)

ABSTRACT: We study the existence and stability of the solitary waves associated to a family of vectorial nonlinear Schrodinger equations with magnetic selfinteraction.

(+) On leave from Fakultat fur Physik/ Universitat Bielefeld/ Postfach 8640 D-4800 Bielefeld 1 / Bundesrepublik Deutschland.
(*) Partially supported by the Swiss National Science Foundation.
(**) Supported by U.S.-Spain Joint Committee for Scientific and Technological Cooperation under grant CCB-8509/001.

I. INTRODUCTION

Nonlinear Schrodinger equations for scalar functions are widely studied in the literature. The existence of solitary waves and their stability properties, the Cauchy problem and the long time behaviour are well established [1,2,3].

Such equations arise in the mathematical description of many physical phenomena, e.g. propagation of thermal pulses in a solid [4], behaviours of nonideal Bose gas with weak interaction among the particles [5], propagation of narrow electromagnetic beams in nonlinear media [6,7] and electromagnetic (Langmuir) waves in a plasma [8,9].

A more precise description of electromagnetic Langmuir waves leads to a vector valued nonlinear Schrodinger equation for the slowly varying envelope of the highly oscillatory electric field. An interesting physical phenomenon is the spontaneous excitation of magnetic fields in such Langmuir plasmas [10] It can be shown that the Zakharov system describing the electrical field, the plasma density and the selfgenerated magnetic field reduces in the static limit (by neglecting also the skin effect) to a generalized nonlinear Schrodinger equation for the field vector $\mathcal{E}$ which contains a magnetic self-interaction of the form $\mathcal{E} \wedge (\mathcal{E} \wedge \mathcal{E}^*)$ where '$\wedge$' denotes the usual cross product operator for three component vectors

$$i \mathcal{E}_t + \Delta \mathcal{E} + |\mathcal{E}|^2 \mathcal{E} + \eta \mathcal{E} \wedge (\mathcal{E} \wedge \mathcal{E}^*) = 0 \qquad \text{on } \mathbb{R}^3$$

In the physical situation described in [10] we have $\eta := (V_{th}/c)^2$ with V_{th} the termal velocity of the electrons. Since the electric field $\mathcal{E}$ is assumed to be potential, it turns out that the selfgenerated magnetic field B_S is of the form $B_S \sim i \mathcal{E} \wedge \mathcal{E}^*$ and the nonlinearity $\mathcal{E} \wedge (\mathcal{E} \wedge \mathcal{E}^*)$ is just the Poynting vector of the electromagnetic field.

In the present contribution we study the existence and stability of standing waves

$$\mathcal{E}(x,t) = \mathcal{E}_\omega(x)\, e^{i\omega t} \tag{1}$$

of the following nonlinear Schrodinger equations with the magnetic field effect

$$\text{(NLS)} \qquad i \mathcal{E}_t + \Delta \mathcal{E} + \lambda |\mathcal{E}|^{2\sigma} \mathcal{E} + \eta \mathcal{E} \wedge (\mathcal{E} \wedge \mathcal{E}^*) = 0$$
$$x \in \mathbb{R}^3, \quad t \in \mathbb{R}_0$$

with $0 < \sigma < 2$

where λ and η are real constants. The bound on σ in the three dimensional case is related to the validity of a certain Sobolev embedding and is therefore important for the well definetness of some physical quantities as will be seen below.

Before stating our main results, let us describe the conservation laws of the NLS. First of all we note that the Schrodinger equation is invariant under the Galilei group which leads to the conservation of the energy, momentum and angular momentum. The energy associated with the NLS is given by

$$H(\mathcal{E}) = \frac{1}{2} \int \sum_{j=1}^{3} |\nabla \mathcal{E}_j|^2 - \frac{\lambda}{\sigma+1} |\mathcal{E}|^{2\sigma+2} - \frac{\eta}{2} |\mathcal{E} \wedge \mathcal{E}^*|^2 \tag{2}$$

For scalar equations there is an additional internal U(1) symmetry (global gauge symmetry) which leads to the charge (particle number) conservation. For the vector-valued equation NLS the situation is different. If there is no magnetic field effect (η =0) then the NLS admits an internal U(3) symmetry which generates nine conservation laws of the form $\int_{\mathbb{R}^d} \mathcal{E}_j \mathcal{E}_k^* \, dx$. For j=k this gives the conservation of the single particle numbers associated to each polarization component

$$Q_K(\mathcal{E}) = \int_{\mathbb{R}^d} |\mathcal{E}_K|^2 dx \qquad K = 1, 2, 3 \tag{3}$$

If a magnetic field effect is present ($\eta \neq 0$), then the U(3) symmetry is broken and we have only four additional conserved quantities, the total particle number

(4) $$Q(\mathcal{E}) = \int_{\mathbb{R}^d} |\mathcal{E}|^2 dx$$

and the average of the selfgenerated magnetic field

(5) $$B(\mathcal{E}) = -i \int_{\mathbb{R}^d} \mathcal{E} \wedge \mathcal{E}^* dx$$

Let us remark that the single particle numbers cannot be conserved any more since the magnetic field effect mixes the polarization components. The conservation of the magnetic field $B(\mathcal{E})$ is also obvious from the physical understanding of the problem since in the limit leading to the NLS, only potential fields $\mathcal{E}$ are considered.

Our main results can be summarized as follows. In terms of the parameters λ, η and σ we give necessary and sufficient conditions for the existence of standing waves of the form (1) heaving minimal energy (ground states) (Theorem 1). Then we show that the stability of these standing waves depends on the behaviour of the charge considered as a function of the frequency ω by using its variational characterization as a ground state [11,12]. If $Q(\mathcal{E}_\omega)$ is increasing in ω we have stability and otherwise instability (Theorem 2). In particular this result implies that for the cubic equation considered in [10] ($\sigma = 1$) the selfgenerated magnetic field does not stabilize the solitary wave.

II. EXISTENCE OF GROUND STATES

We solve the following minimization problem

(6) $$I(\omega) = \operatorname{Inf}\left\{ T(\mathcal{E}) \mid \mathcal{E} \in (H^1(\mathbb{R}^3))^3 ,\ V_\omega(\mathcal{E}) = 1 \right\}$$

where T and V_ω are C^1 functionals on $X \equiv (H^1(\mathbb{R}^3))^3$ given by

(7a) $$T(\mathcal{E}) = \frac{1}{2} \int \sum_{j=1}^{3} |\nabla \mathcal{E}_j|^2$$

(7b) $$V_\omega(\mathcal{E}) = \frac{1}{2} \int -\omega |\mathcal{E}|^2 + \frac{1}{\sigma+1} |\mathcal{E}|^{2\sigma+2} + \frac{\eta}{2} |\mathcal{E} \wedge \mathcal{E}^*|^2$$

By a theorem of Brezis and Lieb [13] $I(\omega)$ admits a nontrivial solution which can be rescaled to a ground state solution $\mathcal{E}_\omega$ of the stationary equation

(8) $$-\Delta \mathcal{E} = -\omega \mathcal{E} + \lambda |\mathcal{E}|^{2\sigma}\mathcal{E} + \eta\, \mathcal{E}\wedge(\mathcal{E}\wedge\mathcal{E}^*)$$

However, much more can be said about the structure of $\mathcal{E}_\omega$. In fact, if we define for fixed $\mathcal{E}_0 \in \mathbb{C}^3$, $|\mathcal{E}_0| = 1$ the following mapping from X to X

(9) $$\Pi_{\mathcal{E}_0}(\mathcal{E}) = \mathcal{E}_0 \cdot |\mathcal{E}|^*$$

where '*' denotes the usual Schwarz symmetrization, then we have

(10a) $$T(\Pi_{\mathcal{E}_0}(\mathcal{E})) \leq T(\mathcal{E})$$

for all $\mathcal{E}_0$, and

(10b) $$V_\omega(\Pi_{\mathcal{E}_0}(\mathcal{E})) \geq V_\omega(\mathcal{E})$$

provided the following condition on $\mathcal{E}_0$ is satisfied

(11a) $$\mathcal{E}_0 \wedge \mathcal{E}_0^* = 0 \qquad \text{if } \eta < 0$$

(11b) $$\mathcal{E}_0 \cdot \mathcal{E}_0 = 0 \qquad \text{if } \eta > 0$$

No condition on $\mathcal{E}_0$ is imposed if η =0. The condition (11a) can be rewritten as $|\mathcal{E}_0 \cdot \mathcal{E}_0| = 1$

Hence we have the following theorem:

Theorem 1. Let be $\omega > 0$ and let either

(a) $\lambda > 0$ $\quad$ η arbitrary

(b) $\lambda = 0$ $\quad$ $\eta > 0$

(c) $\lambda < 0$ $\quad$ $\eta > 0$ with the additional restrictions

$\eta + \lambda > 0$ if $\sigma = 1$

and $0 < \omega < \omega^*$ if $\sigma > 1$

where $\omega^* = \frac{\eta}{2}\,\frac{\sigma - 1}{\sigma}\left(-\frac{\eta}{2}\,\frac{\sigma+1}{\lambda\sigma}\right)^{1/\sigma-1}$

then the stationary equation (8) admits a family of spherically symmetric ground sate solutions of the form

(12) $$\mathcal{E}^0_\omega(x) = \varphi_\omega(x)\,\mathcal{E}_0$$

where $\varphi_\omega(x)$ is a positive real valued function decreasing in $|x|$ and $\mathcal{E}_0$ is a complex vector of length one.
In particular, if $\eta < 0$ then $\mathcal{E}_0$ satisfies (11a) and φ_ω is the ground state solution of the scalar equation

(13a) $$-\Delta\varphi = -\omega\varphi + \lambda|\varphi|^{2\sigma}\varphi$$

On the other hand, if $\eta > 0$, then $\mathcal{E}_0$ satisfies (11b) and φ_ω is the ground state of the scalar equation

(13b) $$-\Delta\varphi = -\omega\varphi + \lambda|\varphi|^{2\sigma}\varphi + \eta|\varphi|^2\varphi$$

If none of the coefficients η and λ is positive then no ground state exists.

Remark 2.1 The last part of the Theorem 1 follows from the fact that the condition (7b) is also necessary for the existence of nontrivial solution, otherwise it cannot be sufficient if η, λ are non positive.

Remark 2.2 Solutions of (13a) correspond to electric fields which do not generate a magnetic field, while solutions of (13b) have a selfgenerated magnetic field of strength $|B(\mathcal{E})| = Q(\mathcal{E})$. In particular, Theorem 1 states that in the attractive case ($\eta > 0$) the ground states have a nonzero field B. Solutions with zero magnetic field may also exist (if (13a) admits nontrivial solutions) but they cannot be be ground states.

Remark 2.3 Theorem 1 is also true in two space dimensions (now $0 < \sigma < \infty$), but we have to solve a different variational problem since (6) does not have a solution [1]. In one space dimension the solitary waves can be constructed by ODE techniques. However, in general they cannot be characterized by a variational principle.

Remark 2.4 There are also solitary waves of the NLS with internal rotations. If $\eta = 0$, the U(3) symmetry of the NLS generates a large family of solitary waves which are not of the form (1). Their existence will be studied in a subsequent publication.

II. STABILITY OF THE GROUND STATES

To prove either the stability or instability of the ground states we apply a general technique using the variational characterization of the solution obtained in Theorem 1 [11,12]. As it was shown in these papers the useful criteria are given by the function $d(\omega)$ which is twice differentiable and we have

(14a) $$d'(\omega) = Q_\omega(\mathcal{E}_\omega^0)$$

(14b) $$d''(\omega) = \frac{d}{d\omega} Q_\omega(\mathcal{E}_\omega^0)$$

We have the following theorem (see [11,12,14])

Theorem 2. Let $d''(\omega_0) \neq 0$. Then the following statements are equivalent.
(a) $d''(\omega_0) > 0$

(b) $\mathcal{E}_{\omega_0}^0$ is a local minimum of the energy subject to fixed charge Q.

(c) $\mathcal{E}_{\omega_0}^0$ is orbitally stable.

Example: If $\sigma = 1$ we can compute $d(\omega)$ explicitily to obtain

(15) $$d(\omega) = \omega^{1/2} d_0$$

Hence all ground states are unstable as predicted in [10] which means that in this case the magnetic field effect does not stabilize the solitary wave.

Stability of other solitary waves:

(a) The case $\eta > 0$
Let $\mathcal{E}_\omega(x) = \varphi_\omega(x)\,\mathcal{E}_0$ be a solution of equation (8) where $\mathcal{E}_0$ satisfies the condition $\mathcal{E}_0 \wedge \mathcal{E}_0^* = 0$ and φ_ω is a ground state solution of the scalar equation (13a). It can be easily checked that $B(\mathcal{E}_\omega) = 0$ and that $\mathcal{E}_\omega$ cannot be a local minimum of the energy subject to constant charge. Therefore we assume $\mathcal{E}_\omega$ to be unstable.

(b) The case $\eta < 0$
Let $\mathcal{E}_\omega(x) = \varphi_\omega(x)\,\mathcal{E}_0$ be a solution of the stationary equation with $\mathcal{E}_0$ satisfying the condition $\mathcal{E}_0 \cdot \mathcal{E}_0 = 0$. Let be φ_ω a ground state of the scalar equation (13b). We have $|B(\mathcal{E}_\omega)| = Q(\mathcal{E}_\omega)$. Again this $\mathcal{E}_\omega(x)$ cannot be a local minimum of the energy subject to fixed charge. But now $\mathcal{E}_\omega$ is a local minimum of the energy if we keep also $|B(\mathcal{E})|$ (the strength of the selfgenerated magnetic field) fixed provided $Q(\mathcal{E}_\omega)$ is increasing in ω . To prove this we note that if $Q(\mathcal{E}) = |B(\mathcal{E})|$ then $\mathcal{E}(x)$ satisfies $\mathcal{E}(x) \cdot \mathcal{E}(x) = 0$ for (almost) all x. Indeed, we have

$$|\mathcal{E} \wedge \mathcal{E}^*| = \left(|\mathcal{E}|^4 - (\mathcal{E} \cdot \mathcal{E})\right)^{1/2} \leq |\mathcal{E}|^2$$ with equality iff. $\mathcal{E} \cdot \mathcal{E} = 0$

Whithin this class $\mathcal{E}_\omega$ is a local minimum of the energy with respect to the constraint $Q(\mathcal{E}) = |B(\mathcal{E})| = Q(\mathcal{E}_\omega)$ if $Q(\mathcal{E}_\omega)$ is increasing in ω . Hence $\mathcal{E}_\omega$ is stable.

ACKNOWLEDGMENTS

We thank to the organizers P.Lochak, M.Balabane and C.Sulem of the Conference "Integrable Systems and Applications" for the stimulating and interdisciplinary atmosphere of the Meeting.

REFERENCES

1. H.Berestycki and P.L.Lions, Arch.Rat.Mech.Anal.82, 313 (1983).

2. M.Weinstein, Comm.Pure Appl.Math. 39, 51 (1986).

3. J.Ginibre and G.Velo, Journ.Funct.Anal. 32, 1 (1979).

4. F.Tappert and C.M.Varma, Phys.Rev.Lett. 25, 1108 (1970).

5. H.Kuratsuyi, Prog.Theor.Phys. 74, 433 (1985).

6. R.Y.Chiao,E.Garmin and C.H.Townes,Phys.Rev.Lett.13, 479 (1964).

7. P.L.Kelley, Phys.Rev.Lett. 15, 1005 (1965).

8. V.E.Zakharov, Sov.Phys.JETP 36, 908 (1972).

9. M.V.Goldman, D.R.Nicholson and J.C.Weatherall, Phys.Fluids 24, 668 (1981).

10.M.Kono,M.M.Skoric and D.Ter Haar, J.Plasma Phys. 26, 123 (1981).

11.J.Shatah, Comm.Math.Phys. 91, 313 (1983).

12.J.Stubbe, SFB preprint Nr.14, SFB-237, Ruhruniversitat Bochum (1988).

13.H.Brezis and E.Lieb, Comm.Math.Phys. 96, 97 (1984).

14.J.Shatah and W.Strauss, Comm.Math.Phys. 100, 173 (1985).

Lecture Notes in Mathematics

Vol. 1236: Stochastic Partial Differential Equations and Applications. Proceedings, 1985. Edited by G. Da Prato and L. Tubaro. V, 257 pages. 1987.

Vol. 1237: Rational Approximation and its Applications in Mathematics and Physics. Proceedings, 1985. Edited by J. Gilewicz, M. Pindor and W. Siemaszko. XII, 350 pages. 1987.

Vol. 1250: Stochastic Processes – Mathematics and Physics II. Proceedings 1985. Edited by S. Albeverio, Ph. Blanchard and L. Streit. VI, 359 pages. 1987.

Vol. 1251: Differential Geometric Methods in Mathematical Physics. Proceedings, 1985. Edited by P.L. García and A. Pérez-Rendón. VII, 300 pages. 1987.

Vol. 1255: Differential Geometry and Differential Equations. Proceedings, 1985. Edited by C. Gu, M. Berger and R.L. Bryant. XII, 243 pages. 1987.

Vol. 1256: Pseudo-Differential Operators. Proceedings, 1986. Edited by H.O. Cordes, B. Gramsch and H. Widom. X, 479 pages. 1987.

Vol. 1258: J. Weidmann, Spectral Theory of Ordinary Differential Operators. VI, 303 pages. 1987.

Vol. 1260: N.H. Pavel, Nonlinear Evolution Operators and Semigroups. VI, 285 pages. 1987.

Vol. 1263: V.L. Hansen (Ed.), Differential Geometry. Proceedings, 1985. XI, 288 pages. 1987.

Vol. 1265: W. Van Assche, Asymptotics for Orthogonal Polynomials. VI, 201 pages. 1987.

Vol. 1267: J. Lindenstrauss, V.D. Milman (Eds.), Geometrical Aspects of Functional Analysis. Seminar. VII, 212 pages. 1987.

Vol. 1269: M. Shiota, Nash Manifolds. VI, 223 pages. 1987.

Vol. 1270: C. Carasso, P.-A. Raviart, D. Serre (Eds.), Nonlinear Hyperbolic Problems. Proceedings, 1986. XV, 341 pages. 1987.

Vol. 1272: M.S. Livšic, L.L. Waksman, Commuting Nonselfadjoint Operators in Hilbert Space. III, 115 pages. 1987.

Vol. 1273: G.-M. Greuel, G. Trautmann (Eds.), Singularities, Representation of Algebras, and Vector Bundles. Proceedings, 1985. XIV, 383 pages. 1987.

Vol. 1275: C.A. Berenstein (Ed.), Complex Analysis I. Proceedings, 1985–86. XV, 331 pages. 1987.

Vol. 1276: C.A. Berenstein (Ed.), Complex Analysis II. Proceedings, 1985–86. IX, 320 pages. 1987.

Vol. 1277: C.A. Berenstein (Ed.), Complex Analysis III. Proceedings, 1985–86. X, 350 pages. 1987.

Vol. 1283: S. Mardešić, J. Segal (Eds.), Geometric Topology and Shape Theory. Proceedings, 1986. V, 261 pages. 1987.

Vol. 1285: I.W. Knowles, Y. Saitō (Eds.), Differential Equations and Mathematical Physics. Proceedings, 1986. XVI, 499 pages. 1987.

Vol. 1287: E.B. Saff (Ed.), Approximation Theory, Tampa. Proceedings, 1985–1986. V, 228 pages. 1987.

Vol. 1288: Yu. L. Rodin, Generalized Analytic Functions on Riemann Surfaces. V, 128 pages, 1987.

Vol. 1294: M. Queffélec, Substitution Dynamical Systems – Spectral Analysis. XIII, 240 pages. 1987.

Vol. 1299: S. Watanabe, Yu.V. Prokhorov (Eds.), Probability Theory and Mathematical Statistics. Proceedings, 1986. VIII, 589 pages. 1988.

Vol. 1300: G.B. Seligman, Constructions of Lie Algebras and their Modules. VI, 190 pages. 1988.

Vol. 1302: M. Cwikel, J. Peetre, Y. Sagher, H. Wallin (Eds.), Function Spaces and Applications. Proceedings, 1986. VI, 445 pages. 1988.

Vol. 1303: L. Accardi, W. von Waldenfels (Eds.), Quantum Probability and Applications III. Proceedings, 1987. VI, 373 pages. 1988.

Lecture Notes in Physics

Vol. 318: B. Mercier, An Introduction to the Numerical Analysis of Spectral Methods. V, 154 pages. 1989.

Vol. 319: L. Garrido (Ed.), Far from Equilibrium Phase Transitions. Proceedings, 1988. VIII, 340 pages. 1988.

Vol. 320: D. Coles (Ed.), Perspectives in Fluid Mechanics. Proceedings, 1985. VII, 207 pages. 1988.

Vol. 321: J. Pitowsky, Quantum Probability – Quantum Logic. IX, 209 pages. 1989.

Vol. 322: M. Schlichenmaier, An Introduction to Riemann Surfaces, Algebraic Curves and Moduli Spaces. XIII, 148 pages. 1989.

Vol. 323: D.L. Dwoyer, M.Y. Hussaini, R.G. Voigt (Eds.), 11th International Conference on Numerical Methods in Fluid Dynamics. XIII, 622 pages. 1989.

Vol. 324: P. Exner, P. Šeba (Eds.), Applications of Self-Adjoint Extensions in Quantum Physics. Proceedings, 1987. VIII, 273 pages. 1989.

Vol. 325: E. Brändas, N. Elander (Eds.), Resonances, Proceedings, 1987. XVIII, 564 pages. 1989.

Vol. 326: A. Grauel, Feldtheoretische Beschreibung der Thermodynamik für Grenzflächen. IX, 317 Seiten. 1989.

Vol. 327: K. Meisenheimer, H.-J. Röser (Eds.), Hot Spots in Extragalactic Radio Source. Proceedings, 1988, XII, 301 pages. 1989.

Vol. 328: G. Wegner (Ed.), White Dwarfs. Proceedings, 1988. XIV, 524 pages. 1989.

Vol. 329: A Heck, F. Murtagh (Eds.), Knowledge Based Systems in Astronomy. IV, 280 pages. 1989.

Vol. 330: J.M. Moran, J.N. Hewitt, K.Y. Lo (Eds.), Gravitational Lenses. Proceedings, 1988. XIV, 238 pages. 1989.

Vol. 331: G. Winnewisser, J.T. Armstrong (Eds.), The Physics and Chemistry of Interstellar Molecular Clouds mm and Sub-mm Observations in Astrophysics. Proceedings, 1988. XVIII, 463 pages. 1989.

Vol. 332: P. Flin, H.W. Duerbeck (Eds.), Morphological Cosmology. Proceedings, 1988. VII, 438 pages. 1989.

Vol. 333: I. Appenzeller, H.J. Habing, P. Léna (Eds.), Evolution of Galaxies – Astronomical Observations. Proceedings, 1988. X, 391 pages. 1989.

Vol. 334: L. Maraschi, T. Maccacaro, M.-H. Ulrich (Eds.), BL Lac Objects. XIII, 497 pages. 1989.

Vol. 335: A. Lakhtakia, V.K. Varadan, V.V. Varadan, Time-Harmonic Electromagnetic Fields in Chiral Media. VII, 121 pages. 1989.

Vol. 336: M. Müller, Consistent Classical Supergravity Theories. VI, 125 pages. 1989.

Vol. 337: A.P. Maclin, T.L. Gill, W.W. Zachary (Eds.), Magnetic Phenomena. Proceedings, 1988. VI, 142 pages. 1989.

Vol. 338: V. Privman, N.M. Švrakić, Directed Models of Polymers, Interfaces, and Clusters: Scaling and Finite-Size Properties. VI, 120 pages. 1989.

Vol. 339: F. Ehlotzky (Ed.), Fundamentals of Laser Interactions II. Proceedings, 1989. XI, 317 pages. 1989.

Vol. 340: M. Peshkin, A. Tonomura, The Aharonov-Bohm Effect. VI, 152 pages. 1989.

Vol. 341: E.F. Milone (Ed.), Infrared Extinction and Standardization. Proceedings, 1988. III, 79 pages. 1989.

Vol. 342: M. Balabane, P. Lochak, C. Sulem (Eds.), Integrable Systems and Applications. Proceedings, 1988. VII, 342 pages. 1989.